Human Factors in the Design and Evaluation of Central Control Room Operations

Human Factors in the Design and Evaluation of Central Control Room Operations

Neville A. Stanton
Paul Salmon
Daniel Jenkins
Guy Walker

CRC Press
Taylor & Francis Group
Boca Raton London New York

CRC Press is an imprint of the
Taylor & Francis Group, an **informa** business

CRC Press
Taylor & Francis Group
6000 Broken Sound Parkway NW, Suite 300
Boca Raton, FL 33487-2742

International Standard Book Number: 978-1-4398-0991-4 (Hardback)

Library of Congress Cataloging-in-Publication Data

Human factors in the design and evaluation of central control room operations / authors, Neville A. Stanton ... [et al.].
p. cm.
Includes bibliographical references and index.
ISBN 978-1-4398-0991-4 (alk. paper)
1. Human engineering. 2. Control rooms--Design and construction. I. Stanton, Neville, 1960- II. Title.

TA166.H793 2009
620.8'2--dc22 2009018069

Visit the Taylor & Francis Web site at
http://www.taylorandfrancis.com

and the CRC Press Web site at
http://www.crcpress.com

Contents

List of Figures

List of Tables

Preface

Central control room operations may be found in many domains, such as aviation, chemical, energy production and distribution, defense, manufacturing, medicine, nuclear, oil and gas, and security. A feature of these control rooms is that the people operating them are often remote from the processes they are controlling and the operations continue 24 hours a day, 7 days a week. The twin demands of remote and continuous operation place special demands on the design of central control rooms. This book deals with the human factors concerns of those operations. As the contents of this book show, this is a complex area, and some 14 human factors topics are presented, covering competencies, training, procedures, communications, workload, automation, supervision, shift patterns, control room layout, SCADA interfaces, alarms, control room environment, human error, and safety culture. This list of topics indicates that the scope of human factors is broad and far reaching. Each of these topics is worthy of a book, but the purpose of this book is to bring them together under one cover with a focus on central control room operations. The bibliography at the end of each chapter can guide the interested reader deeper into the literature.

This book began life when the authors were asked to review human factors pertinent to control room operations. The organizations involved were about to upgrade their control rooms and wanted to get the latest human factors advice and guidance as well as to comply with relevant standards. The Health and Safety Executive in the United Kingdom also wanted to ensure that the safety case approach for the control room upgrade included human factors considerations. The project began with a review of contemporary standards, guidance, and the human factors literature. Because the study was undertaken in 2008, anyone coming to this book for the first time would be well advised to check the latest standards and stay abreast of developments in the human factors and ergonomics literature and methodologies. We hope that the material in this book will serve as a good starting point for such investigations as well as a template for analysis of control rooms. The project served as a good model for collaboration between the Health and Safety Executive, the university sector, and commercial organizations. The resultant material served as a basis for this book, although additional material has been added and the examples have been sanitized to protect the identities of those involved.

This book can be used in many ways. An overview of each chapter is presented in the first chapter. Each of the chapters is self-contained, so those wanting to explore any of the topics can choose to read only the chapter relevant to their needs. Each of the chapters has been written in a modular structure, so readers can identify the parts relevant to them; for example, readers who simply want to find out about the standards can just go to that section (bearing in mind that standards are subject to development and change). We feel that this book will be of most use to those who are involved in developing and upgrading central control rooms. This book shows them how to address the human factors, with methods and examples. The book can also be used as a model of what their safety case might look like. The book should also be useful for those learning about human factors and ergonomics for postgraduate degrees and training courses. They can apply the methods to their own area of study and use the material within the book to prepare case studies, coursework, and report assignments. This is a fascinating time for human factors and ergonomics, particularly as its study brings so many topics together to focus on design improvements for human–system performance in a complex domain. The drawing together of the 14 topic areas is addressed briefly in the final chapter on safety culture. To this end, we hope to have provided an integrated approach to human factors and ergonomics when addressing central control room operations.

The Authors

Neville A. Stanton, PhD, holds a chair in human factors in the School of Civil Engineering and the Environment at the University of Southampton. He has published over 140 peer-reviewed journal papers and 14 books on human factors and ergonomics. In 1998, he was awarded the Institution of Electrical Engineers Divisional Premium Award for a coauthored paper on engineering psychology and system safety. The Ergonomics Society awarded him the Otto Edholm Medal in 2001 and the President's Medal in 2008 for his contribution to basic and applied ergonomics research. In 2007, the Royal Aeronautical Society awarded him the Hodgson Medal and Bronze Award with colleagues for their work on flight deck safety. Dr. Stanton is an editor of the journal *Ergonomics* and is on the editorial boards of *Theoretical Issues in Ergonomics Science* and the *International Journal of Human–Computer Interaction*. He is a fellow and chartered occupational psychologist registered with the British Psychological Society, a fellow of the Ergonomics Society, and a member of the Institution of Engineering and Technology. He has a BSc (Hons) in occupational psychology from the University of Hull, an MPhil in applied psychology, and a PhD in human factors from Aston University in Birmingham.

Paul Salmon, PhD, is a senior research fellow in the Human Factors Group at Monash University; he holds a BSc in sports science and an MSc in applied ergonomics, both from the University of Sunderland. Dr. Salmon has over 6 years of experience in applied human factors research in a number of domains, including the military, civil and general aviation, and rail and road transport, and he has previously worked on a variety of research projects in these areas. This has led to Dr. Salmon gaining expertise in a broad range of areas, including human error, situation awareness, and the application of human factors methods, including human error identification, situation awareness measurement, teamwork assessment, task analysis, and cognitive task analysis methods. His current research interests include the areas of situation awareness in command and control, human error, and the application of human factors methods in sport. Dr. Salmon has authored and coauthored various scientific journal articles, conference articles, book chapters, and books and was recently awarded the Royal Aeronautical Society Hodgson Prize for a coauthored paper in the society's *Aeronautical* journal.

Daniel Jenkins, PhD, graduated in 2004 from Brunel University with an MEng (Hons) in mechanical engineering and design, receiving the University Prize for top student in the department. He has several years of experience as a design engineer in the automotive industry and has worked in a number of roles throughout the world. This wide range of placements has provided experience encompassing design, engineering, project management, and commercial awareness. Dr. Jenkins returned to Brunel in 2005 to become a research fellow in the Ergonomics Research Group, working primarily on the Human Factors Integration Defence Technology Centre project. Dr. Jenkins earned his PhD in human factors and interaction design in 2008 and has maintained a strong focus on customer-orientated design, design for inclusion, and human factors. He has authored and coauthored numerous journal papers, conference articles, book chapters, and books.

Guy Walker, PhD, read for a BSc (Hons) in psychology at Southampton University specializing in engineering psychology, statistics, and psychophysics. During his undergraduate studies he also undertook work in the auditory perception laboratories at Essex University and the applied psychology unit at Cambridge University. After graduating in 1999, he moved to Brunel University, where he earned his PhD in human factors in 2002. His research focused on driver performance, situational awareness, and the role of feedback in vehicles. Since this time, Dr. Walker has worked for a human factors consultancy on a project funded by the Rail Safety and Standards Board that examines driver behavior in relation to warning systems and alarms fitted in train cabs. Currently, he is a senior research fellow within the Human Factors Integration Defence Technology Centre at the University of Southampton, engaged primarily in work on future C4i systems. He is also the author of numerous journal articles and book contributions.

1 Introduction to Human Factors and Ergonomics

Human factors and ergonomics have a 100-year history in the United Kingdom and United States, from humble beginnings at the turn of the last century to the current day. A detailed account of the historical developments in both the United States and United Kingdom may be found in Meister (1999), which even covers the prehistory of the discipline. A recent special issue of the journal *Ergonomics* covered the development of this discipline within the United Kingdom (Waterson and Sell, 2006). To cut a very long story short, the discipline emerged out of the recognition that analysis of the interaction between people and their working environment revealed how work could be designed to reduce errors, improve performance, improve quality of work, and increase the work satisfaction of the workers themselves. Two figures stand out at the early beginnings of the discipline in the 1900s: Frank and Lillian Gilbreth (Stanton, 2006). The Gilbreths sought to discover more efficient ways to perform tasks. By way of a famous example of their work (Gilbreth, 1911), they observed that bricklayers tended to use different methods of working. With the aim of seeking the best way to perform the task, they developed innovative tools, job aids, and work procedures. The resultant effect of these changes to the work meant that the laying of a brick had been reduced dramatically from approximately 18 movements by the bricklayer down to some 4 movements. Thus, the task was performed much more efficiently. This analysis, among others, led to the discovery of "laws of work," or "Ergo-nomics," as it was called (Oborne, 1982). Although the discipline has become much more sophisticated in the way it analyzes work (as indicated in the next section), the general aims to improve system performance and the quality of working life remain the principal goals. *Human factors* and *ergonomics* have been defined variously as "the scientific study of the relationship between man and his working environment" (Murrell, 1965), "a study of man's behaviour in relation to his work" (Grandjean, 1980), "the study of how humans accomplish work-related tasks in the context of human–machine systems" (Meister, 1989), "applied information about human behaviour, abilities, limitations and other characteristics to the design of tools, machines, tasks, jobs and environments" (Sanders and McCormick, 1993), and "that branch of science which seeks to turn human–machine antagonism into human–machine synergy" (Hancock, 1997). From these definitions, it may be gathered that the discipline of human factors and ergonomics is concerned with human capabilities and limitations, human–machine interaction, teamwork, tools, machines and material design, environments, and work and organizational design. The definitions also place some implied emphasis on system performance, efficiency, effectiveness, safety, and well-being. These remain important aims for the discipline.

The term "human factors" seems to have come from the United States and has been used to encompass any aspect of design, operation, maintenance, and disposal that has bearing on human performance as part of a technological system. The terms "human factors" and "ergonomics" are often used interchangeably or together. In the United Kingdom, "ergonomics" is mostly used to describe physiological, physical, behavioral, and environmental aspects of human performance, whereas "human factors" is mostly used to describe cognitive, social, and organizational aspects of human performance. Human factors and ergonomics encompass a wide range of topics in system design, including manpower, personnel, training, communications media, procedures, team structure, task design, allocation of function, workload assessment, equipment design, system safety, and health hazards.

The term "human factors" will be used throughout this book, although this may also mean ergonomics. Modern-day human factors study focuses on integration with other aspects of system engineering. According to the U.K. Ministry of Defence, human factors integration is about "providing a balanced development of both the technical and human aspects of equipment provision. It provides a process that ensures the application of scientific knowledge about human characteristics through the specification, design and evaluation of systems" (Ministry of Defence, 2000, p. 6).

Human factors methods have been developed over the past century to help design new systems and evaluate existing systems. This book focuses on the examination of a central control room in the energy distribution industry and provides an analysis of human factors that relate to 24/7 operation of the control room and the implications for control room staff. It is intended to help organizations and people plan for the design or evaluation of a control room.

Specific topics under analysis are the competencies for control room work, training and training needs analysis, development and maintenance of procedures, communication processes within and without the control room, manning of the control room, automation and allocation of function, supervision of staff, and shift patterns. Finally, control room layout (e.g., positioning and spacing of equipment), environment (e.g., lighting, heating, cooling, and noise), human error potential, and safety culture are assessed against contemporary human factors and ergonomics guidelines and principles. The examples presented represent a mixture of evaluations of existing control rooms and proposals for changes in communication structures, shift patterns, and so on.

Following is a brief overview of the chapters:

- *Competencies* (Chapter 2)—Competencies have been defined as the ability to undertake responsibilities and perform activities to a regular standard on a regular basis. Competencies are comprised of a combination of practical and cognitive skills, experience, and knowledge to conduct work in accordance with standards, rules, and procedures. Different facets of competencies are acknowledged, including knowledge (e.g., knowing how equipment works and the limits of its performance), skills (e.g., the ability to perform tasks and operate equipment to a given standard), and attitudes (e.g., motivation to perform tasks safely and approach to team work).

- *Training* (Chapter 3)—Training is the systematic acquisition of knowledge, skills, and attitudes that enable people to perform work to a required standard. A systematic approach to training development and delivery requires identification of the training needs, designing the training content and media, delivery of the training, and evaluation of the effectiveness of the training. Thus, it is a cyclical process, with continual revision and refinement with changes in the audience, tasks, and organizational requirements. The basis on which all training depends is identification of the training need from the perspective of the organization (to identify what training is required), the tasks (to identify what knowledge, skills, and attitudes are required), and the person (to identify who requires training). This analysis forms the foundation of all of the subsequent training design processes.
- *Procedures* (Chapter 4)—Procedures are the written narrative, checklist, flowchart, or other descriptions of the sequence of steps required to be followed to perform a given task or set of tasks. The analysis of a quality management system (QMS) for the development and maintenance of procedures is undertaken as well as a review of example procedures produced through the application of the QMS. The evaluation of the sampled procedures uses nine criteria: format, comprehensiveness, consistence, accuracy, detail, clarity, layout, ease of use, and review date.
- *Communication Process* (Chapter 5)—Communication process refers to sending and receiving of information between people in any medium available (e.g., via direct voice, telephone, paper notes, e-mail) as well as information communicated between people and computers (e.g., via information displays, controlled actions from keyboards and mice). An analysis of the communications networks within the 24/7 control room and among the management, support staff, field engineers, and external plant equipment is undertaken. The network analysis approach reveals key staff in the 24/7 control room through which communications must pass.
- *Manning and Workload* (Chapter 6)—Manning and workload are defined as the number of personnel working together at any time and the demands placed upon them to perform their tasks. The analysis of workload is based on the time spent in different activities and a comparison of the changes in tasks over the past 10 years. Although the data need to be treated with caution, as they are averaged over many shifts, they do indicate that the tasks are manageable in very general terms. The shift handover time is greater in the winter than in the summer, as there is far more activity on the network. Most of the other times are similar except for the monitor and control task, which is almost double in the winter due to greater network usage.
- *Automation* (Chapter 7)—Automation addresses the dichotomy of scheduling tasks between humans and machines. An automated device is, by definition, able to operate without human input, but automation does tend only to work effectively within prescribed system limits. Often this relegates the human role to one of monitoring the automated systems, intervening only if they fail to cope. Allocation of function methods have been developed to assist in the appropriate design of automation. A sociotechnical approach

is advocated that employs automation in a manner that designs meaningful work for people. This optimizes the performance of both the technical and social aspects of the system, as well as their interaction.

- *Supervision* (Chapter 8)—Supervision involves controlling, influencing, and leading a team and includes activities such as directing the work of others, allocating tasks and workload, and planning and scheduling tasks. Supervisory activities imply authority, responsibility, and accountability (such as maintaining team discipline), but they also imply fostering team spirit and developing the capability of team members. The Health and Safety Executive (HSE)-approved method has been applied to the analysis of the supervisory arrangements in the 24/7 control room team, support team, and management. These arrangements appear to have three main facets. Within normal office hours, the supervisory arrangements have the traditional formal hierarchical line management structure. Within the control room coaching and mentoring also occur. Outside of normal office hours, there is an additional self-managed team aspect to the 24/7 control room operation. The analysis also undertakes to identify the tasks and responsibilities of the various supervisors.
- *Shift Patterns* (Chapter 9)—Shift patterns refer to the scheduling of work activity outside the standard office hours, such as in 24/7 control room operations. These operations assume continuous staffing and a handover of duty from one team to another. The example shifts are assessed against human factors best practice. Comparison of the two proposed shift cycles (i.e., summer and winter) with a notional ideal on the HSE risk and fatigue indices reveals similar results.
- *Control Room Layout* (Chapter 10)—Control room layout refers to the functional and physical structure of the room and associated workstations (including desks, chairs, computers, peripherals, and storage cabinets) where operators are stationed to conduct centralized monitoring and control activities. Evaluation of the functional grouping and control room layout reveals the planned arrangement of workstations in the example.
- *Supervisory Control and Data Acquisition (SCADA)* (Chapter 11)—This chapter presents a review of the SCADA interface by assessing compliance with the Engineering Equipment and Materials Users' Association (EEMUA) 201 guidelines. EEMUA 201 is accepted as the industry best practice guidelines for design and operation control desk human–computer interfaces. Ideally, the SCADA interface should be designed to allow the operators to conduct their tasks effectively and efficiently. This means that the design of the human–computer interface conforms to their expectations and allows them to find information and perform operations in a straightforward manner.
- *Alarms* (Chapter 12)—This chapter presents an assessment of alarm systems. EEMUA 191 is a guide to the design, management, and procurement of alarm systems. It covers all aspects of human factors in the design of alarm systems for control centers, such as alarm system philosophy, key design principles, and key performance metrics. The approaches taken are

multifaceted. When a review of the control system alarm interface was undertaken in the control room, alarm questionnaires were completed by operators, and data from the alarm system over a 24-hour period have been analyzed.

- *Environment* (Chapter 13)—Control room environment refers to the prevailing ambient conditions, such as temperature, air quality, lighting, acoustics, vibration, interior design, and aesthetics, that are likely to have an effect on human performance in 24/7 operations. The example shows how measurements were taken on all six of these environmental factors at the workstations in a control room to see if they conformed to BS/EN/ISO 11064-6 (Ergonomic Design of Control Centers: Environmental Requirements for Control Centers).
- *Human Error* (Chapter 14)—Human error is defined as the performance of an inappropriate action or the failure to perform an appropriate action, commonly referred to as errors of commission and errors of omission. Human error identification techniques are used to identify the error potential within systems. One such technique is the Systematic Human Error Reduction and Prediction Approach (SHERPA). SHERPA classifies human activity into one of five basic types: (1) action (e.g., pressing a mouse button, typing a command), (2) retrieval (e.g., getting information from a screen or procedure database), (3) checking (e.g., conducting a procedural check), (4) communication (e.g., talking to another engineer), and (5) selection (e.g., choosing one alternative over another). This approach is used to analyze the error potential of the roles of Control Room Engineers presented in the examples. Each of the errors identified is classified in terms of its likelihood (where high means that it has been known to happen before) and criticality (where high means that it has a negative effect on the system). Error reduction strategies are classified into training, automation, communication, supervision, workload, competencies, SCADA interface design, procedures, safety culture, shift patterns, and environment.
- *Safety Culture* (Chapter 15)—Safety culture is defined as the product of group values, attitudes, and perceptions about the patterns of behavior that determine an organization's health and safety management. In the example provided, the safety culture is analyzed by using a questionnaire that was developed in the energy distribution industry. The questionnaire has eight main factors, although only seven were used in this study: work pressure, incident investigation, adequacy of procedures, communication and training, relationships, spares, and safety. A mean summed score below 5 is indicative of an area of safety culture that warrants further investigation.

The book is divided into chapters covering these 14 topic areas, each of which addresses the main areas of analysis—namely, competencies for control room work, training needs analysis, communication processes within and without the control room, manning of the control room, automation and allocation of function, supervision of staff, shift patterns, control room layout, SCADA interface, alarms, environment, human error identification, and safety culture. Within each of the subject areas are eight common sections: Outline, Key Issues, Standards and Guidelines, Human Factors Literature, Guidance and Recommendations, Assessment and Evaluation, Example, and Bibliography. The contents of these sections are described briefly below:

- *Outline*—Definition of the subject area and the scope of the assessment
- *Key Issues*—Highlight of the high-level topics within the subject area
- *Standards and Guidelines*—Relevant standards and guidelines
- *Human Factors Literature*—Review of the key ideas from the literature pertinent to the subject of the assessment
- *Guidance and Recommendations*—Summary of the main guidance and recommendations for the subject area
- *Assessment and Evaluation*—Assessment of the example for the subject area with a human factors evaluation
- *Example*—A brief example applying a human factors method to a case study in the control room
- *Bibliography*—List of key references cited and works used in the course of the assessments

BIBLIOGRAPHY

Gilbreth, F.B. (1911). *Motion Study*. Princeton, NJ: Van Nostrand.

Grandjean, E. (1980). *Fitting the Task to the Man*. London: Taylor & Francis.

Hancock, P.A. (1997). *Essays on the Future of Human–Machine Systems*. Minneapolis, MN: Banta.

Meister, D. (1989). *Conceptual Aspects of Human Factors*. Baltimore, MD: The Johns Hopkins University Press.

Meister, D. (1999). *A History of Human Factors and Ergonomics*. London: Lawrence Erlbaum.

Ministry of Defence. (2000). *Human Factors Integration: An Introductory Guide*. London: HMSO.

Murrell, K.F.H. (1965). *Human Performance in Industry*. New York: Reinhold.

Oborne, D.J. (1982). *Ergonomics at Work*. Chichester: John Wiley & Sons.

Sanders, M.S. and McCormick, E.J. (1993). *Human Factors in Engineering and Design*. New York: McGraw-Hill.

Stanton, N.A. (2006). Hierarchical task analysis: developments, applications and extensions. *Applied Ergonomics*, 37(1), 55–79 (invited paper for special issue on fundamental reviews of ergonomics).

Waterson, P. and Sell, R. (2006). Recurrent themes and developments in the history of the Ergonomics Society. *Ergonomics*, 49(8), 743–799.

2 Competencies

2.1 OUTLINE

The safe and efficient operation of process control rooms is dependent upon the competence of the operators working within them. Further, the design of process control rooms should be based on a thorough understanding of human operator competencies. The term *competency* is generally used to describe individual attributes associated with the quality of work performance (Hedge and Borman, 2006) but can also refer to a team's job-related attributes and also to an overall organization's attributes. The identification, training, maintenance, monitoring, and assessment of competencies is critical for ensuring that staff possess the appropriate competence levels required for safe and efficient work performance.

The Health and Safety Executive (HSE) defined *competence* as "the ability to undertake responsibilities and to perform activities to a regular standard on a regular basis" (HSE, 2007b, p. 2). Similarly, Wright, Turner, and Horbury (2003, p. 7) defined *competence* as "the ability to perform activities within an occupation or function to the standards expected in employment." Whiddett and Hollyforde (2007, p. 7) defined competencies as the "behaviours that individuals demonstrate when undertaking job relevant tasks effectively within a given organisational context." According to the HSE, competence is comprised of a combination of practical and thinking skills, experience, and knowledge and can also include a willingness to undertake work activities in accordance with standards, rules, and procedures (HSE, 2007b).

In addition to reducing performance efficiency, the employment of operators with substandard competence levels within safety-critical systems can have grave consequences. For example, a review of major accidents in the hazardous industries indicated that a lack of operator skills and/or knowledge led to the errors that contributed to the accidents analyzed (Wright et al., 2003). Wright et al. (2003) also referred to the Southall rail crash, the Piper Alpha oil rig disaster, the Longford Esso Gas plant explosion, and the Hickson & Welch fire as examples of incidents where staff incompetence played a part and where staff competency levels were not sufficiently tested by the organizations involved.

In recent times, legislation has led to the requirement for hazardous plants to test and demonstrate their staff's competence. Wright et al. (2003), for example, pointed out that the introduction of the Control of Major Accident Hazard (COMAH) regulations means that it is now no longer acceptable for organizations to make untested assumptions about the competence levels of their staff. The HSE *Safety Report Assessment Manual* also states that safety reports should demonstrate that the operator in question has in place a system for providing and maintaining appropriate levels of management and employee competence.

Ensuring that process control operators possess appropriate competence levels is therefore critical; however, it is notable that there is much more to this than simply providing operators with appropriate levels of training. Organizations must clearly identify specifically what competencies are required for each job or role in order to operate their plant safely and efficiently. They must also develop training programs that enable staff to be trained in these competencies and also systems to monitor and maintain staff competence levels. Further, organizations require valid and reliable approaches for assessing staff competence levels and also procedures to deal with instances in which staff competence levels are found to be inadequate.

Competencies can also be discussed with regard to teams and organizations. *Team competencies* have been defined as what team members need to know, how they need to behave, and what attitudes they need to hold (Salas, 2005) and are typically expressed as the knowledge, skills, and attitudes (KSAs) required for efficient task performance. *Organizational competencies* are defined as the "things that organisations are best at" (Whiddett and Hollyforde, 2007, p. 8) and refer to technical competencies rather than behavioral ones.

2.2 KEY ISSUES

The key issues surrounding the concept of operator competencies are related to the identification of the competencies that are required for each of the different jobs and roles within an organization, the training and maintenance of staff competence levels, and the assessment of staff competence levels. Collectively, this process is known as *competence management*. A brief summary of the key issues surrounding each area is presented below.

2.2.1 Individual Operator Competencies

Operator competence refers to the ability to undertake responsibilities and to perform activities to a regular standard on a regular basis (HSE, 2007b, p. 2). Organizations should have a clear understanding of what the competencies are for each job or role and what their current staff competence levels are.

2.2.2 Team Competencies

Team competencies include what team members need to know, how they need to behave, and what attitudes they need to hold (Salas, 2005). Team competencies therefore include all of the KSAs that are necessary for effective team performance. Organizations should have a clear understanding of what team competencies are required by the teams working for them.

2.2.3 Competence Identification

Organizations should have a clear understanding of exactly what the competencies are for each job or role. This allows the appropriate staff to be recruited, permits the development of appropriate staff training programs, and also informs the monitoring and assessment of operator competencies. Typically, task analysis approaches such as hierarchical task analysis (HTA) (Annett, 2005; Stanton, 2006) are used to inform competency identification within complex sociotechnical systems, and the competencies identified are expressed as the KSAs required for safe and efficient task performance.

2.2.4 Competency-Based Training

When a clear understanding of the competencies associated with a particular job or role has been acquired, appropriate training programs designed to train staff members in these competencies can be developed. These training programs can also be used to maintain staff competency levels or to retrain competencies in the event of staff competency levels failing to meet the required levels. According to Whiddett and Hollyforde (2007), competencies add the behavioral dimension to training and will contribute to the identification of training needs, the structuring of training plans and programs, the design and selection of training events and activities, and the evaluation of training program success.

2.2.5 Competence Assessment

Competence assessment involves ensuring that operators are capable of carrying out tasks safely and efficiently by assessing their skills and knowledge. As outlined above, it is no longer acceptable for organizations to assume that their staff are competent based on their exposure to training and experience. Organizations therefore need to regularly undertake competence assessments of all staff in order to ensure that they can safely and efficiently fulfill their jobs or roles. Despite this, Wright et al. (2003) reported that, according to HSE Hazardous Installations Directorate inspectors, few COMAH-duty holders have developed appropriate competence assessment methods. According to Wright et al. (2003), competence assessment is intended to improve accident prevention via the following:

- Identifying and filling gaps in individuals' competencies before they can contribute to major incidents
- Developing standards that will provide individuals with a clear understanding of what competencies they need and thus should encourage the development of competencies
- The setting of performance standards by duty holders which will prompt the provision of appropriate training and highlight unrealistic staff performance expectations
- Revealing substandard competence
- Including competence assessment within management systems that should reduce the likelihood of substandard staff performance being overlooked
- Setting standards and associated assessment of competence to mitigate pressure for staff to work beyond their personal limits

Appropriate competence assessment requires the collection of outcome-based evidence regarding an individual's ability to perform as opposed to simply assuming that exposure to training and experience ensure competence (Wright et al., 2003).

2.2.6 Competence Management Systems

So-called competence management systems (CMSs) are used by organizations to ensure that their staff possess and maintain the appropriate competencies required for efficient system performance and safety. According to the HSE (2007b, p. 2), the aim of a CMS is to ensure that individuals:

- Clearly understand the competencies expected of them.
- Receive appropriate training, development, and assessment.
- Have appropriate experience.
- Maintain or improve their competence over time.

2.3 STANDARDS AND GUIDELINES

A summary of the relevant standards and guidelines is presented below.

2.3.1 Managing Competence for Safety-Related Systems

The HSE's *Managing Competence for Safety-Related Systems* guide provides guidance on the management of competence (for all industry sectors) to ensure that workers are suitably qualified and experienced for their work activities, roles, and responsibilities (HSE, 2007a). The guidance is aimed at those responsible for managing and ensuring the competence of individuals and teams working within safety-related organizations and describes the core requirements for a competence management system, including a breakdown of the different phases and component steps. The aim of the guide is to enable organizations to meet the U.K. legal requirements for competence in safety-related systems. The guide also contains a list of the regulations that contain requirements concerning competence for health and safety.

2.3.2 Developing and Maintaining Staff Competence

The HSE *Developing and Maintaining Staff Competence* guide is aimed at those responsible for managing the competence of individuals working within the railway and other guided transport system industries (HSE, 2007b). Essentially, the guide presents an overview of the areas of competence, competence management, and CMSs. It begins with a general introduction to the area of staff competence and then presents an overview of a generic CMS framework. The overall objectives of the guide include describing the principles and factors that should be considered in any CMS and explaining how organizations should go about ensuring that the competence of individuals and teams working for them satisfies the requirements of existing legislation. This document is likely to be particularly useful for any organization wishing to develop and implement a new CMS or redevelop an existing CMS.

2.3.3 Competence Assessment for the Hazardous Industries

The *Competence Assessment for the Hazardous Industries* report (HSE, 2003) presents an extensive overview of the area of competence assessment in the hazardous industries, including an introduction to the area of competence assessment, a review of current practice (including case studies), guidance on competence assessment for major accident prevention, and, finally, examples of competence assessment (including process plant maintenance fitter, emergency response manager, operations team leader, and field operator examples). The report also presents a summary of the different methods available for conducting competence assessments and an overall competence assessment framework. This document is likely to be particularly useful for any organization wishing to develop a new process for undertaking competence assessments.

Various domain-specific guidelines and standards state that organizations should identify the level of competence required by their employees and that their employees should be competent in their roles or jobs. For example, guidance for the 1998 Provision and Use of Work Equipment Regulations states that employers should identify the level of employee competence required to operate work equipment (Wright et al., 2003). Likewise, the 1998 Control of Substances Hazardous to Health Regulations state that employees must be competent in the use and handling of equipment and substances. National vocational standards are also used in many cases to assess operator competence (Wright et al., 2003), whereby a person meeting the required standards is deemed competent.

2.3.4 Control of Major Accident Hazard Regulations

The 1999 COMAH regulations were designed to prevent and mitigate major accidents involving dangerous substances (e.g., chlorine, petroleum gas, arsenic pentoxide) that can cause harm to people or the environment (HSE, 2007a). Wright et al. (2003) pointed out that, as a result of the introduction of the COMAH regulations, it is no longer acceptable for organizations to make untested assumptions about the competencies of their staff.

2.4 HUMAN FACTORS LITERATURE

Within the human factors literature, the concept of competencies has received most attention from the point of view of team competencies, particularly within the aviation domain (i.e., flight crew competencies) with regard to the development of flight crew training programs. A brief overview of the human factors literature on competencies is presented below.

2.4.1 Competence

The HSE (2007a) suggests that for individuals to be competent, they need the qualifications, experience, and qualities appropriate to their duties, including:

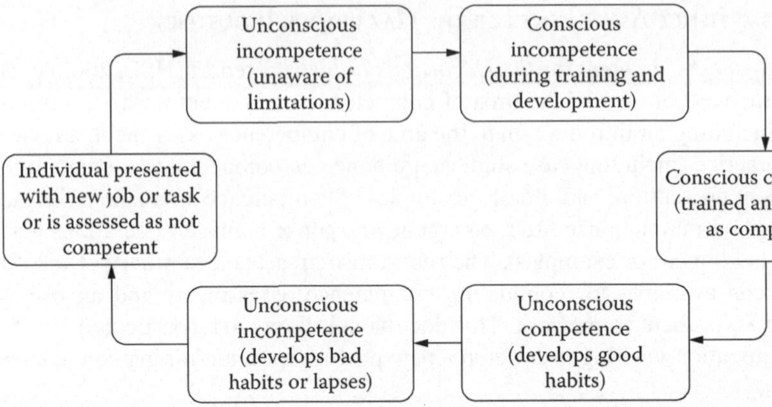

FIGURE 2.1 Stages of individual competence. (Adapted from HSE, *Managing Competence for Safety-Related Systems. Part 1. Key Guidance on Setting Up a Competency Management System Using 15 Guiding Principles*, Health and Safety Executive, London, 2007.)

- Knowledge necessary for the tasks they are required to perform
- Knowledge of the hazards and failures of the equipment for which they are responsible
- Knowledge and understanding of the working practices used in the organization for which they work
- Ability to communicate effectively with their peers and with any staff in the organization for which they work
- An appreciation of their own limitations and constraints, whether of knowledge, experience, facilities, resources, etc., and a willingness to point these out

Figure 2.1 presents a model of the stages of individual competence that describes how individuals, from the initial point of being given a new task or job, progress through different stages of competence based on the training that is given to them.

According to the model, individuals presented with a new job or task are typically unaware of what they can and cannot do, of what their abilities and limitations are (labeled in the model as *unconscious incompetence*). Eventually, as a result of training, individuals become cognizant of what it is they are incapable of doing and what they do not know or cannot do (labeled in the model as *conscious competence*). When they have been exposed to an appropriate form and level of training and have subsequently learned the task appropriately, individuals become *consciously competent.* When consciously competent, individuals can perform their tasks efficiently but they have to consciously think about it and what is required of them in order to do so. Over time, with increased exposure to the task, the individual becomes well practiced in the task and behavior can even become automatic. At this point, the individual's competence level becomes one of *unconscious competence.* In some situations, individuals can unknowingly regress and develop unconscious incompetence, where they develop bad habits. At this point, the model suggests that operators can then become unaware of their limitations once again (due to factors such as new technologies, new practices, or bad habits). The goal of any competence management system should be to avoid this state or detect this state and, in the event of operators reaching this state, to provide the means necessary to address it.

In terms of what actually comprises individual operator competencies, the majority of the human factors literature focuses on competencies as comprising the KSAs required for task performance. Moving into the process control domain, Pascual, Mills, and Henderson (2001) proposed the following set of generic, hypothetical control room operator competencies:

- Good knowledge of drills and procedures
- Ability to operate control equipment accurately
- An understanding of the theory and application of the control process
- An understanding of the safety aspects of control room operations
- The ability to make appropriate decisions
- The ability to assess situations accurately
- The ability to deal with stress and pressure
- The ability to ensure that team goals, roles, and responsibilities are understood
- The ability to anticipate colleagues requirements
- The ability to pass the correct information to colleagues at the right time
- The ability to notice overloaded colleagues and support them appropriately
- The ability to think ahead and develop contingencies
- The ability to ensure that colleagues maintain a shared understanding

Brabazon and Conlin (2001) proposed the following key performance requirements of process control operators:

- Be able to follow the condition of the process, anticipate its behavior, and select appropriate control strategies
- Be in a fit state to monitor the process (e.g., be awake and attentive)
- Be willing to take action when necessary
- Be able to take action reliably and within the necessary time frame

They also suggested that key team performance requirements include:

- The ability to collect and share critical information about the process and control actions
- The ability to coordinate actions

A generic example of how individual competencies can be described is presented in Figure 2.2.

Ivergard (1989) suggested that, in modern control rooms, operators generally face the following tasks:

- Start up/stop the system.
- Control, maneuver, and regulate.
- Check, monitor, and act (only when there is a fault).

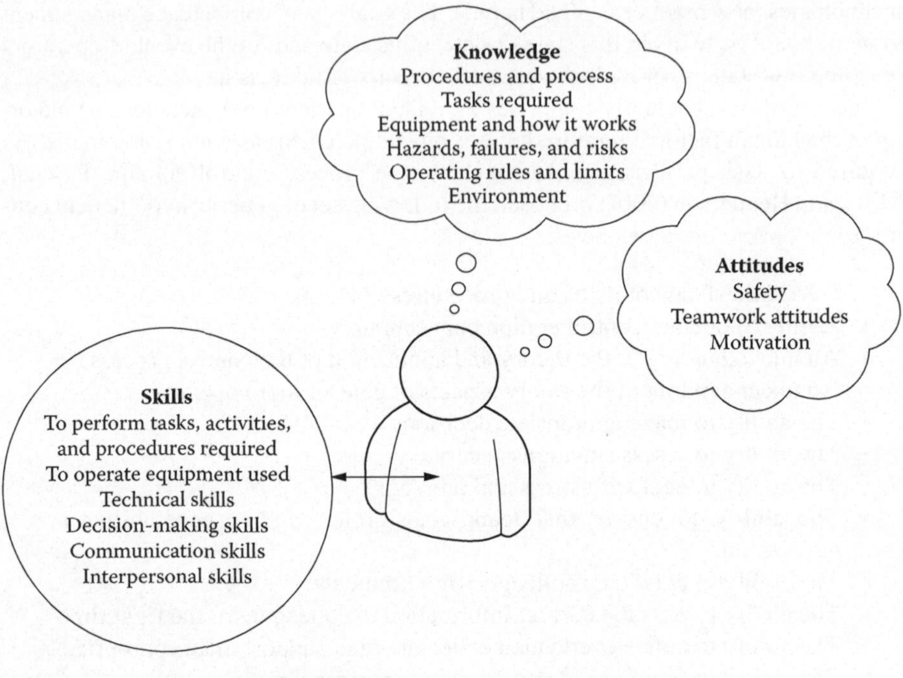

FIGURE 2.2 Operator competencies.

- Keep records and report.
- Repair and maintain.
- Plan, program, and analyze.

According to Ivergard (1989), the different types of knowledge and skills required for these tasks can be decomposed into the following seven main areas: perceptive skills, cognitive/thinking ability, vigilance ability, diagnostic skill, sensorimotor skills, motor abilities, and control skills. These are summarized below:

1. *Perceptive skills*—These are skills related to the ability to notice and distinguish signals from complex patterns. These typically include light or sound signals. Examples of perceptual tasks include radar work and inspection tasks, such as the supervision of electronic circuit room work.
2. *Cognitive/thinking ability*—This ability is related to supervisory, planning, and analysis tasks. The demands on different forms of cognitive and thinking abilities vary considerably in different types of control room.
3. *Vigilance ability*—Typically, control room operators are required to engage in lengthy, low-stimulation monitoring tasks and because of this understimulation can occur. These situations require a vigilance ability on behalf of the operator, which refers to the ability to maintain a form of alertness despite the soporific nature of the work.
4. *Diagnostic skill*—Diagnostic skill is the ability to find suitable guidelines, the ability to interpret guidelines in relation to the situation at hand, and the ability to select effective testing and searching strategies.

5. *Sensorimotor skills*—Various sensorimotor skills are required in control room operations in response to required changes to the system state. Sensorimotor skills pair perception (e.g., visual identification of changes in the system state) to action (e.g., manual inputs to the system to achieve the desired system state). These skills become automated in experienced persons.
6. *Motor ability*—Motor ability is required for the various manual tasks involved in control room operation. Although control room operations require both sensorimotor and motor ability, the motor requirements are relatively limited and will include the movements required to make gross and fine adjustments to the system state, such as pressing keys, changing values, moving the mouse onto targets, clicking the mouse button to select objects, and so on.
7. *Control skills*—People possess varying degrees of control skills depending on training, experience, and other factors.

Individual operator competencies are also often discussed in relation to the Skill, Rule, and Knowledge (SRK) framework. Vicente (1999) described three hierarchical levels of human behavior: skill-, rule-, and knowledge-based behaviors. Each of the levels within the SRK framework defines a different level of cognitive control or human action. Skill-based behavior occurs in routine situations that require highly practiced and automatic behavior and where there is only small conscious control on behalf of the operator. Skill-based behavior consists of smooth, automated, and highly integrated patterns of action that are performed without conscious attention.

The second level of behavior, rule-based behavior, occurs when the situation deviates from the normal but can be dealt with by the operator applying rules that either are stored in memory or are readily available—for example, emergency procedures. Rule-based behavior consists of stored rules derived from procedures, experience, instruction, or previous problem-solving activities.

The third and highest level of behavior is knowledge-based behavior, which typically occurs in nonroutine situations (i.e., emergency scenarios), where the operator has no known rules to apply and has to use problem-solving skills and knowledge of the system characteristics and mechanics in order to achieve task performance. Knowledge-based behavior consists of deliberate searching based on an explicit representation of the goal and a mental model of the functional properties of the environment. Further, knowledge-based behavior is slow, serial, and effortful, as it requires conscious, focused attention. The SRK framework is presented in Figure 2.3.

The SRK framework is used with the cognitive work analysis approach to conduct the worker competencies phase. Vicente (1999) presented a worker competencies analysis of the Dual Reservoir System Simulation (DURESS) II process control microworld. Here, the worker competencies are expressed in terms of the skill-, rule-, and knowledge-based behaviors required. Most of the demands associated with industrial process control are cognitive rather than perceptual–motor in nature, and only limited tasks require skill-based behavior. Vicente suggested that operators should have basic perceptual skills and should be able to recognize and keep track of the labels for the various components within DURESS II. He also suggested that operators should be able to directly interact with displays rather than have to

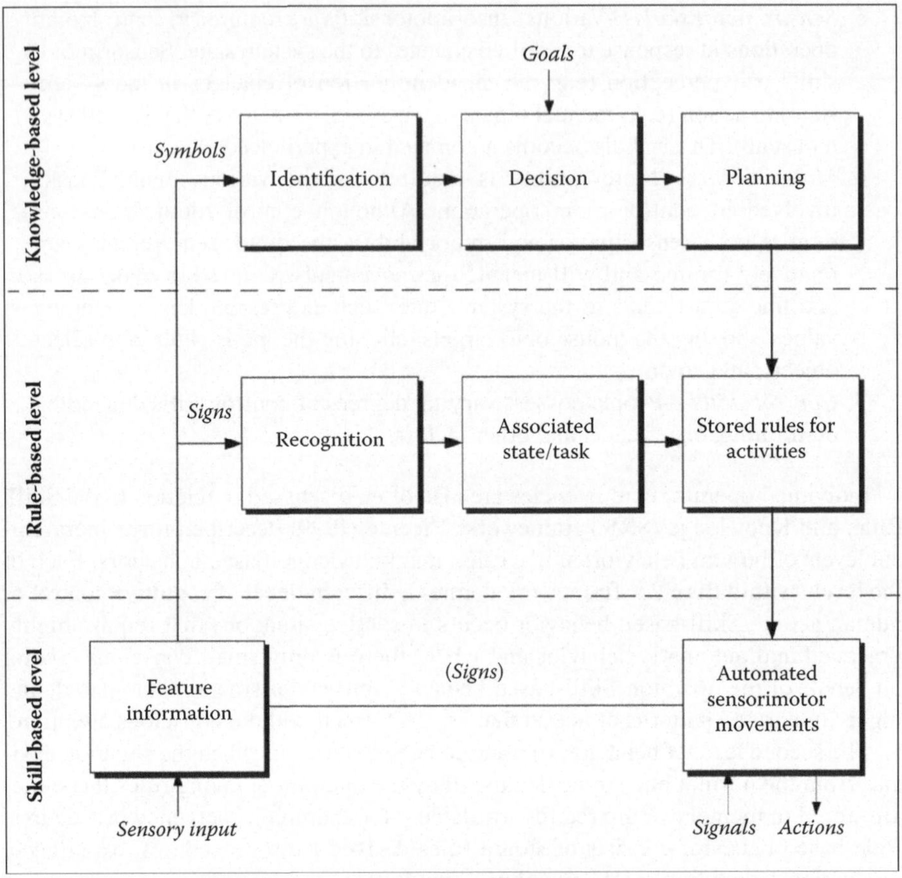

FIGURE 2.3 Skill, Rule, and Knowledge (SRK) behavioral classification scheme.

communicate their intentions via some arbitrary command language. Finally, at the skill-based level, Vicente suggested that operators should be sensitive to the dynamics of the work domain. At the rule-based behavior level, Vicente divided the worker competencies into two categories: the creation of salient perceptual cues to support rule-based behavior and awareness of the shortcuts and strategies that can be used to control the process. Finally, Vicente pointed out that, even though training and system design can be used to encourage lower levels of cognitive control, workers will still need to engage in the knowledge-based behavior level.

2.4.2 Team Competencies

Team competencies have been defined as what team members need to know, how they need to behave, and what attitudes they need to hold (Salas, 2005) and include all of the KSAs that are necessary for effective team performance. Salas and Cannon-Bowers (2001) defined *team competencies* as resources that team members draw

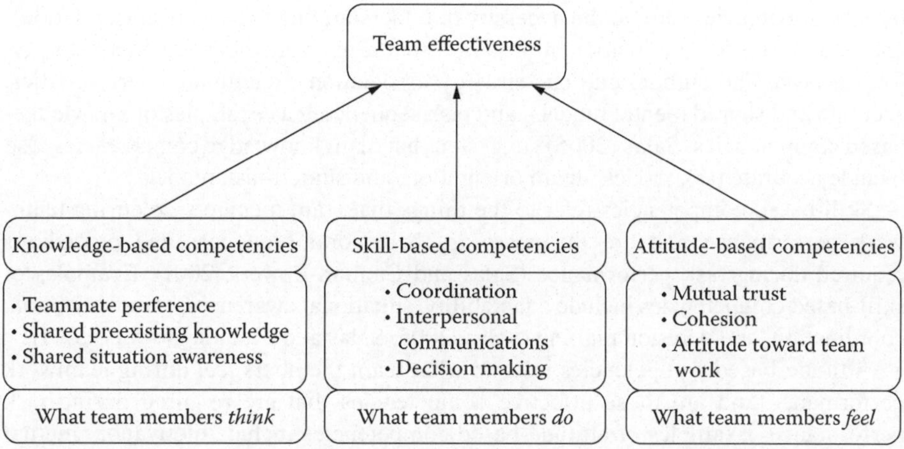

FIGURE 2.4 (Adapted from Guzzo, R.A. and Salas, E., *Team Effectiveness and Decision Making in Organizations*, Pfeiffer, New York, 1995. Reproduced with permission of John Wiley & Sons.)

from in order to function. Salas (2005) made a distinction between teamwork competencies and taskwork competencies. Taskwork competencies refer to the "knowledge, skills, and attitudes that the individual needs to accomplish his or her portion of the team task," whereas teamwork competencies refer to what team members need to know, how they need to behave, and what attitudes they need to hold (Salas, 2005, pp. 44–42). Teamwork competencies include the technical KSAs required for task performance and also the important teamwork processes that comprise effective teamwork. Cannon-Bowers and Salas (2006) suggested that, although taskwork competencies are necessary, they are not sufficient for effective team performance. Team members need to be proficient in their individual tasks, but they also require competencies at the team level. A model of team competencies developed by Cannon-Bowers and Salas (2006) is presented in Figure 2.4.

Most human factors practitioners now agree that team competencies can be decomposed into the KSAs required for efficient task performance; for example, Cannon-Bowers, Tannenbuam, Salas, and Volpe (1995) suggested that teamwork competencies are comprised of three components:

1. The requisite knowledge, principles, and concepts underlying the team's effective task performance
2. The repertoire of required skills and behaviors necessary to perform the team task effectively
3. The appropriate attitudes on the part of the team members (about themselves and the team) that foster effective team performance

Knowledge-based competencies refer to what team members think during teamwork performance. According to Salas and Cannon-Bowers (2001), knowledge-based competencies refer to the necessity of understanding facts, concepts, relations, and underlying foundations of information that a team member must have to perform a task. The authors cite cue-strategy associations, teammate characteristics, accurate and shared mental models, and task sequencing as examples of knowledge-based competencies. Salas (2005) suggested that team knowledge competencies also include accurate task models, team orientation, and shared-task models.

Skill-based competencies refer to the things that team members *do* during teamwork performance and are the necessary behavioral sequences and procedures required during task performance (Salas and Cannon-Bowers, 2001). Examples of skill-based competencies include adaptability, situational awareness, communication, coordination, and decision making (Salas, 2005; Salas and Cannon-Bowers, 2001).

Attitude-based competencies refer to what team members *feel* during teamwork performance and are those affective competencies that are required during task performance. Examples of attitude-based competencies include motivation, mutual trust, shared vision, collective orientation, team orientation, teamwork efficacy, and team cohesion (Salas, 2005; Salas and Cannon-Bowers, 2001). A breakdown of team competency classifications is presented in Table 2.1.

The majority of human-factors-related research into team competencies has been conducted within the area of team training. *Team training* is defined as "a set of tools and methods that form an instructional strategy in combination with requisite competencies and training objectives" (Salas and Cannon-Bowers, 1997). Further, Salas and Cannon-Bowers pointed out that team training consists of strategies derived from integrating a set of outputs from tools such as team task analysis, a delivery method of some sort, and content (e.g., knowledge, skills, and abilities). For team training to be effective, an understanding of the nature of team performance is required. Salas, Fowlkes, Stout, Milanovich, and Prince (1999) suggested that team training should focus on the team competencies identified for specific types of team's organizations. Salas and Cannon-Bowers (1997) suggested that an understanding of the KSAs that a team must possess for effective performance is required and that team training interventions should be designed on the basis of the KSAs identified. The primary goal of team training is to develop competencies to allow effective synchronization, coordination, and communication among team members; the competencies or KSAs provide the foundation for training goals and objectives (Salas, 2005). Four critical elements of team training identified by Salas (2005) are (1) determining the team training requirements, (2) delineating required KSAs, (3) selecting the appropriate instructional delivery method, and (4) designing the team training strategy.

Cannon-Bowers et al. (1995) proposed a framework for identifying the appropriate competencies to be included within team training programs. Team competencies can be specific to a particular task or team, or they may be generic in that they can be applied to across-team and across-task types. In the framework presented in Paris, Salas, and Cannon-Bowers (2000), task-specific competencies (e.g., interactions required for a task, knowledge of specific role responsibilities) involve performing teamwork behaviors for a specific task or situation, whereas team-specific competencies (e.g., knowledge of teammates' characteristics, team cohesion) relate to a specific team and influence the performance of that team only. Task-generic competencies (e.g., planning skills, interpersonal skills), on the other hand, are transportable and can be used for other tasks, and team-generic competencies (e.g., communication skills) are transportable from one team to another and can influence the performance of any team on which an individual serves.

Team-competency-related research has taken place mainly within the realm of civil aviation with a view to identifying and training the competencies required for safe and efficient flight-crew performance. This form of competency-based training is known as Crew Resource Management (CRM) training. CRM training programs were developed in response to analyses of aviation accident data which demonstrated that many accidents are caused not by lack of technical proficiency but by lack of coordination among team members or underutilization of team members' resources (Salas et al., 1999a). *Crew Resource Management* is formally defined as "using all available resources—information, equipment, and people—to achieve safe and efficient flight operations" (Lauber, 1984). According to Helmreich and Foushee (1993), CRM includes "optimising not only the person–machine interface and the acquisition of timely, appropriate information, but also interpersonal activities including leadership, effective team formation and maintenance, problem solving, decision making, and maintaining situation awareness." Salas et al. (1999b) defined CRM as "a set of teamwork competencies that allow the crew to cope with the situational demands that would overwhelm any individual crew member."

Inherent within CRM programs is a focus on enhancing collaborative activity, or teamwork, to improve performance. CRM programs have been applied successfully in the civil and military aviation domains. Salas et al. (1999a) reported that studies have consistently demonstrated that CRM training can improve performance by 6 to 20% and that CRM training is a viable strategy for improving teamwork in the cockpit. Despite their origin within the aviation domain, CRM training programs have also been applied in a number of additional domains, including offshore oil (O'Connor and Flin, 2003), health care (Howard, Gaba, Fish, Yang, and Sarnquist, 1992), helicopter rescue (Schmeiser, Bömmel, and Bühren, 2000), and air traffic control (Smith-Jentsch, Baker, Salas, and Cannon-Bowers, 2001).

Prince and Salas (1993) outlined seven different behaviors to be trained within CRM training programs, including:

1. *Communication*—Communication skills ensure that information about plans, problems, and observations is communicated clearly. When communication is unclear, clarification is requested. Communication is acknowledged.
2. *Assertiveness*—Assertiveness is the ability to communicate opinions, make suggestions, raise questions, and advocate for a position in an effective way, regardless of the relative seniority of those involved in the interaction.
3. *Mission organizing and planning*—Mission planning is the ability to develop plans and strategies for the mission, assign tasks, and review plans throughout the flight.
4. *Decision-making*—Decision-making requires the ability to gather information, evaluate alternative actions, and anticipate consequences.
5. *Leadership*—Leadership skills are used to assign tasks, establish procedures, communicate plans, review plans, and reassign tasks dynamically.

TABLE 2.1
Team Competencies

Nature of Team Competency	Description of Team Competency	Knowledge	Skills	Attitudes
Context driven	Team-specific Task-specific	Cue-strategy associations Task-specific teammate characteristics and responsibilities Shared task models Team mission objectives, norms, resources	Task organization Mutual performance monitoring Shared problem-model development Flexibility Compensatory behavior Information exchange Dynamic reallocation of functions Task structuring Task interaction Motivation of others	Team orientation (morale) Collective efficacy Shared vision
Team contingent	Task-specific Task-generic	Team-mate characteristics Team mission, objectives, norms, resources Relationship to larger organization	Conflict resolution Motivation of others Information exchange Intra-team feedback Compensatory behavior Assertiveness Planning Flexibility Morale building Cooperation	Team cohesion Interpersonal relations Mutual trust

Task contingent	Team-generic Task-specific	Task-specific role responsibilities Team role and interaction patterns Procedures for task accomplishment Accurate task models Accurate problem models Boundary spanning role Cue-strategy associations	Task structuring Mission analysis Task sequencing Mutual performance monitoring Compensatory behavior Information exchange Intra-team feedback Assertiveness Flexibility Planning Task interaction Situational awareness	Task-specific teamwork attitudes
Transportable	Team-generic Task-generic	Teamwork skills	Morale building Conflict resolution Information exchange Task motivation Cooperation Consulting with others Assertiveness	Collective orientation Belief in importance of teamwork

Source: Salas, E. et al., *International Journal of Management Reviews*, 2(4), 339–356, 2000. Used with permission of John Wiley & Sons.

The leader should establish an environment in which feedback and opinions are elicited from all members.

6. *Adaptability*—Adaptability is important if plans must be changed because of changed circumstances. It involves the ability to be receptive to other ideas and to flexibly change behavior.
7. *Situation awareness*—Situation awareness refers to the capacity to extract relevant information from the environment, understand it, and use it to anticipate future events. Teams should have shared situation awareness.

Cannon-Bowers et al. (1995) outlined a core set of teamwork skills that should be considered when designing CRM training programs: adaptability, shared situational awareness, performance monitoring and feedback, leadership, interpersonal skills, coordination skills, communication, and decision making.

2.4.3 Identifying Competencies

Various ways of identifying competencies have been described within the literature (e.g., McIntyre and Salas, 1995). These include observation, interviews, questionnaires, verbal protocol analysis, and team task analysis approaches. Salas and Cannon-Bowers (1997) recommended that the development of team training programs should involve identification of the skills required for team task performance and identification of the component KSAs required. Following this, methods for measuring the teamwork skills should be selected and then methods for training the KSAs required should be identified. Cannon-Bowers et al. (1995) also suggested that defining team training requirements involves identifying the competencies required by the team in question. Salas et al. (1999b) proposed a series of guidelines for the design and delivery of CRM training. These guidelines are summarized in Table 2.2.

The application of CRM training programs has not been limited to the aviation domain. Indeed, there is no theoretical reason why CRM training cannot be applied in domains other than aviation (Helmreich, Wiener, and Kanki, 1993). As generic teamwork skills are fundamental to CRM programs, it is feasible that CRM training could potentially be applied in any domain involving collaborative or teamwork activity. In the past, CRM training programs have been applied across a wide range of domains, including offshore oil (O'Connor and Flin, 2003), medicine (Howard, Gaba, Fish, Yang, and Sarnquist, 1992), helicopter mountain rescue (Schmeiser, Bömmel, and Bühren, 2000), air traffic control (Smith-Jentsch et al., 2001), maritime operations (Barnett, Gatfield, and Pekcan, 2004; Byrdorf, 1998), nuclear power (Flin and O'Connor, 2001), and the rail safety domain.

2.4.4 Competence Management Systems

As noted earlier, the process of identifying, training, assessing, and maintaining staff competence is known as *competence management.* Competence management systems are used by organizations to ensure that their staff possess and maintain the appropriate competencies required for efficient system performance and safety. The HSE guide for managing competence in safety-related systems suggests that to

TABLE 2.2
Methodology for the Development and Delivery of a CRM Training Program

Step	Procedure
Identify operational requirement.	Review existing training materials, including course material lists, instructor guides, standard operating procedures. Interview subject matter experts. Observe crews performing missions. Review mishaps/accident reports.
Assess team training needs and coordination demand.	Using material from previous step, specify all tasks involving teamwork and identify deficiencies in existing team training procedures.
Identify teamwork competencies and KSAs.	Link training needs to competencies required for performing the team tasks identified in previous step.
Determine team training objectives.	For each KSA, determine objectives that can be empirically evaluated.
Determine instructional delivery method.	Determine delivery method (i.e., information, demonstration, or practice) and feedback.
Design scenario exercises and create opportunities for practice.	Design scenarios to give trainees an opportunity to demonstrate KSAs.
Develop performance assessment/measurement tools.	Develop reliable and valid measures that can assess whether each of the KSAs was demonstrated.
Design and tailor tools for feedback.	Design tools to be used during debriefing to measure team behaviors exhibited and also to diagnose poor performance and provide guidance.
Evaluate the extent of improved teamwork in the cockpit.	Design experiments to assess effectiveness of the training.

Source: Adapted from Salas, E. et al., *Human Factors*, 41, 161–172, 1999.

ensure an effective and consistent standard of competence in individuals and teams a continuous process of improvement is required. In their guidance for developing and maintaining staff competence document, the HSE presents the CMS cycle as shown in Figure 2.5.

Flin (2005) conducted a case study of competence assurance in the civil aviation, nuclear power generation, and offshore oil production domains. Using interviews to generate descriptions of the licensing, competence assurance, and revalidation procedures in these domains, she found that the Civil Aviation Authority (CAA) has a well-developed competence assurance system for pilots and aviation engineers that is based on training, qualifications, and licensing. Among other things, the pilots have competency checks every 6 months which involve an assessment of their flying skills using an observation and rating procedure. Standards for performance are clearly stipulated, and these form the foundation for training and assessment. The focus of these standards is on both nontechnical (i.e., cognitive and social) and technical skills, and there is a range of safety monitoring systems in place that complement and strengthen the competence assurance approach. Within the nuclear power generation

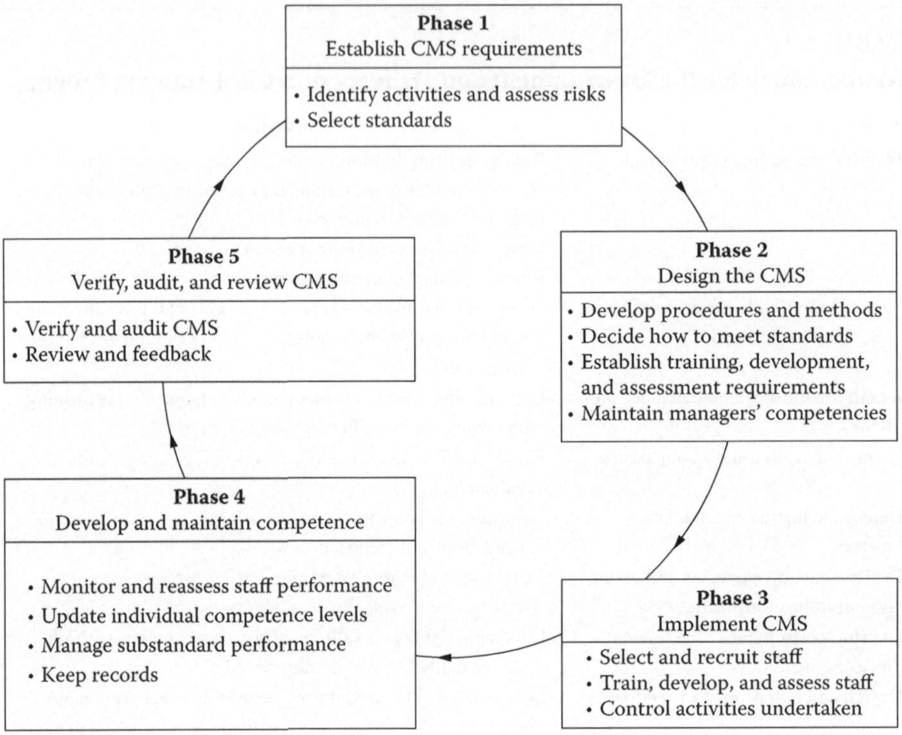

FIGURE 2.5 Competence management cycle. (Adapted from Health and Safety Executive, *Developing and Maintaining Staff Competence*, 2nd ed., HSE Books, London, 2007.)

domain, Flin found that it is the operators' responsibility to design and implement competence assurance systems for safety-critical roles. Unit desk managers undertake a formal competence check involving an observation of performance of a simulated task and an interview every 2 years. Competence assurance and authorizations are built into an organization's safety management systems and are linked to task analyses and risk calculations. Again, the focus is on both nontechnical and technical skills, and safety and job performance monitoring systems are used to strengthen the competence assurance system in place. For the offshore oil industry, responsibility for competence assurance is given to the operator in question. The competence assessment and authorizations are also built into an organization's safety management systems and are linked to task analyses and risk calculations. Competence assessment varies between companies, but Flin found general adherence to industry guidelines. One company's approach was for the offshore installation manager to take control of an emergency scenario in a simulator and be assessed by two assessors, an exercise that takes place on a 3-year basis. Safety and job performance monitoring systems are used to strengthen the competence assurance system in place.

Based on this review, Flin (2005) specified nine general principles for competence assessment that may be useful for licensing and revalidation in the healthcare profession:

1. *Independent regulator*—In the three domains analyzed, the function of awarding and revalidating licenses to operate sites is managed by an independent regulatory body (HSE for nuclear and oil domains and CAA for the aviation industry).
2. *Regular, formal proficiency checks*—Each of the three domains analyzed uses regular, formal proficiency checks, with periods ranging from 6 months to 3 years. Attention is given to maintaining competence between assessments.
3. *Clearly defined standards of competence*—Competence assessments in all three domains are based on detailed standards of competence derived from task analyses.
4. *Trained and accredited assessors*—Within the aviation industry, assurance procedures are also used to check the competency of the authorized examiners, and it is recognized that assessment programs should consider the reliability and validity of the evaluation process. This means that assessors need to be trained and calibrated against the standards in question.
5. *Nontechnical skills*—The aviation and nuclear power companies are concerned with the training and assessment of nontechnical (e.g., cognitive and social) skills, including situation awareness, decision making, leadership, teamwork, and understanding personal limitations.
6. *Failure of competence assessment*—Within the three domains, clear procedures are in place for reassessment and retraining in the event that an individual fails a competence check.
7. *Simulators*—In all three domains, simulators were used as part of the competence assessment process.
8. *Health checks*—In all three organizations analyzed systems were in place to ensure physical and mental fitness for duty.
9. *Linkage of competence assessment to safety management*—Within the oil and nuclear domains the competence assessment processes were integrated into overall competence assurance systems (e.g., selection, training, employee support).

2.5 GUIDANCE AND RECOMMENDATIONS

The effective management of staff competence levels is inherently based on the organization in question having a clear understanding of the competencies required for the various jobs and roles involved in the operation of their plant. This clear and specific breakdown of the competencies required for each job or role can then be used to inform the recruitment of appropriate staff, the development of staff training programs, the assessment of competence levels, and the development of remedies for substandard competence levels. An approach to competence management is presented in Figure 2.6.

Within the approach presented in Figure 2.6, it is recommended that organizations begin by developing a specific breakdown of the competencies that individuals and teams require to undertake each job or role efficiently. It is recommended that task analysis procedures, such as hierarchical task analysis (HTA), are used to identify

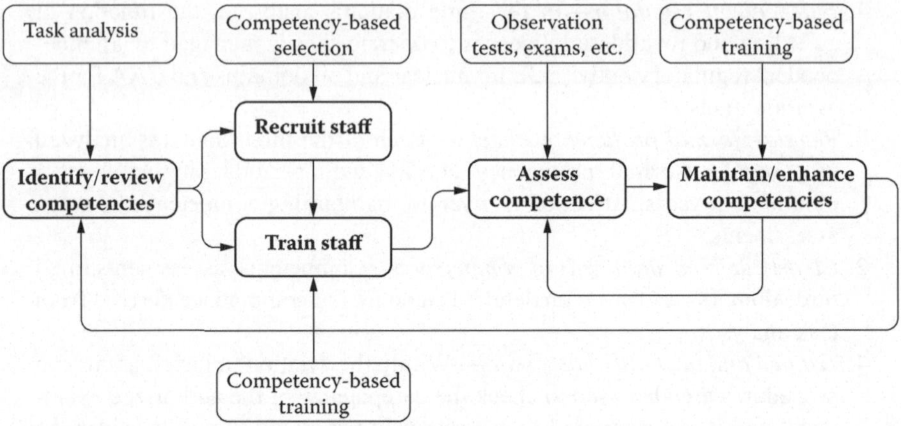

FIGURE 2.6 Competence management approach.

the competencies associated with each job or role (see Assessment and Evaluation section for procedure and advice on HTA). The competencies for each job or role should be extracted from the associated HTA in the form of the KSAs required for efficient task performance. Once the organizations have a clear description and understanding of the competencies required for each job or role they can use this information to recruit staff with appropriate competence levels or develop suitable training programs for existing staff members. Staff competence levels should then be regularly assessed via methods such as observation, aptitude tests, and written and oral exams (see assessment and evaluation section) and maintained appropriately. In the event of substandard levels of competence, measures should be taken to bring staff competence to the required levels (e.g., retraining). Finally, also of importance within a competence management process is the continual assessment of the competencies required for each job or role. The introduction of new technologies or procedures may serve to change the level of staff competence required for a particular task, so it is important that organizations continually assess the competence levels required for each job or role in addition to the competence levels of the staff undertaking them.

2.6 ASSESSMENT AND EVALUATION

2.6.1 Competence Identification

Most organizations use task analysis procedures to identify the competencies required for different jobs and roles. Hierarchical task analysis (Annett, 2005; Stanton, 2006) is the most commonly used task analysis approach and is essentially a descriptive task analysis technique that is used to describe systems, goals, and tasks. HTA works by decomposing activities into a hierarchy of goals, subordinate goals, operations, and plans that allows systems and tasks to be described exhaustively. HTA outputs therefore specify the overall goal of a particular system, the subgoals required to achieve this goal, and the operations required to achieve each of the subgoals specified.

The HTA process is relatively straightforward, involving collecting data about the task or system under analysis—through techniques such as observation, questionnaires, interviews with subject matter experts (SMEs), walkthroughs, user trials, and documentation review, to name but a few—and then using these data to decompose and describe the goals, subgoals, and tasks involved. The HTA procedure is presented in Figure 2.7, and an example HTA output is presented in Figure 2.8 For the identification of competencies, each subordinate goal within the HTA can be decomposed based on the knowledge, skills, and attitudes required for its successful completion. From Figure 2.8, for example, Subgoal 1.5, "Discuss planned/requested operations that may affect system control," could be decomposed into the following KSAs:

- Knowledge of planned/requested operations that may affect system control
- Knowledge of system control process
- Ability to determine how planned/requested operations might affect system control
- Ability to articulate the effects of planned/requested operations on system control
- Safety orientation
- Team cohesion
- Mutual trust

2.6.2 Competence Assessment

Competence assessment is used to ensure that individuals are capable of carrying out tasks safely and efficiently. Fletcher (2000) presented a standard approach to competence assessment (see Figure 2.9) that involves first identifying the competence levels required by defining performance criteria of some sort and then collecting evidence of the outcomes of an individual's task performance and matching those outcomes with the specified outcomes (performance criteria). On the basis of this comparison, a judgment should then be made regarding achievement of the required performance outcomes, and a "competent" or "not yet competent" rating should be allocated.

It is also important to consider the competence of the assessors who are undertaking the competence assessments. Wright et al. (2003) highlighted the importance of assessors understanding the tasks and functions, as well as the assessment process, and providing impartial opinions. There are various methods available to organizations wishing to undertake competence assessments. Wright et al. (2003) reviewed the methods available, including observational assessment, task performance assessment, oral questioning, written exams, multiple choice exams, aptitude tests, psychometric tests and assessment, and development centers. A summary of their findings is presented in Table 2.3. In developing an appropriate competence assessment procedure, it is necessary to identify the following factors (Wright et al., 2003):

- What are the competencies that must be assessed?
- What assessment criteria and competence standards are required?
- What methods of assessment and reassessment are required to acquire evidence of operator competence?
- What qualifications and competence do the assessors need to possess?

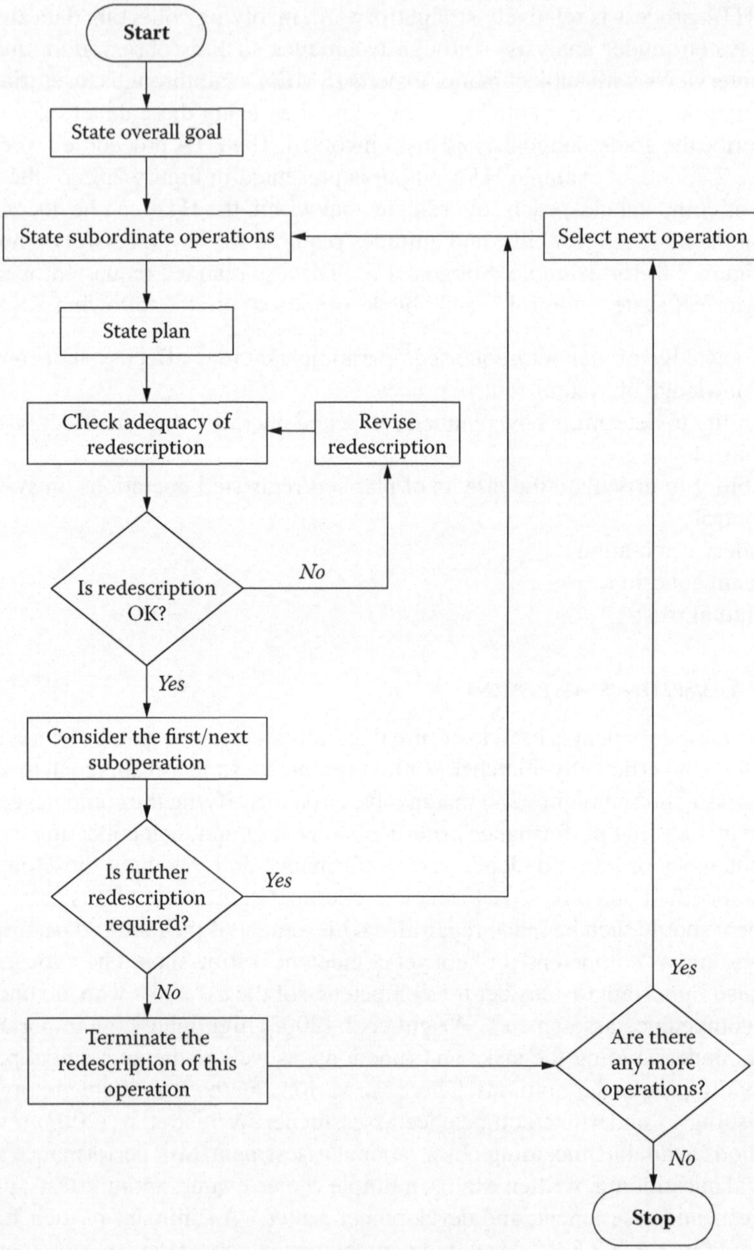

FIGURE 2.7 Hierarchical task analysis (HTA) procedure.

- How often should competence assessment take place?
- What is to be done in the event of operator incompetence?

A flowchart depicting a generic competence assessment procedure is presented in Figure 2.10. A brief overview of the procedure is given below.

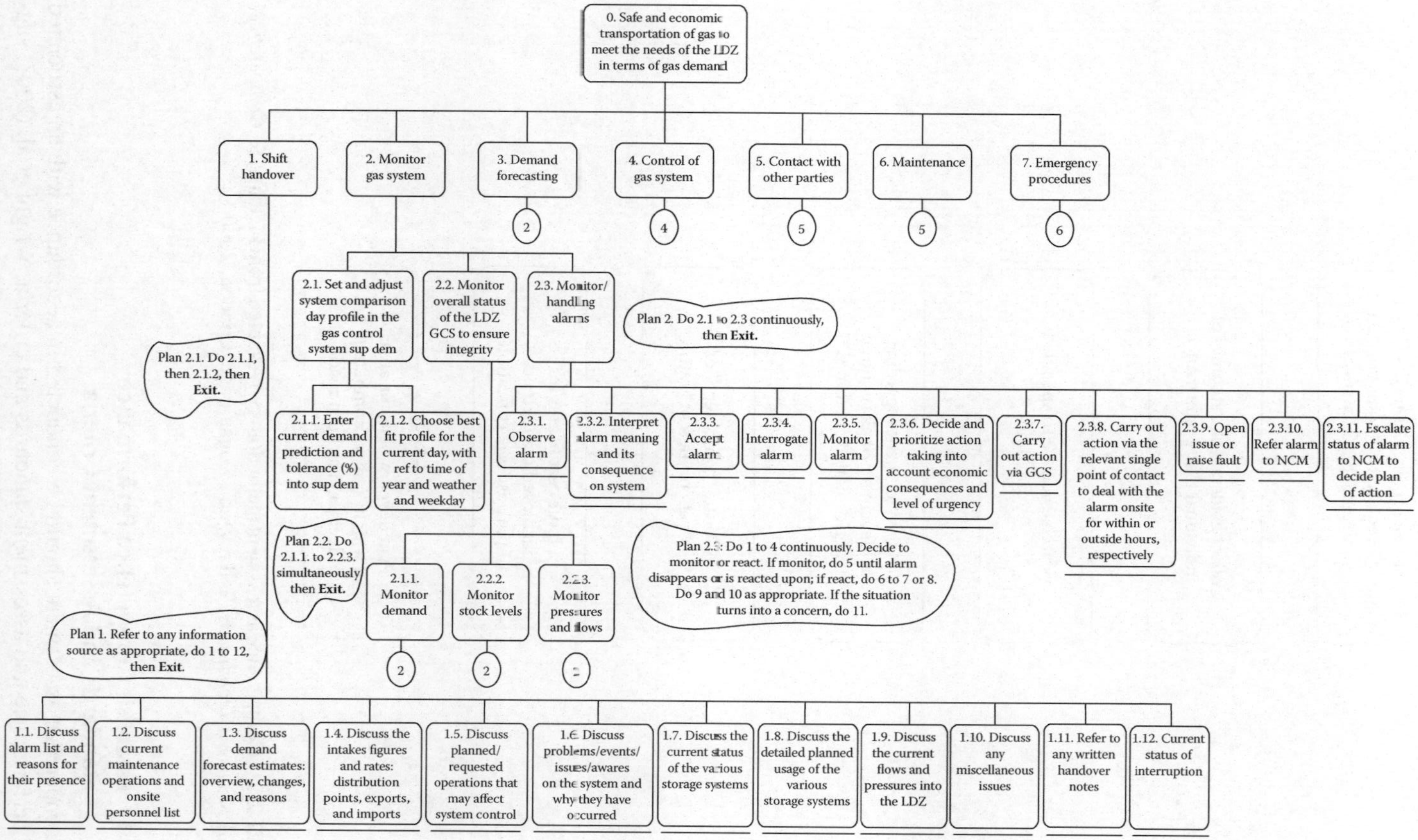

FIGURE 2.8 Control Engineer HTA extract.

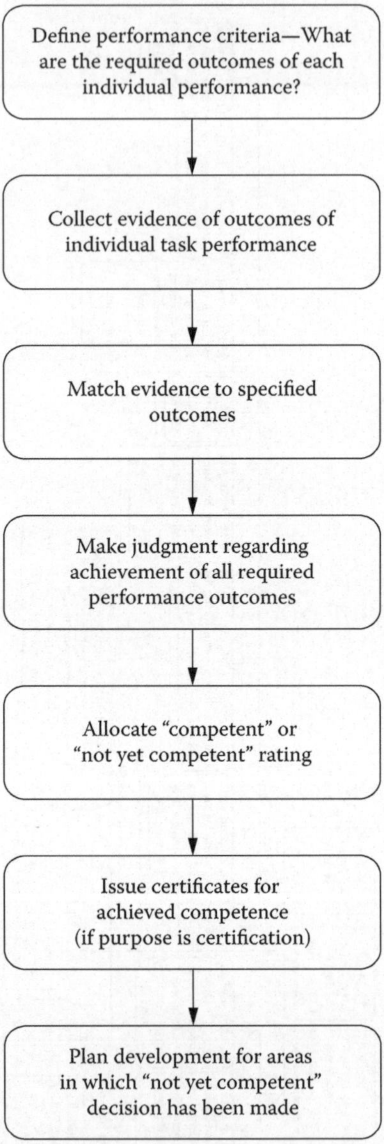

FIGURE 2.9 Competence assessment methodology. (Adapted from Fletcher, S., *Competence-Based Assessment Techniques*, 2nd rev. ed., Kogan Page, London, 2000.)

2.6.2.1 Establish Competence Performance Standards and Assessment Criteria

Any competence assessment should be conducted in accordance with an associated set of performance and assessment standards and criteria. Wright et al. (2003) suggested that a competence standard should provide a testable statement or description

TABLE 2.3
Competence Assessment Methods

Method	Main Strengths	Main Weaknesses
Direct observation: Involves observing individuals undertaking their jobs	Direct evidence of task performance is provided. Assessment can be undertaken as part of the job. Individuals are familiar with assessment method. Continuous assessment is possible. Evidence is produced, regardless of assessment.	Opportunity to demonstrate full range of competencies is limited. Assessors may become overloaded. Assessor/assessee relationships can affect results. Observation may be disruptive to the workforce. Observation can be resource intensive (e.g., expensive and time consuming). Trained assessors are required.
Specially set tasks: Includes skills, tests, simulations, projects, and assignments; typically used when observing normal work activities is not possible	Provided appropriate tasks are used, such tasks generate evidence of staff competence. Tasks can be undertaken offsite (e.g., in a simulator), thus avoiding the working (and perhaps disruptive) environment. Test conditions can be standardized for skill sets. Time for testing can be effectively allocated.	It is questionable whether or not the tasks are representative of real working conditions. Individuals may act differently in a test environment. It is difficult to predict what type of evidence will be generated.
Oral questioning: Can be used to supplement task performance evidence by querying respondent about performance in a difference context	Evidence on the full range of job activities can be collected. An individual's underpinning knowledge can be determined. Questioning can be rigorous and standardized.	Assessors may often ask their own questions. Questioning alone is insufficient to demonstrate operator competence. Questioning is less likely to reflect or represent real working conditions. Knowledge of task requirements is very different from performing a task efficiently.

(continued)

TABLE 2.3 (continued)
Competence Assessment Methods

Method	Main Strengths	Main Weaknesses
Written exam	Knowledge-based activities can be evaluated. Well-structured exams can elicit key areas of knowledge and understanding. Evidence on the full range of job activities can be collected. An individual's underpinning knowledge can be determined.	Exam reflects an individual's ability to write and construct written materials. Skilled assessors for marking are required. Exam can lack validity. This is not a quick approach, due to the time required to take the examination and grade it. Written exams alone are insufficient to demonstrate operator competence. Written exams are less likely to reflect or represent real working conditions.
Multiple-choice examination	Questions can be standardized. Questions can quickly elicit key knowledge and understanding. Knowledge-based activities can be evaluated. An individual's underpinning knowledge can be determined.	Guessing may be an issue. Careful design is necessary. Time away from the job is required.
Aptitude tests	Tests are quick and easy to administer. Well-designed questions can be standardized.	Guessing may be an issue. Careful design is necessary. Aptitude tests do not provide direct evidence of job performance. Time away from the job is required. Individuals may prepare for the tests and may also behave differently during the test.

Psychometric tests	Tests are quick and easy to administer. Tests are especially useful for well-established tasks with well-established performance criteria.	Tests must be appropriate. Tests may lack validity. Tests must be designed carefully. Psychometric tests measure characteristics that influence behavior but not how a person behaves when using these characteristics. Time away from the job is required. Individuals may give false answers to pass.
Assessment and development centers	High-quality results are obtained. Many different approaches can be used.	Centers are costly and resource intensive (e.g., time spent at center). Significant time away from the job is required. It is necessary to use various different assessment approaches, each with its own flaws.

Source: Adapted from Wright, M. et al., *Competence Assessment for the Hazardous Industries*, Research Report 086, HSE Books, London, 2003.

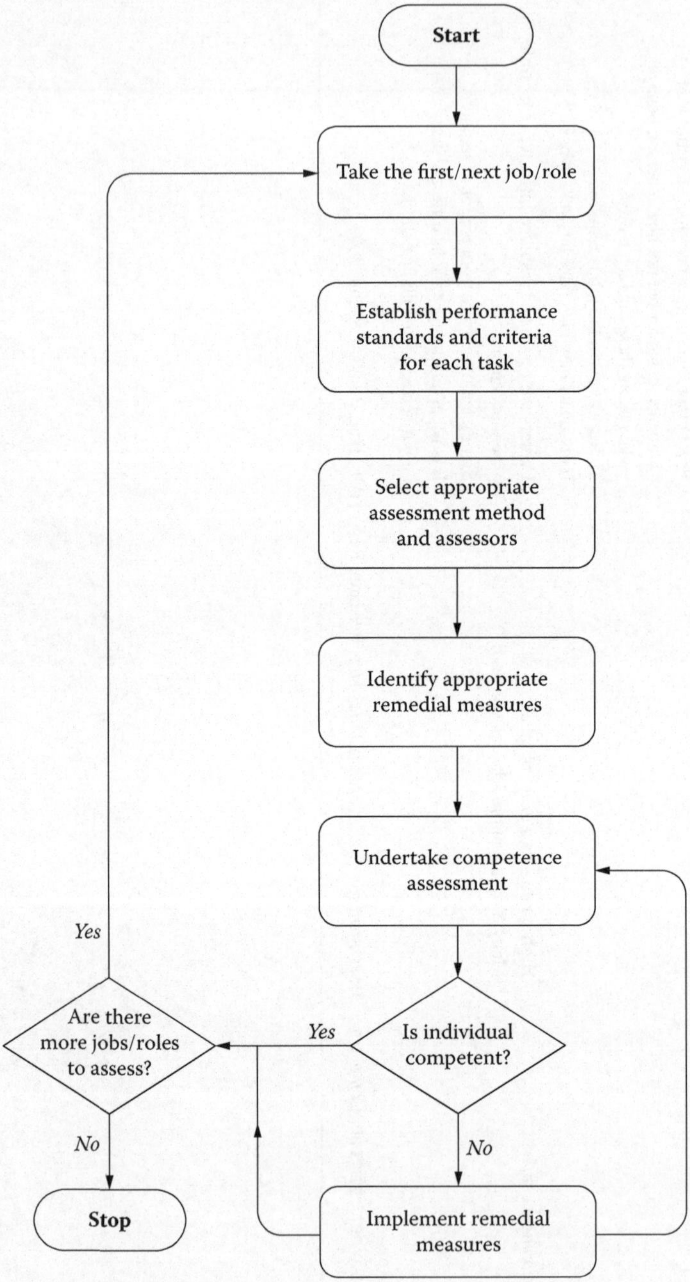

FIGURE 2.10 Competence assessment flowchart.

of competence, define measurable criteria for judging performance, specify what level of performance evidence is required, and reflect the type of competence in question (e.g., skills vs. knowledge vs. behaviors). Competence standards tend to cover skills (i.e., the ability to perform a certain procedure), knowledge (i.e., an understanding of a task or its associated hazards), and safety behaviors and attitudes. A number of examples of competence standards and assessment criteria are given below (Wright et al., 2003):

- Process control room supervisors need to successfully manage a simulated emergency response in three tests (out of a possible set of ten scenarios) and display appropriate behaviors such as task delegation and effective communication.
- Safety engineers must be able to correctly interpret a piping and instrumentation diagram, identify all (contrived) engineering defects, and specify safety devices and engineering modifications as noted in company standards.
- A maintenance technician should be able to recollect all key safety actions required in the isolation of a hydrocarbon pump, its dismantling, and its restoration.

2.6.2.2 Select Appropriate Competence Assessment Methods

The next step involves selecting appropriate methods for assessing each competency identified. Wright et al. (2003) provided a guide for the most appropriate assessment methods to use for each competency type. An overview of this is presented in Table 2.4.

2.6.2.3 Select Remedial Measures and Undertake Competence Assessments

Once appropriate competence assessment methods are selected, remedial measures (e.g., training) should be developed for use in the event of inadequate staff competence levels. Following this, competence assessments should be undertaken using the approaches selected.

2.7 EXAMPLE

The following example is taken from an analysis of the competencies required for a Control Engineer in our example domain. Based on observational study and interviews with subject matters experts, hierarchical task analysis of the Control Engineer job was conducted (see Figure 2.8). The HTA was then decomposed into the KSAs required for each role. Extracts of the KSAs required for each task and a recommendation on the assessment methods to use during competence assessments are presented in Table 2.5 and Table 2.6.

TABLE 2.4
Competence Assessment Method Selection

Competence Type	Assessment Methods	Examples
Physical and sensory motor skills (e.g., visual, auditory, touch)	Practical "show me" tests Simulated tasks or mock-ups Peer review of quality of work Evidence of prior experience	National Vocational Qualification (NVQ)-based assessment Driving skill road tests Workshop-based tests of welding ability on a mocked-up item of equipment
Ability to carry out a prescribed procedure of work	Simulation exercises Pen and paper tests Talk throughs Shadowed work Peer-assessed decision making Post-task briefing (verbalization of decisions)	Classroom verbal test of candidate's recollection of a work procedure Talking through the correct procedure of isolating a hydrocarbon pump using a piping and instrumentation diagram Observation of a fitter following the procedure for installing a seal on a hydrocarbon pump, checking operability, and advising that it is safe to restart
Cognitive skills	Simulation exercises Pen and paper tests Shadowed work Peer-assessed decision making Post-task briefing (verbalization of decisions) Peer observation and feedback Psychometric tests	Performance of a sample of tasks on a control room simulator Talking through the interpretation of a set of alarms Completion of SHL aptitude tests

Knowledge of equipment, plant and processes	Talk throughs Post-task briefing (verbalization of decisions) Verbal knowledge tests by experts	Verbal or written examination of an individual's knowledge of safety functions and various items of equipment, including formal qualifications or tests Explanation of how a chemical reaction may go exothermic due to process deviations
Interpersonal skills	Shadowed work Peer observation and feedback Group exercises Self-assessment questionnaires and psychometric tests	Observation of behavior using behavioral markers in real or simulated activities Self-completion of psychometric tests
Team management skills	Peer observation in real or simulated tasks Self-assessment questionnaires and psychometric tests	Observation of behavior using behavioral markers in real or simulated activities Self-completion of psychometric tests
Safety attitudes and behaviors	Peer observation in real or simulated tasks Verbal tests	Observation of behavior using behavioral markers in real or simulated activities Statement about the appropriate way of responding to, for example, conflicting operational and safety requirements Verbal examination of supervisors' understanding of how their behavior influences the safety climate

TABLE 2.5
Control Engineer Competence Assessment Extract, Part 1

Task	Knowledge	Skills	Attitudes	Recommended Assessment Method
Shift handover objective: Provide a snapshot overview of the system and prepare the next shift for any ramifications from the last shift.	Knowledge of the shift handover procedure Knowledge of the different items contained within the shift handover procedure Knowledge of the current system status	Ability to undertake shift handover procedure Ability to identify status of system and shift handover items Ability to articulate current system status in a form amenable for the shift handover	Safety Team cohesion Mutual trust	Observation Simulated exercises Shadowed work Talk throughs

TABLE 2.6
Control Engineer Competence Assessment Extract, Part 2

	Task	Knowledge	Skills	Attitudes	Recommended Assessment Method
1.1	Discuss the alarm list and reasons for their presence.	Knowledge of plant alarms and possible reasons for their presence Knowledge of different plant states	Ability to discriminate among various alarm types and states Ability to identify reason for presence of alarms	Safety Team cohesion Mutual trust Collective orientation Teamwork	Talk throughs Observation Verbal knowledge tests by experts Simulated exercises Pen and paper tests Shadowed work
1.2	Discuss current maintenance operations and onsite personnel list.	Knowledge of current maintenance operations Knowledge of maintenance procedures Knowledge of onsite personnel list	Ability to identify current maintenance operations Ability to read onsite personnel list	Safety Team cohesion Mutual trust Collective orientation Teamwork	Talk throughs Observation Verbal knowledge tests by experts Post-task debriefing (verbalization of decisions)

1.3	Discuss the demand forecast estimates: overview, changes, and reasons.	Knowledge of demand forecast estimates Knowledge of demand forecast changes and reasons for their changes	Ability to read demand forecast estimates Ability to identify demand forecast estimate changes Ability to provide an overview of demand forecast estimates	Safety Team cohesion Mutual trust Collective orientation Teamwork	Talk throughs Observation Verbal knowledge tests by experts Post-task debriefing (verbalization of decisions)
1.4	Discuss the intake figures and rates: distribution points and exports/imports.	Knowledge of intake figures and rates Knowledge of distribution points Knowledge of exports/imports	Ability to read intake figures Ability to read distribution points Ability to read exports/imports Ability to discuss intake figures and rates: distribution points, exports, and imports	Safety Team cohesion Mutual trust Collective orientation Teamwork	Talk throughs Observation Verbal knowledge tests by experts Post-task debriefing (verbalization of decisions) Simulated exercises Shadowed work
1.5	Discuss planned and requested operations that may affect system control (e.g., NROs).	Knowledge of planned and requested operations that may affect system control Knowledge of system control process	Ability to determine how planned and requested operations may affect system control Ability to articulate the effects of planned and requested operations on system control	Safety Team cohesion Mutual trust Collective orientation Teamwork	Talk throughs Observation Verbal knowledge tests by experts Post-task debriefing (verbalization of decisions)
1.6	Discuss problems, events, and issues regarding the system and why they have occurred.	Knowledge of different problems, events, and issues and how they might affect the system Knowledge of problems, events, and issues regarding the system	Ability to determine how problems, events, and issues might affect the system Ability to articulate the effects of problems, events, and issues on the system	Safety Team cohesion Mutual trust Collective orientation Teamwork	Talk throughs Observation Verbal knowledge tests by experts Post-task debriefing (verbalization of decisions)

BIBLIOGRAPHY

Annett, J. (2005). Hierarchical task analysis. In N.A. Stanton, A. Hedge, K. Brookhuis, E. Salas, and H. Hendrick (Eds.), *Handbook of Human Factors and Ergonomics Methods* (pp. 33.1–33.7). Boca Raton, FL: CRC Press.

Barnett, M., Gatfield, D., and Pekcan, C. (2004). A Research Agenda in Maritime Crew Resource Management. Paper presented at the National Maritime Museum Conference on Safer Ships, Safer Lives, London, March 2004 (http://www.he-alert.com/documents/published/HE00240.pdf).

Brabazon, P. and Conlin, H. (2001). *Assessing the Safety of Staffing Arrangements for Process Operations in the Chemical and Allied Industries*. London: HSE Books.

Byrdorf, P. (1998). Human Factors and Crew Resource Management: An Example of Successfully Applying the Experience from CRM Programmes in the Aviation World to the Maritime World. Paper presented at the 23rd Conference of the European Association for Aviation Psychology, Vienna, September 1998.

Cannon-Bowers, J.A. and Salas, E. (2006). Team effectiveness and competencies. In W. Karwowski (Ed.), *International Encyclopedia of Ergonomics and Human Factors*, Vol. 2, 2nd ed. (pp. 2379–2383). Boca Raton, FL: Taylor & Francis.

Cannon-Bowers, J.A., Tannenbaum, S.I., Salas, E., and Volpe, C.E. (1995). Defining competencies and establishing team training requirements. In R.A. Guzzo and E. Salas (Eds.), *Team Effectiveness and Decision Making in Organizations* (pp. 333–380). San Francisco, CA: Jossey-Bass.

Fletcher, S. (2000). *Competence-Based Assessment Techniques*, 2nd rev. ed., London: Kogan Page.

Flin, R. (2005). *Safe in Their Hands? Licensing and Competence Assurance for Safety-Critical Roles in High Risk Industries*. London: Department of Health.

Flin, R. and O'Connor, P. (2001). Applying crew resource management on offshore oil platforms. In E. Salas, C.A. Bowers, and E. Edens (Eds.), *Improving Teamwork in Organizations: Applications of Resource Management Training* (pp. 217–233). Mahwah, NJ: Lawrence Erlbaum.

Health and Safety Executive (HSE). (2002). *Developing and Maintaining Staff Competence*, Railway Safety Publication 1. London: HSE Books.

Health and Safety Executive (HSE). (2003). *Competence Assessment for the Hazardous Industries*, RR086. London: Health and Safety Executive.

Health and Safety Executive (HSE). (2007a). *Managing Competence for Safety-Related Systems*. Part 1. *Key Guidance on Setting Up a Competency Management System Using 15 Guiding Principles*. London: Health and Safety Executive.

Health and Safety Executive (HSE). (2007b). *Developing and Maintaining Staff Competence*, 2nd ed., Railway Safety Publication 1. London: HSE Books.

Hedge, J.W. and Borman, W.C. (2006). Personnel selection. In G. Salvendy (Ed.), *Handbook of Human Factors and Ergonomics* (pp. 458–471). Upper Saddle River, NJ: John Wiley & Sons.

Helmreich, R.L. and Foushee, H.C. (1993). Why crew resource management? Empirical and theoretical bases of human factors training in aviation. In E.L. Wiener, B.G. Kanki, and R.L. Helmreich (Eds.), *Cockpit Resource Management* (pp. 3–45). San Diego, CA: Academic Press.

Helmreich, R.L., Wiener, E.L., and Kanki, B.G. (1993). The future of Crew Resource Management in the cockpit and elsewhere. In E. Wiener, B. Kanki, and R. Helmreich (Eds.), *Cockpit Resource Management* (pp. 479–501). San Diego, CA: Academic Press.

Howard, S.K., Gaba, D.M., Fish, K.J., Yang, G., and Sarnquist, F.H. (1992). Anaesthesia crisis resource management training: teaching anaesthesiologists to handle critical incidents. *Aviation, Space, and Environmental Medicine*, 63, 763–770.

Ivergard, T. (1989). *Handbook of Control Room Design and Ergonomics*. London: Taylor & Francis.

Lauber, J.K. (1984). Resource management in the cockpit. *Air Line Pilot*, 53, 20–23.

McIntyre, R.M. and Salas, E. (1995). Measuring and managing for team performance: emerging principles from complex environments. In R. Guzzo and E. Salas (Eds.), *Team Effectiveness and Decision Making in Organizations* (pp. 9–45). San Francisco, CA: Jossey-Bass.

O'Connor, P. and Flin, R. (2003). Crew resource management training for offshore oil production teams. *Safety Science*, 41, 591–609.

Paris, C.R., Salas, E., and Cannon-Bowers, J.A. (2000). Teamwork in multi-person systems: a review and analysis. *Ergonomics*, 43(8), 1052–1075.

Pascual, R., Mills, M., and Henderson, S. (2001). Teamworking. In M. Bransby and J.M. Noyes (Eds.), *People in Control: Human Factors in Control Room Design* (pp. 69–78). Stevenage, U.K.: Institution of Engineering and Technology.

Prince, C. and Salas, E. (1993). Training and research for teamwork in the military aircrew. In E. Wiener, B.G. Kanki, and R.L. Helmreich (Eds.), *Cockpit Resource Management* (pp. 337–366). San Diego, CA: Academic Press.

Salas, E. (2005). Team methods. In N.A. Stanton, A. Hedge, K. Brookhuis, E. Salas, and H. Hendrick (Eds.), *Handbook of Human Factors and Ergonomics Methods* (pp. 43-1–43-4). Boca Raton, FL: CRC Press.

Salas, E. and Cannon-Bowers, J.A. (1997). Methods, tools, and strategies for team training. In M.A. Quinones and A. Ehrenstein (Eds.), *Training for a Rapidly Changing Workplace: Applications of Psychological Research* (pp. 249–279). Washington, D.C.: American Psychological Association.

Salas, E. and Cannon-Bowers, J.A. (2001). Teamwork and team training. In J.J. Smelser and P.B. Baltes (Eds.), *International Encyclopedia of the Social and Behavioural Sciences* (pp. 15487–15492). Oxford, U.K.: Pergamon.

Salas, E., Burke, C.S., and Cannon-Bowers, J.A. (2000). Teamwork: emerging principles. *International Journal of Management Reviews*, 2(4), 339–356.

Salas, E., Fowlkes, J.E., Stout, R.J., Milanovich, D.M., and Prince, C. (1999a). Does CRM training improve teamwork skills in the cockpit? Two evaluation studies. *Human Factors*, 41, 326–343.

Salas, E., Prince, C., Bowers, C.A., Stout, R.J., Oser, R.L., and Cannon-Bowers, J.A. (1999b). A methodology for enhancing crew resource management training. *Human Factors*, 41, 161–172.

Schmeiser, G., Bömmel, T.V., and Bühren, V. (2000). Crew Resource Management (CRM): cooperation of the police helicopter crew at mountain missions. *Air Medical Journal*, 19, 118.

Smith-Jentsch, K.A., Baker, D.P., Salas, E., and Cannon-Bowers, J.A. (2001). Uncovering differences in team competency requirements: the case of air traffic control teams. In E. Salas, C.A. Bowers and E. Edens (Eds.), *Improving Teamwork in Organizations: Applications of Resource Management Training* (pp. 31–54). Mahwah, NJ: Lawrence Erlbaum.

Stanton, N.A. (2006). Hierarchical task analysis: developments, applications and extensions. *Applied Ergonomics*, 37, 55–79.

Vicente, K.J. (1999). *Cognitive Work Analysis: Towards Safe, Productive, and Healthy Computer-Based Work*. Mahwah, NJ: Lawrence Erlbaum.

Whiddett, S. and Hollyforde, S. (2007). *A Practical Guide to Competencies: How to Enhance Individual and Organisational Performance*. London: Chartered Institute of Personal Development.

Wright, M.S, Turner, D., and Horbury, C. (2003). *Competence Assessment for the Hazardous Industries*, Research Report 086. London: HSE Books.

3 Training

3.1 OUTLINE

Training is a critical factor in safe and efficient organizational performance. A skilled and prepared workforce can yield higher productivity and better service, improved quality, higher motivation and commitment, fewer errors, increased safety, and greater levels of teamwork and morale (Salas, Wilson, Priest, and Guthrie, 2006). Perhaps indicative of its importance, research suggests that in the United States alone, organizations spend an estimated $54.2 to $200 million annually on training and developing employees (Salas and Cannon-Bowers, 2001).

Training is formally defined as the "systematic acquisition of knowledge (i.e., what we need to know), skills (i.e., what we need to do), and attitudes (i.e., what we need to feel) that together lead to improved performance in a particular environment" (Salas et al., 2006, p. 473). On a simpler level, training is about "people getting the right competencies to do a job" (Salas et al., 2006, p. 473).

3.2 KEY ISSUES

Far more complex than it first appears, the area of training involves much more than merely designing and delivering training programs; notwithstanding the importance of these aspects, dealing with training includes ensuring that appropriate preconditions for learning are in place, undertaking training needs analyses (i.e., identifying training requirements, including who and what needs training), determining training program content, selecting appropriate methods of delivery, designing training programs, delivering training programs, setting appropriate conditions for effective transfer of training, and evaluating training program outcomes and the effectiveness of the training delivered. Organizations wishing to deliver successful training programs therefore need to carefully consider all aspects of training program design, delivery, and evaluation. A summary of the key issues involved is presented below.

3.2.1 Training Needs Analysis

The most first and most important step when designing training programs involves identifying the training needs of the organization. This is achieved by conducting so-called *training needs analysis* (TNA). TNA involves analyzing the organization, task, and trainees in order to identify exactly where training is required, who needs to be trained, and, in terms of the knowledge, skills, and attitudes (KSAs), what should be trained by a particular training program.

3.2.2 Training Preconditions

The conditions present within an organization before training delivery are of critical importance. Salas and Cannon-Bowers (2001), for example, suggested that in some cases the events that occur before training can be more important than those events that occur during and after training. The relevant preconditions include what trainees bring to a training setting (e.g., individual characteristics such as cognitive ability and self efficacy), variables that engage trainees to learn and participate (e.g., level of motivation), and how training is prepared (Salas and Cannon-Bowers, 2001).

3.2.3 Training Design

Designing training programs is an important element and involves more than simply providing trainees with information regarding the task or job in question. Training program design should take into account learning objectives, trainee characteristics, knowledge regarding learning processes, and such practical considerations as organizational constraints and resources available (Tannenbaum and Yukl, 1992). Particularly important is the selection of appropriate training content and training delivery methods.

3.2.4 Training Content

One of the most important issues faced when designing training programs is determining what the content will actually be. This is informed by the TNA component, which uses methods such as task analysis (e.g., HTA) and cognitive task analysis (e.g., CDM) to identify what it is that must be trained. However, translating the outputs of the TNA into actual training content is more difficult and involves determining the nature and amount of the training program content and also the structure and sequencing of the training content (Patrick, 1992).

3.2.5 Training Delivery Methods

Another important consideration during the training design process is the selection of appropriate training delivery methods. Many methods are available for delivering training programs within complex sociotechnical systems. Salas (2004) categorized the methods available into information-based methods (e.g., lectures, handbooks and presentations), demonstration-based methods (e.g., video demonstrations, multimedia systems), and practice-based methods (role-play exercises, high-fidelity simulation). The form of training method used is a key factor in the success of a training program, and careful consideration should be given to the training methods adopted.

3.2.6 Team Training

Due to the increased presence of teams in complex sociotechnical systems (Fiore, Salas, Cuevas, and Bowers, 2003), there is an increased requirement for effective team training strategies. *Team training* is defined as "a set of tools and methods that form an instructional strategy in combination with requisite competencies and training objectives" (Salas and Cannon-Bowers, 1997). Team training programs are used to develop the KSAs required for effective team performance; their primary goal is to develop competencies to allow effective synchronization, coordination, and communication between team members (Salas, 2004).

3.2.7 Transfer of Training

Transfer of training refers to the extent to which the KSAs trained by a particular training program are applied, generalized, and maintained over time within the job environment in question (Baldwin and Ford, 1988; Salas and Cannon-Bowers, 2001). It occurs when there is "effective and continuing application, by trainees to their jobs, of the knowledge and skills gained in training, both on and off the job" (Broad and Newstrom, 1992, p. 6). The area has received significant attention from training practitioners, and various transfer of training models exist (e.g., Baldwin and Ford, 1988; Kozlowski and Salas, 1997; Machin, 2002). Key issues relating to the concept of training transfer include how to encourage training transfer and how to measure the levels of training transfer achieved.

3.2.8 Training Evaluation

Due to the high level of resources invested, organizations need to justify their training programs in terms of improved organizational performance, profit, safety, and reduced errors (Salas and Cannon-Bowers, 2001). How organizations evaluate their training programs is therefore a key issue: Do the training programs used by organizations in the safety critical domains actually work? How do they determine whether or not they work? The Health and Safety Executive (HSE, 2008) stated that training should be validated (i.e., did it deliver what it was supposed to?) and evaluated (i.e., is this the right kind of training for our needs?). Although ostensibly an important question given the cost and importance of developing and delivering effective training programs, it appears that it is one that is often ignored by organizations, and the efficiency of training programs is often assumed rather than tested scientifically. Further, training evaluation is typically difficult, resource intensive, and expensive and often generates negative results (Salas et al., 2006).

3.3 STANDARDS AND GUIDELINES

A number of standards and guidelines regarding the competence levels of employees in the safety critical domains are available. A summary of selected examples is presented below.

3.3.1 *Managing Competence for Safety-Related Systems*

The Health and Safety Executive's *Managing Competence for Safety-Related Systems* guide provides guidance on the management of competence (for all industry sectors) to ensure that workers are suitably qualified and experienced for their work activities, roles, and responsibilities (HSE, 2007a). The guidance is aimed at those responsible for managing and ensuring the competence of individuals and teams working within safety-related organizations and describes the core requirements for a competence management system, including a breakdown of the different phases and component steps. The aim of the guide is to enable organizations to meet the U.K. legal requirements for competence in safety-related systems. The guide also includes a list of the regulations that contain requirements concerning competence for health and safety.

3.3.2 Developing and Maintaining Staff Competence

The Health and Safety Executive *Developing and Maintaining Staff Competence* guide is aimed at those responsible for managing the competence of individuals working within the railway and other guided transport system industries (HSE, 2007b). Essentially, the guide presents an overview of the areas of competence, competence management, and competence management systems (CMSs). It begins with a general introduction to the area of staff competence and then presents an overview of a generic CMS framework. The overall objectives of the guide include describing the principles and factors that should be considered in any CMS and explaining how organizations should go about ensuring that the competence of individuals and teams working for them satisfy the requirements of existing legislation. This document is likely to be particularly useful for any organization wishing to develop and implement a new CMS or redevelop an existing CMS.

3.3.3 Competence Assessment for Hazardous Industries

This report (HSE, 2003) presents an extensive overview of the area of competence assessment in the hazardous industries, including an introduction to the area of competence assessment, a review of current practice (including case studies), guidance on competence assessment for major accident prevention, and examples of competence assessment (including process plant maintenance fitter, emergency response manager, operations team leader, and field operator examples). The report also presents a summary of the different methods available for conducting competence assessments and an overall competence assessment framework. This document is likely to be particularly useful for any organization wishing to develop a new process for undertaking competence assessments.

Various domain-specific guidelines and standards state that organizations should identify the level of competence required by their employees and that employees should be competent in their role or job. For example, guidance for the 1998 Provision and Use of Work Equipment Regulations state that employers should identify the level of employee competence required to operate work equipment (Wright, Turner, and Horbury, 2003). Likewise, the 1998 Control of Substances Hazardous to Health Regulations state that employees must be competent in the use and handling of equipment and substances. National vocational standards are also used in many cases to assess operator competence (Wright et al., 2003), whereby a person meeting the required standards is deemed competent.

3.3.4 Control of Major Accident Hazard Regulations

The Control of Major Accident Hazard (COMAH) regulations passed in 1999 were designed to prevent and mitigate major accidents involving dangerous substances (e.g., chlorine, petroleum gas, arsenic pentoxide) that can cause harm to people or the environment (HSE, 2007a). Wright et al. (2003) pointed out that, as a result of the introduction of the COMAH regulations, it is no longer acceptable for organizations to make untested assumptions about the competencies of their staff.

3.4 HUMAN FACTORS LITERATURE

3.4.1 Definitions

Most training texts present the Department of Employment's definition of training, which describes training as "the systematic development of the attitudes/knowledge/skill behavior pattern required by an individual in order to perform adequately a given task or job" (Department of Employment, 1971). However, various definitions also exist within the human factors literature. Goldstein, for example, formally defined *training* as "the acquisition of skills, concepts or attitudes that result in improved performance in an on-the-job situation" (Goldstein, 1980, p. 230). More recently, Salas et al. defined *training* as the "systematic acquisition of knowledge (i.e., what we need to know), skills (i.e., what we need to do), and attitudes (i.e., what we need to feel) [knowledge, skills, and attitudes] that together lead to improved performance in a particular environment" (Salas et al., 2006, p. 473).

Training is therefore all about ensuring that trainees acquire the KSAs required to perform a particular job to a predefined standard; however, the definitions presented do not do justice to the breadth of the area. Training is clearly more than simply giving trainees KSAs for a particular job; it involves analyzing training needs and preconditions (e.g., organizational and trainee characteristics), designing training programs appropriately (e.g., ensuring that the content and delivery method are appropriate), delivering training, supporting and ensuring the application of the KSAs trained on the job, and evaluating training program effectiveness.

Operator competence is closely linked to the area of training; the HSE defines *competence* as the ability to undertake responsibilities and to perform activities to a regular standard on a regular basis (HSE, 2007b, p. 2). Wright et al. (2003) defined *competence* as "the ability to perform activities within an occupation or function to the standards expected in employment" (p. 7). According to the HSE, *competence* is comprised of a combination of practical and thinking skills, experience, and knowledge and can also include a willingness to undertake work activities in accordance with standards, rules, and procedures (HSE, 2007b).

3.4.2 Training Theory

Various models of training exist (e.g., Annett, 1968; Kozlowski and Salas, 1997; Tannenbaum, Cannon-Bowers, and Mathieu, 1993). Snow and Swanson (1992) presented a simplistic training framework that includes the initial state (i.e., trainee's existing KSAs and other characteristics), the desired state (i.e., the performance level required after training), the learning process (i.e., the transition from the initial state to the desired state), and an evaluation of the training program design and training effectiveness. Snow and Swanson's model is presented in Figure 3.1.

Based on a review of the literature, Tannenbaum et al. (1993) proposed a more comprehensive model which focuses on training effectiveness and includes the variables that are likely to affect the level of training transfer achieved. The Tannenbaum model is presented in Figure 3.2. This model takes a systems perspective on training and considers various individual, organizational, and situational elements (both before, during, and after training) that are likely to impact the level of training transfer achieved. The model considers several variables, including individual (trainee) variables (e.g., abilities, non-abilities, motivation, reactions to training) and organizational and situational characteristics (e.g., organizational climate, supervisor and peer support policies, practices and organizational constraints). Underpinning the model, Tannenbaum et al. (1993) presented comprehensive taxonomies of each variable category presented within the model (e.g., individual characteristics, organizational and situational characteristics, expectations, motivation). Examples of these variables are presented in Table 3.1.

According to the model, individual variables (e.g., existing abilities, attitudes, and experience) and organizational and situational characteristics (e.g., organizational climate, trainee selection and notification, organizational history) influence trainee expectations, desires, and motivation toward training, as well as their motivation to learn. For TNA purposes, the model suggests that TNA efforts should focus on the characteristics of the individuals who need to be trained and the characteristics of the organization and of the tasks requiring training. TNA outputs should then be used to inform the development of training content and selection of the training method.

Following training, post-training motivation is influenced by organizational and situational characteristics, such as supervisor and coworker support and resource availability, as well as any training maintenance strategies implemented by the organization in question. According to Tannenbaum et al. (1993), the level of training transfer is determined by post-training motivation to transfer and post-training organizational and situational characteristics. Trainees who can perform the task in question during training will perform it successfully on the job only if they want to and if the necessary conditions are in place.

Basing their work on organizational theory (Salas and Cannon-Bowers, 2001), Kozlowski and Salas (1997) developed a model of training implementation and transfer that is presented in Figure 3.3. This model considers training transfer achieved at three different levels: the organizational level, the team level, and the individual level. Various factors at each level are involved in the level of training transfer achieved. The model also focuses on the concept of vertical training transfer, which refers to the "upward propagation of individual-level training outcomes that emerge as team- and organizational-level outcomes" (Salas and Cannon-Bowers, 2001, p. 474).

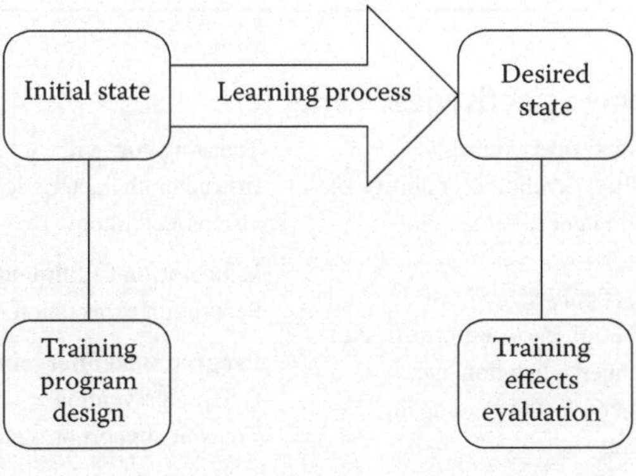

FIGURE 3.1 Snow and Swanson's training model. (Adapted from Koubek, R.J. and Liang, S., in *The Occupational Ergonomics Handbook*, W. Karwowski and W.S. Marras, Eds., CRC Press, Boca Raton, FL, 1999, pp. 1641–1657.)

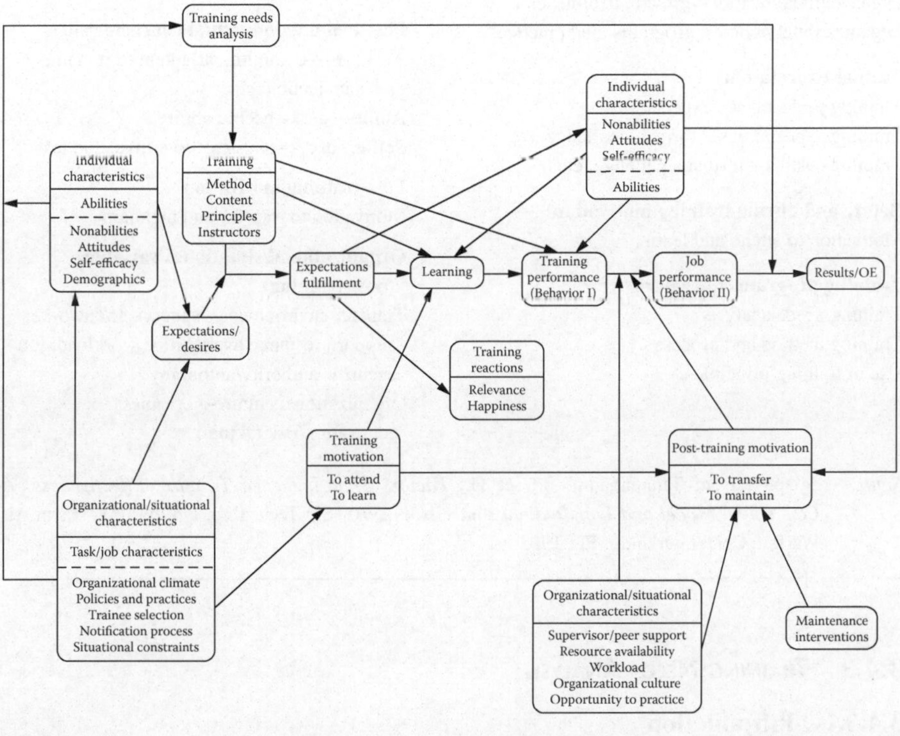

FIGURE 3.2 Training effectiveness model. (Adapted from Tannenbaum, S.I. et al., *Factors That Influence Training Effectiveness: A Conceptual Model and Longitudinal Analysis*, NAWCTSD Tech. Rep. No. 93-011, Naval Air Warfare Center, Orlando, FL, 1993.)

TABLE 3.1
Examples of Training Effectiveness Variables

Individual characteristics (pretraining)
Abilities—cognitive ability, psychomotor ability, etc.
Attitudes—commitment, intent to remain, job satisfaction, etc.
Self-efficacy—physical, cognitive, and task
Personality—locus of control, ego, conformity, etc.
Demographics—age, gender, education, etc.
Experience—tenure, experience with company, experience with task, etc.

Organizational/situational characteristics (pretraining)
Organizational climate
Trainee selection/notification process
Purpose of training
Task characteristics—complexity, type, difficulty, etc.
Organizational history—growth, decline, etc.
Organizational policies, programs, and practices

Trainee expectations
Training performance expectations
Training expectations—format, challenge, etc.
Training desires—format, challenge, etc.

Before and during training motivation
Motivation to attend and learn

Training program characteristics
Training needs analysis
Training method and process
Use of training principles

Training content
Instructor characteristics
Use of technology

Expectation fulfillment
Perception/expectation match

Programmed interventions
Relapse prevention
Transfer support programs

Training effectiveness
Training reactions—relevance, value, affective response, etc.
Learning
Training performance
Job performance
Results/organizational effectiveness

Post-training individual characteristics
Attitudes—commitment, intent to remain, job satisfaction, etc.
Ability—task-specific ability
Self-efficacy—physical, cognitive, and task

Post-training motivation
Motivation to transfer and maintain

Organizational/situational variables (post-training)
Transfer environment—supervisor/coworker support, resource availability, workload, job security, authority/autonomy
Organizational culture—openness to innovation/risk taking

Source: Adapted from Tannenbaum, S.I. et al., *Factors That Influence Training Effectiveness: A Conceptual Model and Longitudinal Analysis*, NAWCTSD Tech. Rep. No. 93-011, Naval Air Warfare Center, Orlando, FL, 1993.

3.4.3 Training Needs Analysis

3.4.3.1 Introduction

The first and most important step when designing training programs is to conduct a TNA to identify where training is needed, exactly what must be trained, and who it is that needs to be trained (Salas et al., 2006). Surprisingly, given its acknowledged importance within the training process, only limited published empirical research

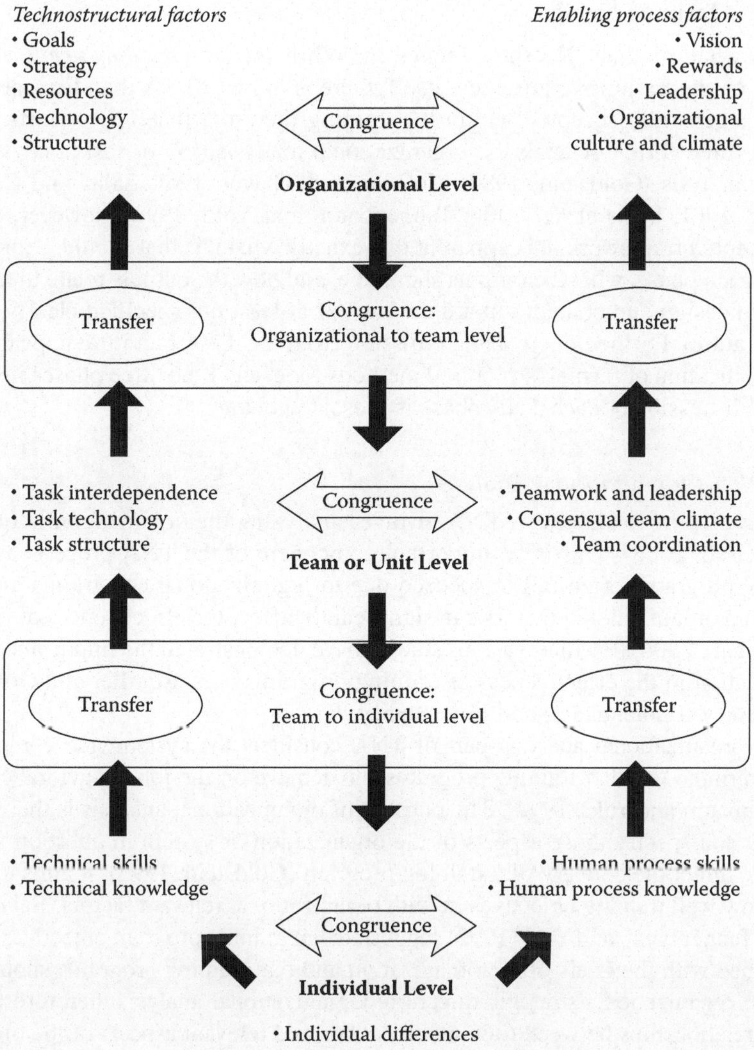

FIGURE 3.3 Model of training implementation and transfer. (From Kozlowski, S.W.J. and Salas, E., in *Improving Training Effectiveness in Work Organizations*, J.K. Ford et al., Eds., Lawrence Erlbaum, Mahwah, NJ, 1996, pp. 247–287. With permission.)

focuses on TNA. Further, structured methods for TNA remain limited; a survey conducted by Saari, Johnson, McLaughlin, and Zimmerle (1988) found that only 27% of those organizations surveyed had TNA procedures in place. Salas and Cannon-Bowers (2001) also pointed out that, despite being the most important component of training program design and delivery, TNA remains largely an art rather than a science.

3.4.3.2 Training Needs

Exactly what is it that TNA should focus on? While the term *training needs* suggests a focus on the training requirement itself, there is more to TNA than first meets the eye. It is generally accepted within the training literature that TNA efforts should include three forms of analysis: organizational analysis, job or task analysis, and person analysis (Goldstein, 1980; McGhee and Thayer, 1961; Salas and Cannon-Bowers, 2001; Salas et al., 2006; Tannenbaum and Yukl, 1992); however, despite this accepted framework, it is apparent that exactly what it is that should be analyzed during each phase, what the outputs should be, and how the outputs relate to training program design and content remain unclear (or at least not specified clearly within the literature). Further, one glaring omission from the TNA literature appears to be the specification of formal, structured methods for each of the three phases involved. A brief discussion of each TNA phase is presented below.

3.4.3.2.1 Organizational Analysis

The first step in conducting a TNA involves analyzing the organization in question (Salas et al., 2006). This is an important component of the TNA process, as many training programs often fail to succeed due to organizational constraints and conflicts, and organizational factors can significantly affect the effectiveness of training (Salas et al., 2006). Further, various studies have demonstrated the importance of the organization in the effectiveness of training programs (e.g., Rouiller and Goldstein, 1993; Tracy, Tannenbaum, and Kavanagh, 1995).

The organizational analysis part of TNA considers the systemwide components that determine whether training programs can achieve on-the-job behavioral changes (Tannenbaum and Yukl, 1992). The purpose of organizational analysis is therefore to identify and specify those aspects of the organization or system in question that are likely to affect the delivery of a training program (Goldstein, 1993); it aims to determine how well training objectives fit with organizational relevant factors (Salas et al., 2006). Tannenbaum and Yukl (1992) suggested that training program objectives should be aligned with the goals of the organization and that training programs should support the organization's strategic direction. Organizational analysis therefore focuses on the relationships between training objectives and relevant aspects of the organization in question, including organizational goals, resources, constraints, and support for training transfer (Salas and Cannon-Bowers, 2001). According to Koubek and Liang (1999), through organizational analysis the organization's goals, resources, and allocation of resources can be identified. Further, psychosocial characteristics, such as organizational climate and manager, supervisor, and peer support, are analyzed.

Despite its importance within the TNA process, it is apparent that more research is needed concerning the development of structured methods for assessing organizational context relative to training (Salas and Cannon-Bowers, 2001). In the authors' opinion, clear guidance for the organizational analysis phase is currently absent from the literature. In particular, the provision of structured methods for undertaking organizational analysis and guidance on exactly what the outputs of an organizational analysis should be are lacking. Further, there is limited discussion of how the outputs of organizational analyses actually impact the content of a particular training program. From the literature reviewed, it appears that organizational analysis efforts should include analyses of the organizations goals and strategic direction, resources and resource allocation, constraints, and the organizational climate. Further, the training goals and objectives of the organization should also be specified.

3.4.3.2.2 *Task Analysis*

When the organization is understood with regard to training, the job or tasks that must be trained should be the next focus. The task analysis part of TNA (also referred to as *job analysis*) involves analyzing the tasks to be trained so learning objectives can be specified (Salas et al., 2006) and training content can be determined. The task analysis phase therefore involves describing the tasks in question and identifying the competencies required (i.e., KSAs) to perform them successfully. Salas and Cannon-Bowers (2001) suggested that the output of a task analysis should include a detailed description of the work functions to be performed, the conditions under which the tasks will be performed, and the KSAs required to perform the tasks involved. Salas et al. (2006) suggested that the task analysis component can be broken down into three subcomponents: task description, task specification, and task requirements. The task description involves describing the work functions and the resources required to undertake the job. The task specification step involves describing the specific tasks that need to be performed and the range of conditions under which they are performed. The task requirements component involves describing the KSAs required by trainees to complete the tasks safely and efficiently. Koubek and Liang (1999) suggested that two key factors should be examined during the task analysis phase: the complexity of the tasks to be learned and similarities between the tasks to be learned and previous tasks.

Due to the popularity of task analysis as a method for system design and evaluation, the task analysis phase of TNA is undoubtedly the most developed of the three phases and has the most literature associated with it. Unlike the organizational analysis phase, various structured human factors methods exist for analyzing tasks. Typically, task analysis approaches, such as hierarchical task analysis (HTA) (Stanton, 2006), and cognitive task analysis approaches, such as the Critical Decision Method (CDM) (Klein, Calderwood, and McGregor, 1989), are used for the task analysis component of TNA. Further, Salas and colleagues (e.g., Salas et al., 2006) have clearly specified how task analysis outputs should be presented (e.g., as KSAs) and how task analysis outputs inform training program design.

3.4.3.2.3 *Person Analysis*

Whereas the first two phases focus on the organization and the task, the third and final TNA phase focuses on the individuals who are to be trained. The person analysis phase involves identifying *who* needs to be trained and *what* training is required by each trainee (Salas et al., 2006). The person analysis component is therefore used to ensure that each trainee receives the right training at the right level. Differing roles, abilities, and levels of experience mean that, within complex sociotechnical systems, different individuals are likely to require different levels and forms of training. Tannenbaum and Yukl (1992) suggested that inadequate person analyses can lead to training that is targeted at an inappropriate level or at the wrong people. The person

3.4.3.3.3 Cognitive Work Analysis

More recently, the Cognitive Work Analysis (CWA) framework (Jenkins, Stanton, Walker, and Salmon, 2009; Vicente, 1999) has been used for TNA purposes. In their conclusion to a review of existing TNA approaches, Naikar and Sanderson (1999) concluded that traditional approaches to TNA are not suited to the analysis of cognitive work because they cannot accommodate adaptive behavior and can only deal with known events and not non-routine ones. For these reasons, they used the work domain analysis (WDA) component of CWA to identify training needs for F/A-18 pilots. Because WDA focuses exclusively on the functional structure of the system in question, it defines the conditions and constraints that shape human behavior, and by identifying the functional structure of a particular work system WDA provides a framework for designing training systems that will support adaptive behavior. The functional structure of the work domain (as described by the WDA) can be used as the basis for defining training needs, as each level of the abstraction hierarchy identifies particular training needs. Training needs are then further defined by conducting a decomposition analysis (Naikar and Sanderson, 1999). The relationship between the WDA and defining training needs is described in Table 3.3. The following training needs can be derived from each cell within the abstraction–decomposition space (ADS) in the WDA (Naikar and Sanderson, 1999):

- *Functional purposes*—Functional purposes cells within the ADS provide the training objectives (i.e., the purpose for training personnel is to fulfill the functional purpose of the system in question). The training system must be designed to satisfy the training objectives of the system in question.
- *Priorities and values*—Priorities and values cells within the ADS provide the measures of performance with which to evaluate trainee performance and/or the effectiveness of the training program. The training system must be capable of collecting data related to the measures of performance.
- *Purpose-related functions*—Purpose-related functions cells within the ADS highlight functions that workers must be competent in executing and coordinating when performing activity within the system. The training system must be capable of generating scenarios for practicing basic training functions.
- *Physical functions*—Physical functions cells within the ADS highlight the functionality of the physical devices that workers must be trained to exploit and operate. The training system must simulate the functionality of physical devices and signify environmental conditions.
- *Physical form*—Physical form cells within the ADS highlight the functionally relevant properties of the physical devices and environmental features within the system. The training system must recreate functionally relevant properties of physical devices and significant features of the environment.

3.4.4 Training Design

Training design is a critical but often overlooked component of training system delivery. Patrick (1992) described the widespread but erroneous belief that, when training content has been identified (via TNA), it is then a straightforward process of giving

TABLE 3.3
Work Domain Analysis and Training Needs Analysis

Functional purpose—Why the system exists	*Training objectives*—Purpose for training workers (i.e., to fulfill the functional purpose of a work domain)	*Design objectives*—Training system must be designed to satisfy training objectives.
Priorities and values—Criteria for ensuring that purpose-related functions satisfy systems objectives	*Measures of training performance*—Criteria for evaluating trainee performance or the effectiveness of training programs	*Data collection*—Training system must be capable of collecting data related to measures of performance.
Purpose-related functions—Functions that must be executed and coordinated	*Basic training functions*—Functions that workers must be competent in executing and coordinating	*Scenario generation*—Training system must be capable of generating scenarios for practicing basic training functions.
Physical functions—Functionality afforded by physical objects	*Physical functionality*—Trainees trained to exploit the functionality of the physical objects or devices and to operate them under various conditions	*Physical functionality*—Training system must simulate the functionality of physical devices and signify environmental conditions.
Physical form—Physical objects comprising the system	*Physical context*—Trainees trained to recognize functionally relevant properties of physical objects or devices and significant environmental features	*Physical attributes*—Training system must recreate functionally relevant properties of physical devices and significant environmental features.

Source: Naikar, N. and Sanderson, P.M., *International Journal of Aviation Psychology*, 9, 271–290, 1999. Reprinted with permission of Taylor & Francis.

the trainee this information. Of course, how the content is structured and presented is extremely important. According to Patrick (1992), training design covers all of the activities associated with translating content into a training program, including providing pretraining, organizing, and sequencing different types of training content, selecting appropriate training methods, designing training sequences, providing support to trainees, ensuring that appropriate learning strategies are activated, and assessing level of performance. Tannenbaum and Yukl (1992) suggested that training design should consider learning objectives, trainee characteristics, knowledge regarding learning processes, and practical considerations such as constraints and costs.

Patrick (1992) decomposed training design into three components: training content, training methods and strategies, and trainees. The first component, training content, involves determining the nature and amount of the training program content and also the structure and sequencing of the training content. Ostensibly, this is heavily informed by the results of an initial TNA. The second component, training methods and strategies, involves defining the training methods and strategies to be used. The third and final component focuses on the trainees themselves; the knowledge and skills of the trainees reflect not only what it is that has to be trained but also how it should be trained and how it is understood and assimilated.

Tannenbaum and Yukl (1992) presented the following guidelines regarding the design of training programs:

- The instructional events and training method should be consistent with the cognitive, physical, and/or psychomotor processes that lead to mastery.
- The learner should be encouraged to produce the capability actively (e.g., via practice behaviors). Greater retention and transfer are achieved through more active capability production.
- Feedback should be timely, accurate, and constructive, and all available sources of feedback should be utilized.
- Instructional processes that enhance trainee self-efficacy and expectations should be adopted (i.e., initial training should focus on simple behaviors that are easily mastered).
- The training methods adopted should be adaptable based on trainee aptitude and prior knowledge.

3.4.5 Pretraining Conditions

It is widely accepted that pretraining conditions and events can influence the effectiveness of training programs (e.g., Tannenbaum and Yukl, 1992; Tannenbaum et al., 1993). Salas and Cannon-Bowers (2001), for example, suggested that events occurring before training can be equally as, if not more, important than those that occur during and after training. Salas and Cannon-Bowers (2001) identified the following key elements:

- Trainee factors are those factors that trainees bring to the training setting, including individual characteristics such as cognitive ability, self-efficacy, and goal orientation, as well as level of training motivation.
- Learning and participation variables are those variables that engage trainees to learn and participate in developmental activities, such as level of motivation regarding training and practice.
- Training preparation factors are the ways in which training can be prepared to maximize the learning experience of trainees, including factors related to the pretraining environment and climate, such as the manner in which the training is framed and trainees' experiences with previous training programs.

Tannenbaum and Yukl (1992) also cited factors such as environmental cues and signals regarding training, involvement of employees in training decisions, and pretraining preparation as important prerequisites for successful training programs.

TABLE 3.4
Training Strategies and Associated Methods

Training Strategy	Training Methods	Definition
Information	Lecture Discussion Reading	Present basic knowledge to support skill acquisition.
Demonstration	Mishap reports Videotapes Audiotapes	Present examples of situations in which the skill or behavior in question occurs.
Practice	Role plays Simulation Group exercises	Apply skills in a realistic setting.
Feedback	Observation Videotape review Verbal review Simulation	Evaluate trainee performance and provide feedback.

Source: Adapted from Beard, R.L. et al., *International Journal of Aviation Psychology*, 5, 131–143, 1995.

3.4.6 Training Delivery Methods

3.4.6.1 Introduction

One of the most important considerations during the training design process is the selection of appropriate training delivery methods for the training content in question. According to Salas and Cannon-Bowers (2001), training program designers require cost-effective, content-valid, easy-to-use, and engaging training delivery methods. Many different methods are available for delivering training programs within complex sociotechnical systems. Salas (2006) categorized the various methods available into information-based methods (e.g., lectures, handbooks, presentations), demonstration-based methods (e.g., video demonstrations, multimedia systems), and practice-based methods (role-play exercises, high-fidelity simulation). Beard, Salas, and Prince (1995) discussed the training delivery methods used in the aviation domain, and a breakdown of these methods is presented in Table 3.4.

Koubek and Liang (1999) addressed Hinrich's categorization of content-oriented (designed to transmit knowledge on a cognitive level), process-oriented (designed to change attitudes, develop awareness, and enhance interpersonal skills), and mixed (combination of content- and process-oriented methods) training delivery methods. They suggested that content-oriented approaches are best suited for enhancing cognitive learning outcomes, process-oriented approaches are the most appropriate for affective learning outcomes, and for psychomotor skills "learning-by-doing" approaches are the most appropriate. The delivery methods most commonly used for each category are presented in Table 3.5.

TABLE 3.5
Training Delivery Methods and Associated Learning Outcomes

Cognitive Domain	Affective Domain	Psychomotor Domain
Manuals and printed instructions	Discussion or conference	On-the-job training
Audiovisual instruction	Tutoring	Practice
Lectures	On-the-job training	Coaching
Discussions or conference	Case study	Simulators
Tutoring	Simulations or games	
On-the-job training	Role play	
Computer-assisted instruction	Behavior modeling	
Case studies		
Simulations or games		

Source: Adapted from Koubek, R.J. and Liang, S., in *The Occupational Ergonomics Handbook*, W. Karwowski and W.S. Marras, Eds., CRC Press, Boca Raton, FL, 1999, pp. 1641–1657.

3.4.6.2 Training Media

Advances in technological capability are impacting the way training is delivered. Simulation, online resources, serious games, and PC-based programs are all currently being used to deliver training. Salas and Cannon-Bowers (2001) even suggested that web training may represent the end for physically going to training programs. A worrisome aspect, though, is that little consideration is being given to the science of training when designing these new training programs, and many issues require attention, including the level of trainee–instructor interaction required, how to address trainee questions, and learner vs. instructor control over training (Salas and Cannon-Bowers, 2001).

3.4.6.2.1 Simulation

Simulation is currently a popular method for delivering training in the complex sociotechnical domains, particularly in the military, aviation (Salas and Burke, 2002), and healthcare domains. Simulator-based training systems vary in terms of cost, fidelity, sophistication, and functionality, but evidence suggests that, on the whole, simulation is an efficient training approach (Salas and Cannon-Bowers, 2001), and there are various advantages associated with using simulators for training purposes. Training occurs in a realistic context, and errors and mistakes can be made without suffering the associated consequences (but with an appreciation of what they will be). Tasks can also be continually practiced until an acceptable level of competence is achieved, and data recording and replay functions permit the accurate measurement of trainee performance and the provision of detailed feedback. On the downside, high-fidelity simulation can be extremely expensive, and designing effective simulation-based training systems is a difficult prospect. Further, as Salas and Cannon-Bowers (2001) pointed out, there is evidence that simulator-based training systems are being designed without consideration of what is known about training program design.

Caution is urged when considering the use of simulation for training; Salas and Cannon-Bowers (2001), for example, observed that exactly why simulators work is not well understood, and there is a general misleading understanding that simulation in itself leads to learning. So, how can the effectiveness of simulation-based training be maximized? Satish and Streufert (2002) proposed that effective training simulation must be built on sound underlying theory, using structured exercises, assessing performance, and providing feedback. In their article, Salas and Burke (2002) discussed the dos and don'ts of using simulation for training; in particular, they offered six core guidelines for maximizing the effectiveness of simulation as a tool for training:

- Instructional features are embedded within the simulation. Training simulations should be designed around key instructional components. One example given is the event-based approach to training (EBAT), which uses a series of embedded events within the training scenario that act as opportunities for trainees to exhibit the competencies being targeted by the training program. In addition, the events provide opportunities for performance measurement and feedback.
- Appropriate scenarios are embedded within the simulation. When using simulation for training, the importance of the scenarios used should be underestimated. The scenarios represent the training curriculum and so should be carefully created using storyboarding and with specific learning outcomes in mind. The authors recommended the use of cognitive task analysis approaches (e.g., CDM) to identify the scenario content.
- Opportunities for assessing and diagnosing individual or team performance are presented to trainers. The assessment of trainee competence is key to the effectiveness of simulation-based training. Accordingly, simulations must provide opportunities for the assessment of trainee competence. One approach is to use prescripted, learner-focused events, as trainers know when they are likely to occur. Further, measurement opportunities must provide the basis for diagnosing skill deficiencies by capturing moment to moment actions and behaviors.
- The learning experience is guided. Trainees who are given unguided practice often learn the wrong thing, fail to focus on practicing the correct behaviors, spend too much time on only one aspect of the training, and may be unable to transfer the skills trained to on-the-job performance. For efficient training transfer, practice must be guided through the use of scenarios and diagnostic feedback.
- Simulation fidelity is matched to training requirements. Simulation fidelity is a key factor in the success of simulation-based training system; however, the much-used adage that more is better is not always true in this case. It has been suggested by many in the field that the physical fidelity of the simulation is less important than the psychological fidelity of the scenarios used. The level of simulation fidelity used is a function of the cognitive and behavioral requirements of the task and training involved.

- There is a reciprocal partnership between subject matter experts (SMEs) and training specialists. The provision of effective training requires a partnership between task SMEs and training specialists, and both have a key contribution to make. SMEs are able to articulate task requirements (e.g., competencies), whereas training specialists create learning environments.

The same general training program design principles apply to the design of simulation-based training systems. Initially, a TNA is required, and careful consideration should be given to the training content in terms of the KSAs trained and the scenarios used to train them.

3.4.7 Team Training

Due to the increased presence of teams within complex sociotechnical systems (Fiore et al., 2003), the concept of team training is currently receiving increased attention from the human factors community. The use of teams has increased significantly over the past three decades (Salas, 2004), primarily due to two factors: (1) work and work procedures are growing in complexity, and (2) appropriately trained and constructed teams can potentially offer a number of advantages over the use of individual operators, including the ability to better perform more difficult and complex tasks, greater productivity and improved decision making (Orasanu and Fischer, 1997), more efficient performance under stress (Salas and Cannon-Bowers, 2000), and a reduction in the number of errors made (Wiener, Kanki, and Helmreich, 1993).

A team is characterized as consisting of two or more people, dealing with multiple information sources, and working to accomplish a shared or common goal of some sort. Salas, Prince, Baker, and Shrestha (1995) defined a *team* as "a distinguishable set of two or more people who interact dynamically, interdependently and adaptively toward a common and valued goal, who have each been assigned specific roles or functions to perform and who have a limited life span of membership." Paris, Salas, and Cannon-Bowers (2000) suggested that team characteristics include multiple sources of information, task interdependencies, coordination among team members, common and valued goals, specialized member roles and responsibilities, task-relevant knowledge, intensive communication, and adaptive strategies for responding to change.

Collaborative work involves two forms of activity: teamwork and taskwork. Teamwork refers to those instances where individuals interact or coordinate behavior to achieve tasks that are important to the team's goals (i.e., behavioral, attitudinal, and cognitive responses coordinated with fellow team members), while taskwork (i.e., task-oriented skills) describes those instances where team members are performing individual tasks separate from their team counterparts (i.e., tasks that do not require interdependent interaction with other team members) (Salas, Cooke, and Rosen, 2008). *Teamwork* was formally defined by Wilson, Salas, Priest, and Andrews (2007) as "a multidimensional, dynamic construct that refers to a set of interrelated cognitions, behaviours and attitudes that occur as team members perform a task that results in a coordinated and synchronised collective action."

Team training programs (see Figure 3.4) are used to develop the knowledge, skills, and abilities required for effective team performance. *Team training* is defined as

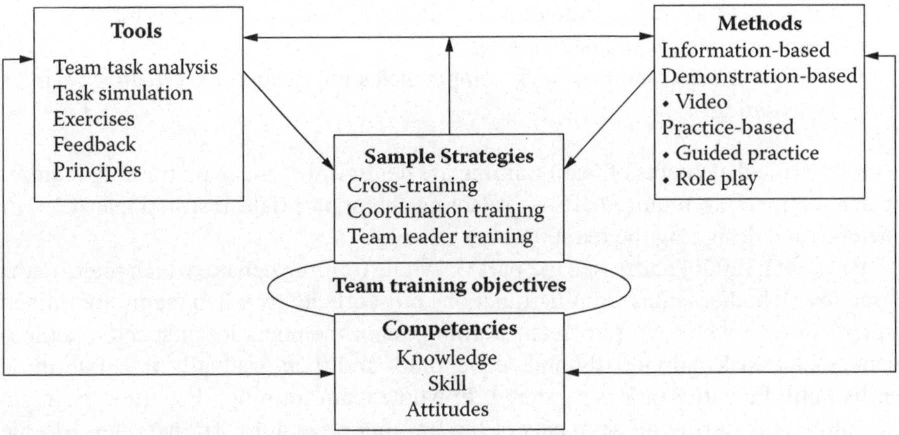

FIGURE 3.4 Structure of team training. (From Guzzo, R.A. and Salas, E., *Team Effectiveness and Decision Making in Organizations*, Pfeiffer, New York, 1995. Reproduced with permission of John Wiley & Sons.)

"a set of tools and methods that form an instructional strategy in combination with requisite competencies and training objectives" (Salas and Cannon-Bowers, 1997). Further, Salas and Cannon-Bowers have suggested that team training consists of strategies derived from integrating a set of outputs from tools such as team task analysis, a delivery method of some sort, and content (e.g., knowledge, skills, and abilities). The primary goal of team training is to develop competencies to allow effective synchronization, coordination, and communication between team members (Salas, 2004). Salas and Cannon-Bowers (2004) described four main team training strategies: team coordination training, cross-training, team self-correction training, and team leadership training. Additional team training strategies include EBAT (Salas and Cannon-Bowers, 2000), stress exposure training, scenario-based training, guided self-correction training, simulation-based training and games, and on-the-job training. Several key principles of team training are summarized below (Salas, 2004):

- Team training should lead to teamwork skill development (e.g., leadership, adaptability, compensatory behavior).
- Both process and outcome measures are needed for teamwork diagnosis.
- Simulations should allow team members to experience some alternative course of action.
- Response strategies provided during training should be linked to cues in the environment and other team members.
- Team members should be provided sufficient opportunities to interact with novel environments (to develop adaptive mechanisms).
- Training should be theoretically based.
- Team training should be more than a "feel good" intervention.
- Guided practice is essential.
- Training should establish a mechanism to foster teamwork.
- Team training is embedded in an organizational system.

- Creation of an appropriate transfer climate for teamwork is essential.
- Team training should be ongoing.
- Both teamwork and taskwork competencies are needed for effective team functioning.

Four critical elements of team training are determining the team training requirements, delineating required KSAs, selecting the appropriate instructional delivery method, and designing the team training strategy.

Paris et al. (2000) addressed the part vs. whole training debate, which refers to the issue over whether teams training functions most efficiently when teams are trained in part or as a whole. In part team training, team members learn specific components of the task individually and sequentially and then gradually integrate those skills until the entire task is mastered. In whole team training, they are exposed to the entire task during all segments of the training procedure. Trainers must decide among four different approaches:

- Provide individual skills training that includes a focus on the interactive requirements of the team task (individual–whole training).
- Provide individual skills training that ignores the team context (individual–part training).
- Provide team skills training with specific focus on each individual's assigned subtask (team–part training).
- Provide team skills training within the context of the team and its coordination and communication requirements (team–whole training).

Paris et al. (2000) suggested that if the task in question is characterized by high complexity and high organization then the most appropriate strategy may be to combine the individual–whole and team–whole approaches.

According to Salas (2005), performance measurement is the key to learning, and embedding performance measurement tools into training allows learning to occur. Salas (2006) described a number of principles for the measurement of team performance during team training: First, any system of performance measurement should have a strong theoretical foundation upon which to draw. Second, the purpose for developing a measurement system should be established from the onset; this can be achieved via the use of team cognitive task analysis to identify the KSAs required. Third, the opportunities for measuring performance during training must be identified and specified clearly, and any measurement approach should be used to determine why something happened as well as what happened. Fourth, any team training performance measurement tool should be capable of diagnosing performance deficiencies so team members can be remediated accordingly. Finally, the tool should not be overly cumbersome to use and should be quick and easy to apply.

3.4.7.1 Crew Resource Management

One of the most successful forms of team training used in the complex sociotechnical domains is Crew Resource Management (CRM) training. Emerging within the civil aviation domain, CRM training programs were first developed in response to accident investigations that indicated that most accidents were caused by a lack of coordination among team members or underutilization of team member resources, rather than by a lack of technical proficiency (Salas et al., 1999b). Accordingly, CRM training focused on the team, rather than individual crewmembers (which was unique at the time), and aimed to promote the use of all available resources during problem solving on the flight deck. CRM programs have since been used in a number of other complex collaborative domains, including offshore oil (O'Connor and Flin, 2003), health care (Howard, Gaba, Fish, Yang, and Sarnquist, 1992), helicopter rescue (Schmeiser, Bömmel, and Bühren, 2000), and air-traffic control (Smith-Jentsch, Baker, Salas, and Cannon-Bowers, 2001).

Crew Resource Management is formally defined as "using all available resources—information, equipment and people—to achieve safe and efficient flight operations" (Lauber, 1984) and includes "optimising not only the person–machine interface and the acquisition of timely, appropriate information, but also interpersonal activities including leadership, effective team formation and maintenance, problem solving, decision making, and maintaining situation awareness" (Helmreich and Foushee, 1993, p. 4). Salas et al. (1999b) noted that CRM involves training teamwork competencies that allow team members to cope with situational demands that overwhelm individuals. CRM programs therefore focus on enhancing the skills required for improving teamwork and performance.

Early CRM training programs, known as Cockpit Resource Management training (the first was introduced by a U.S. airline in the early 1980s), focused on changing attitudes that had been identified as distinguishing between highly skilled and less skilled pilots (Baker, Prince, Shrestha, Oser, and Salas, 1993; Seamster and Kaempf, 2001) and on social interaction within the cockpit (Salas et al., 1999b). Since then, CRM has evolved somewhat; Helmreich, Klinect, and Wilhelm (1999), for example, described five generations through which CRM training programs have evolved. More recently, a sixth generation of CRM training has emerged, the so-called Threat and Error Management (TEM) training programs, which focus on the management of threat and error, including training on the management of errors as they arise and on the management of external threats and errors such as high terrain, poor weather, aircraft system malfunctions, and errors made by other agents in the system, including maintenance personnel and air-traffic controllers (Klinect, Wilhelm, and Helmreich, 1999).

In terms of what CRM training actually is composed of, there is great variation in the content, length of the training provided, and the training methods used (Salas et al., 1999b). The U.S. Federal Aviation Administration (FAA) provides comprehensive guidelines for the methods used and the content of CRM training (Federal Aviation Administration, 2001). They recommend that training should include situation awareness, communication skills, teamwork, task allocation, and decision making. The training should also include awareness, practice, and feedback, with continuing reinforcement of the skills through simulator training and feedback. The main content areas suggested by the FAA include:

1. Communications processes and decision behavior
    - Briefings
    - Inquiry, advocacy, and assertion

and skills gained in training, both on and off the job" (Broad and Newstrom, 1992, p. 6); however, it has previously been estimated that only 10 to 20% of what is learned during training is actually applied on the job (Newstrom, 1986).

Transfer of training can be either positive or negative and occurs both during the learning of new skills and the performance of old ones. Patrick (1992) made the distinction between general and specific transfer of training. General transfer takes place when one task affects learning of a whole range of tasks, whereas specific transfer refers to transfer relating to a specific task. Patrick used the example of trainees "learning how to learn" as a demonstration of general transfer. Further, many in the field have distinguished between horizontal and vertical transfer. Horizontal transfer refers to the transfer of KSAs *between* tasks from one task to another, whereas vertical transfer refers to transfer *within* a task (Patrick, 1992) and involves the propagation of individual-level training outcomes upward to team- and organizational-level outcomes (Salas and Cannon-Bowers, 2001).

Based on a comprehensive review of the training transfer literature, Burke and Hutchins (2007) concluded that three primary factors influence levels of training transfer: learner characteristics (e.g., trainee personality, ability, motivation), intervention design and delivery (e.g., TNA, learning goals, content relevance, instructional strategies and methods adopted, self-management strategies, technological support for transfer), and the work environment (e.g., transfer climate, supervisor and peer support, opportunity to perform, accountability). Tannenbaum et al. (1993) presented a model of training transfer in which the level of training transfer achieved is determined by post-training motivation to transfer and post-training organizational and situational characteristics. They suggested that trainees who can perform the task in question during training will perform it successfully on the job only if they want to and if the necessary conditions are in place. Weiner et al. (2003) presented a useful summary of the various ways in which researchers in the field have suggested that transfer of training can be enhanced through training program design by:

- Ensuring a high level of fidelity of the training setting relative to the work setting
- Teaching general principles (i.e., teaching trainees how to apply principles across a range of different settings)
- Ensuring variability in the stimulus used by using different examples and practice conditions
- Using varying practice conditions
- Developing adaptive expertise
- Using discovery and error-based learning
- Developing trainee metacognitive skills

3.4.8.2 Post-Training Conditions

In addition to the type of training program used and the training content delivered, the importance of post-training conditions is also highlighted as a key factor in the level of training transfer achieved (e.g., Baldwin and Ford, 1988). Factors such as trainee self-efficacy, career motivation, and the opportunity to perform trained tasks on the job (Ford, Quinones, Sego, and Sorra, 1992), as well as the level of support provided by peers, supervisors, and the organization itself, have all been cited as key elements in the level of training transfer achieved. Other factors thought to be important include the time between training and on-the-job performance (Ford et al., 1992), the use of intervention strategies to enhance the probability of transfer (Brinkerhoff and Montesino, 1995), and informal reinforcement of transfer activities (Smith-Jentsch et al., 2001). The organizational context and climate also have influential roles in the transfer of training. Salas et al. (2006) pointed out that context helps determine trainee motivations, expectations, perceptions, and attitudes toward training transfer. The literature also suggests that the organizational climate can have an impact on the extent to which newly trained KSAs are applied on the job.

3.4.9 Training Evaluation

It is essential that organizations evaluate their training programs. Various questions need to be asked by organizations wishing to evaluate their training programs: Do their training programs actually work? Is transfer of training occurring? Do trainees continue to learn after training program attendance? Salas et al. (2008) observed that often huge amounts of expenditure are made by organizations without asking these sorts of questions. The inescapable conclusion appears to be that very few organizations partake in training program evaluations.

Training evaluation is formally defined as "a system for measuring the intended outcomes of training. It is concerned with issues of measurement, design, learning objectives, and the attainment of desired knowledge, skills, and abilities. Ultimately, training evaluation asks 'Did the training work?'" (Salas et al., 2006, p. 494). Goldstein (1986) defined *training evaluation* as the systematic collection of data to determine whether a training program achieved its training objectives and also whether or not job performance improved. Perhaps explaining its limited uptake, training evaluation is typically difficult, resource intensive, and expensive and often generates negative results (Salas and Cannon-Bowers, 2001). Further, evaluation often has to take place in the field during real-world job performance which is difficult to organize. Tannenbaum et al. (1993) observed that training effectiveness is a complex, multidimensional construct. They defined *training effectiveness* as "the extent to which training yields desired or relevant outcomes" (p. 18) and suggested that training effectiveness is normally assessed by comparing post-training performance to a specified criterion or standard. Koubek and Liang (1999) explained that when training results match training goals and objectives the training program can be concluded to be effective. Ultimately, the form of training evaluation used is dependent on the organization, training program, and tasks involved.

The most popular approach used for guiding training evaluation purposes is Kirkpatrick's (1976) typology (see also Salas and Cannon-Bowers, 2001; Salas et al., 2006). Kirkpatrick (1976) proposed a four-level approach to training evaluation, and his model outlines four levels of training outcomes: reaction, learning, behavior, and results. Level one, *reactions*, involves assessing trainee reactions to the training program, including reactions to the training, assessing judgments on the quality and relevance of the training, satisfaction with the training and instructor, and trainee ratings of course materials and effectiveness of training delivery (Salas et al.,

2006). Level two, *learning*, involves assessment of the learning that has taken place during the course of the training, including examining the extent to which trainees have acquired the KSAs being trained. Level three, *behavior*, involves assessing the extent to which the KSAs trained are applied on the job, and level four, *results*, involves assessing the impact that the training has in terms of organizational objectives. Factors assessed include reduced turnover, reduced costs, improved efficiency, and improved quality (Salas et al., 2006).

Tannenbaum et al. (1993) presented an extended version of Kirkpatrick's typology that features six categories of training effectiveness:

- Reactions
- Attitude change
- Learning
- Training performance (behavior I)
- Job performance (behavior II)
- Results/organizational effectiveness

Ford and Sego (1990) proposed a conceptual model for evaluating training effectiveness. The five main purposes of training evaluation are to determine content validity (the relevance between training and job content), training efficiency (quality of the training delivered), training validity (extent of trainees' learning), transfer validity (trainees' job performance after training), and predictive validity (the prediction of job performance from training performance).

3.4.10 Summary

A huge body of literature concerns the area of training program design, delivery, and evaluation, and it is only summarized briefly here. For exhaustive reviews, the reader is referred to Salas and Cannon-Bowers (2001), Salas et al. (2006), Burke and Hutchins (2007), and Tannenbaum and Yukl (1992). Each of the different areas involved is well described in its own right. In particular, extensive literature and exhaustive reviews are available on topics such as TNA, training program design, training transfer, team training (CRM), and training evaluation. It is notable, however, that significant gaps exist in the literature, particularly regarding the specification of formal, structured methodologies for TNA and training program design, delivery, and evaluation, as well as literature describing empirical case studies regarding training program design, delivery, and evaluation. TNA, for example, is one component of the process that is well described in terms of what is required (e.g., organizational, task, and person analyses) but which also leaves much to the imagination in terms of the appropriate methodologies to use, how to use them, and what the outputs should be and how they should be represented by the training program.

3.5 GUIDELINES AND RECOMMENDATIONS

Salas et al. (2006) developed a useful set of guidelines for the design, delivery, and evaluation of training systems. Their guidance is presented in Table 3.7. The process begins (assuming that a requirement for training has arisen) with the conduct of a

TNA, including the organizational analysis, task analysis, and person analysis components. As noted previously, we recommend that the organizational analysis phases should involve conducting a work domain analysis of the organization in question and that the job and person analysis phases should involve the conduct of detailed HTA and cognitive task analysis for the task in question. From both analyses, the KSAs required to perform the tasks in question should be identified and documented. Finally, it is our opinion that the social organization and cooperation analysis phase of CWA should be used for the person analysis component of the TNA (along with the KSAs identified during the task analysis phase).

The next phase involves the design of the training program itself. According to Salas et al. (2006), key tasks here include developing the training program objectives, considering any characteristics of the trainees and organization that might affect the training program, establishing and specifying practice opportunities (when, how many, levels of difficulty), establishing and specifying feedback opportunities (when and at what level feedback will be provided), selecting the most appropriate training methods, and outlining the training program content in terms of structure and sequence.

Next, the training development phase involves developing the practice scenarios and training performance measures. According to Salas et al. (2006), the development of practice scenarios should include developing and scripting various practice scenarios of varying difficulty. Developing performance measures includes identifying and developing an appropriate measurement plan, relevant criteria for success, and appropriate performance measures.

When the training development phase is complete and the training program has been developed, the training implementation phase follows. This involves selecting and preparing an appropriate location for the training to be delivered, selecting and training appropriate instructors to enable them to deliver the training, conducting a pilot test of the training program to identify any issues or concerns and collect feedback from trainees, making any adjustments to the training program, and, finally, when the program is ready, delivering the training program to the appropriate trainees.

Following delivery of the training program, the training evaluation phase involves evaluating the training program through the collection and analysis of appropriate data regarding the training program and post-training job performance. The final transfer of training phase involves establishing a positive post-training environment (e.g., by supporting competencies on the job, rewarding trainees, discouraging contradictory behaviors) and using job aids (e.g., manuals, decision support systems) to facilitate training transfer.

3.6 ASSESSMENT AND EVALUATION

3.6.1 Training Needs Analysis

Training needs analysis involves conducting organizational, task, and person analyses. Although there is general agreement within the training literature that these three forms of analysis are required for TNA purposes, there is limited guidance on the methods that should be used for each phase. Encouragingly, various human factors methods exist that can produce the outputs required for each phase of TNA;

TABLE 3.7
Phases of Training System Design, Delivery, and Evaluation

Phase	Outcome
Training Needs Analysis	
1. Perform organizational analysis.	Identifies: Where training is needed When training is needed Resources and constraints Support for transfer
2. Perform task analysis.	Identifies: Task specifications (e.g., what tasks and under what conditions) Task characteristics (e.g., equipment needed for task) Competencies (KSAs) required for tasks in question
3. Perform person analysis.	Identifies: Who requires training What specifically trainees need to be trained on
Training Design	
4. Develop objectives.	Identifies: Desired outcomes and goals of training Training assumptions Training objectives Competencies (KSAs)
5. Consider individual characteristics.	Identifies: Trainee characteristics that might affect training, including cognitive abilities, self-efficacy, goal orientation, and motivation
6. Consider organizational characteristics.	Identifies: Organizational characteristics that might affect training, including organizational culture, policies and procedures, situational influences, and pre-practice conditions

7. Establish practice opportunities.	Leads to the specification of practice opportunities, including: When they will occur during training The number of opportunities provided Levels of practice difficulty
8. Establish feedback opportunities.	Identifies: When feedback will be provided What level of feedback will be provided
9. Select an instructional strategy.	Involves selecting the most appropriate instructional strategy based on the needs of the organization
10. Outline program content.	Involves identifying and specifying the sequence and structure of the training program
Training Development	
11. Develop practice scenarios.	Involves developing and scripting appropriate practice scenarios; scenarios of varying levels of difficulty should be scripted
12. Develop performance measures.	Involves the identification and development of: Appropriate measurement plan Criteria for success Performance measures and the tools required to support them
Training Implementation	
13. Select the instructional setting/location.	Involves: Identification of an appropriate training site Preparation of the training environment
14. Train instructors.	Involves: Training instructors to enable them to deliver training Ensuring instructors are knowledgeable in terms of training program content and any problems that might arise
Training Implementation	
15. Conduct a pilot test.	Involves: Identifying any issues or concerns associated with the training program Collecting feedback from trainees Making appropriate adjustments (based on issues identified and feedback received) to the training program
16. Conduct the instruction.	Involves using the instructional materials to deliver the training

(continued)

TABLE 3.7 (continued)
Phases of Training System Design, Delivery, and Evaluation

Phase	Outcome
Training Evaluation	
17. Consider evaluation design issues.	Involves developing an experimental plan for the evaluation of the training program, including the experimental design and location of the evaluation
18. Consider evaluation cost.	Involves considering the cost of the evaluation and identifying and exploring low-cost alternatives
19. Evaluate training system (at multiple levels).	Involves: Collecting and analyzing data on the training programs effectiveness Collecting and analyzing data on job performance
Transfer of Training	
20. Establish a positive post-training environment.	Involves: Supporting competencies on the job Creating a continuous learning environment Rewarding trainees Discouraging contradictory behaviors
21. Use job aids.	Involves enhancing job performance through the use of job aids such as manuals and decision support systems

Source: Adapted from Salas, E. et al., in *Handbook of Human Factors and Ergonomics*, 3rd ed., G. Salvendy, Ed., John Wiley & Sons, New York, 2006, pp. 475–512.

for example, it is our opinion that the WDA component of the CWA framework can be used for the organizational phase, that the HTA approach can be used for the task analysis phase, and that the social organization and cooperation analysis phase can be used for the person analysis phase. A brief outline of each approach, including guidance on how to undertake the analyses, is presented below.

3.6.1.1 Organizational Analysis

Work domain analysis involves the construction of an abstraction hierarchy (AH) and an abstraction–decomposition space (ADS), which provides a two-dimensional representation of the system under analysis (Vicente, 1999). The AH consists of five levels of abstraction, ranging from the most abstract level of purposes to the most concrete level of form. A description of each of the five abstraction hierarchy levels is given below (Naikar, Hopcroft, and Moylan, 2005):

- *Functional purpose*—The overall purposes of the system and the external constraints on its operation
- *Abstract function*—The criteria that the system uses for measuring progress toward the functional purposes
- *Generalized function*—The general functions of the work system that are necessary for achieving the functional purposes
- *Physical function*—The functional capabilities and limitations of the physical objects within the system that enable the purpose-related functions
- *Physical form*—The physical objects within the work system that afford the physical functions

The ADS is comprised of a combination of an abstraction hierarchy and a decomposition hierarchy. The decomposition hierarchy uses five levels of resolution, ranging from the broadest level of total system to the finest level of component and comprised of total system, subsystem, function unit, subassembly, and component (Vicente, 1999). Each of the cells within the ADS presents a different representation of, or way of thinking about, the same system; for example, the top left cell describes the functional purposes of the entire system (e.g., why the system exists and what its primary purposes are), while the bottom right cell describes the physical components that comprise the system (the people, objects, tools, and technological artifacts used within the system).

The WDA provides a useful tool for the organizational analysis component of TNA because it models a system, or organization, based on the constraints imposed on activity within it. The purpose of the organizational analysis component of TNA is to identify those aspects of an organization that are likely to affect the delivery of a particular training program. Thus, the constraints identified in the WDA, at each of the different levels of decomposition, can be analyzed for their impact on the design and delivery of training programs. The relationships between organizational goals and resources and the design and delivery of training can also be identified, as the functional purpose and priorities and values levels specify the overall functional purpose of the organization and their measure of performance, whereas the physical objects and functions levels specify the physical resources present in the domain along with their physical functionality. Naikar et al. (2005) presented a nine-step methodology for undertaking WDA and a series of WDA prompts (see Table 3.8).

TABLE 3.8
Example WDA Prompts

	Prompts	Keywords
Functional purposes	*Purposes:* For what reasons does the work system exist? What are the highest level objectives or ultimate purposes of the work system? What role does the work system play in the environment? What has the work system been designed to achieve?	*Purposes:* Reasons, goals, objectives, aims, intentions, mission, ambitions, plans, services, products, roles, targets, aspirations, desires, motives, values, beliefs, views, rationale, philosophy, policies, norms, conventions, attitudes, customs, ethics, morals, principles
	External constraints: What kinds of constraints does the environment impose on the work system? What values does the environment impose on the work system? What laws and regulations does the environment impose on the work system?	*External constraints:* Laws, regulations, guidance, standards, directives, requirements, rules, limits, public opinion, policies, values, beliefs, views, rationale, philosophy, norms, conventions, attitudes, customs, ethics, morals, principles
Values and priority measures	What criteria can be used to judge whether the work system is achieving its purposes? What criteria can be used to judge whether the work system is satisfying its external constraints? What criteria can be used to compare the results or effects of the purpose-related functions on the functional purposes? What are the performance requirements of various functions in the work system? How is the performance of various functions in the work system measured or evaluated and compared?	Criteria, measures, benchmarks, tests, assessments, appraisals, calculations, evaluations, estimations, judgments, scales, yardsticks, budgets, schedules, outcomes, results, targets, figures, limits Measures of effectiveness, efficiency, reliability, risk, resources, time, quality, quantity, probability, economy, consistency, frequency, success Values, including laws, regulations, guidance, standards, directives, requirements, rules, limits, public opinion, policies, beliefs, views, rationale, philosophy, norms, conventions, attitudes, customs, ethics, morals, principles

Purpose-related functions	What functions are required to achieve the purposes of the work system? What functions are required to satisfy the external constraints on the work system? What functions are performed in the work system? What are the functions of individuals, teams, and departments in the work system?	Functions, roles, responsibilities, purposes, tasks, jobs, duties, occupations, positions, activities, operations
Object-related processes	What can the physical objects in the work system do or afford? What processes are the physical objects in the work system used for? What are the functional capabilities and limitations of physical objects in the work system?	Processes, functions, purposes, utility, roles, uses, applications, functionality, characteristics, capabilities, limitations, capacity, physical processes, mechanical processes, electrical processes, chemical processes
Physical objects	What are the physical objects or physical resources in the work system—both manmade and natural? What physical objects or physical resources are necessary to enable the processes and functions of the work system?	Manmade and natural objects, including tools, equipment, devices, apparatus, machinery, items, instruments, accessories, appliances, implements, technology, supplies, kit, gear, buildings, facilities, premises, infrastructure, fixtures, fittings, assets, resources, staff, people, personnel, terrain, land, meteorological features

Source: Adapted from Naikar, N. et al., *Work Domain Analysis: Theoretical Concepts and Methodology*, Tech. Report No. DSTO-TR-1665, Australian Government, Department of Defence, Defence Science and Technology Organisation, Canberra, 2005.

3.6.1.2 Task Analysis

The task analysis phase of TNA involves exhaustively describing the tasks in question and then identifying the KSAs required to perform the tasks successfully. For this purpose, it is recommended that HTA, by far the most commonly used task analysis approach, be used. HTA (Annett, 2005; Stanton, 2006) is essentially a descriptive task analysis technique that is used to describe systems, goals, and tasks. HTA works by decomposing activities into a hierarchy of goals, subordinate goals, operations, and plans, thus allowing systems and tasks to be described exhaustively. HTA outputs specify the overall goal of a particular system, the subgoals required to achieve this goal, and the operations required to achieve each of the subgoals specified.

The HTA process is simplistic, involving collecting data about the task or system under analysis (through such techniques as observation, questionnaires, interviews with subject matter experts, walkthroughs, user trials, and documentation review, to name but a few) and then using these data to decompose and describe the goals, subgoals, and tasks involved. The HTA procedure is presented in Figure 3.5.

3.6.1.3 Person Analysis

The person analysis component of TNA involves identifying *who* needs to be trained and *what* training is required by each trainee (Salas et al., 2006). Analogous to the organizational analysis phase, there is limited guidance in the literature regarding what method or approach should be used to elicit this information during TNA efforts; however, a skills gap analysis, focusing on what KSAs the workers currently have in relation to what KSAs are required for a particular role, would be useful for this purpose. Such an analysis identifies the training that various workers need to become appropriately qualified for a particular role that is the focus of the TNA.

3.7 EXAMPLE TNA

3.7.1 Training Needs Analysis

An example TNA for the organization and example domain is presented here, including organizational analysis, task analysis, and person analysis in the example domain.

3.7.1.1 Organizational Analysis

The first step involves analyzing the organization in question with a view to identifying organizational factors that might affect training program design and delivery (Goldstein, 1993). Salas et al. (2006), for example, suggested that organizational factors may significantly affect the effectiveness of training. Organizational analysis therefore aims to determine how well training objectives fit with organizational relevant factors (Salas et al., 2006). To demonstrate how organizational analysis might be undertaken, a WDA was conducted for the example organization (see Figure 3.6). The functional purpose of the organization in question is to make a profit (dividend) for shareholders, and the abstract functions include the safe and efficient movement of gas and the supply of gas. The training objectives should therefore be to train workers in the functions and processes required to fulfill the functional purpose of

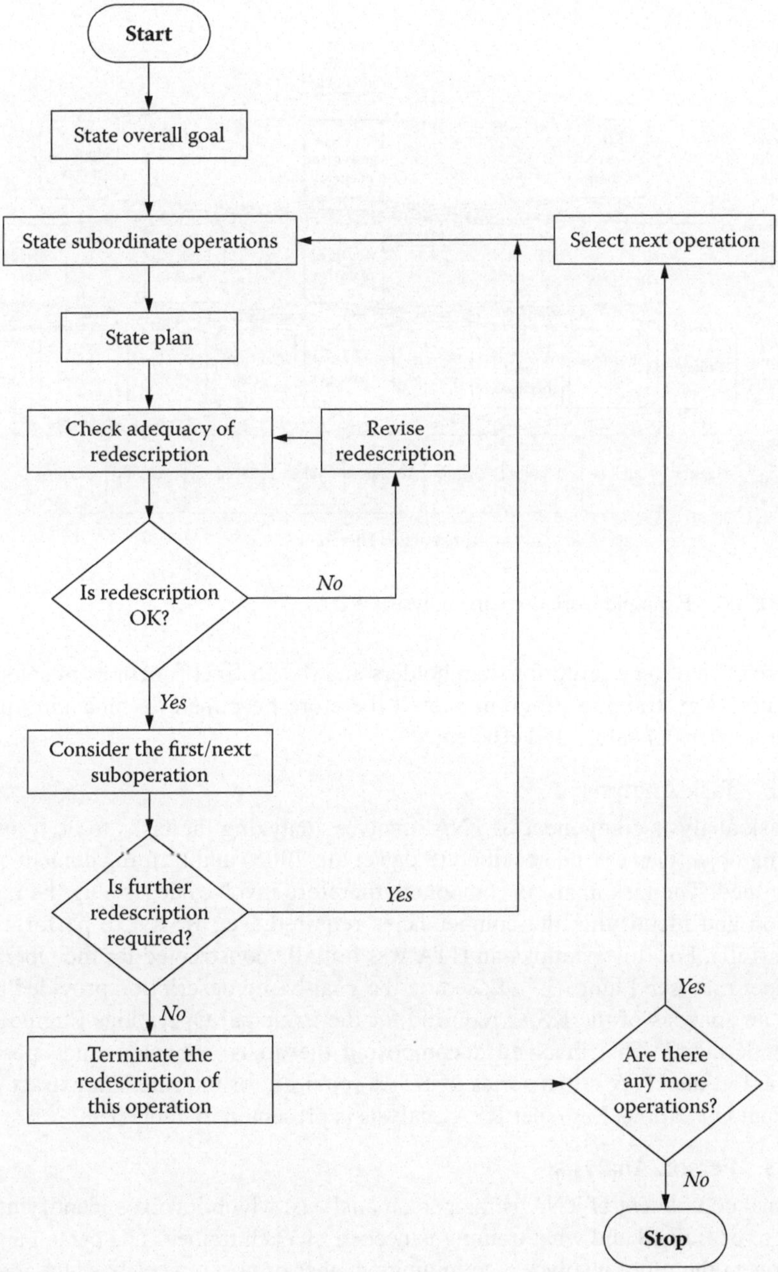

FIGURE 3.5 Hierarchical task analysis (HTA) procedure.

making a profit, whereas the measures of training performance should be safety, efficiency, and maintenance of supply. The WDA suggests that the organizational factors that are likely to affect training program design and delivery are the functional

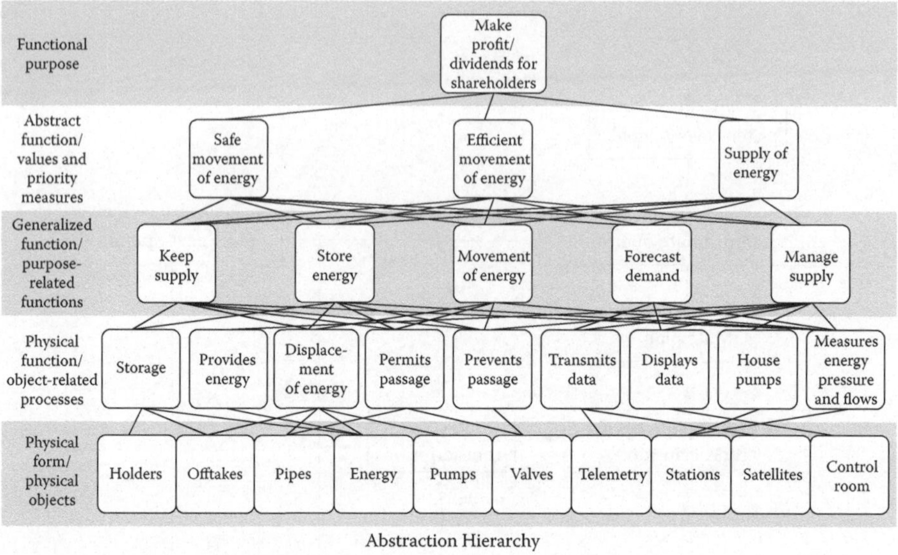

FIGURE 3.6 Example work domain analysis (WDA).

purpose of making a profit for shareholders and the abstract functions of safety and efficiency. Any training program should therefore be aimed at enhancing performance in terms of safety and efficiency.

3.7.1.2 Task Analysis

The task analysis component of TNA involves analyzing the tasks to be trained so learning objectives can be specified (Salas et al., 2006) and training content can be determined. The task analysis component therefore involves describing the tasks in question and identifying the competencies required (i.e., KSAs) to perform them successfully. For this example, an HTA was initially constructed for the Operations Engineer role (see Figure 3.7). Based on the goal-based description provided by the HTA, an analysis of the KSAs required for the Principal Operations Engineer role was undertaken. This involved decomposing the goals, subgoals, and operations specified in the HTA into a series of KSAs required for the role. An extract of the Principal Operations Engineer KSA analysis is presented in Table 3.9

3.7.1.3 Person Analysis

The final component of TNA is the person analysis, which involves identifying who needs to be trained and what training is needed by each trainee. The person analysis component therefore involves determining whether or not trainees have the requisite competencies (i.e., KSAs) for the task or job in question. For this purpose, a gap analysis was undertaken using the Principal Operations Engineer HTA and KSA analyses and the Operations Engineer HTA and KSA analyses. The aim of the gap analysis was to identify the training required for an Operations Engineer to assume the Principal Operations Engineer role. An extract of the gap analysis is presented in Table 3.10.

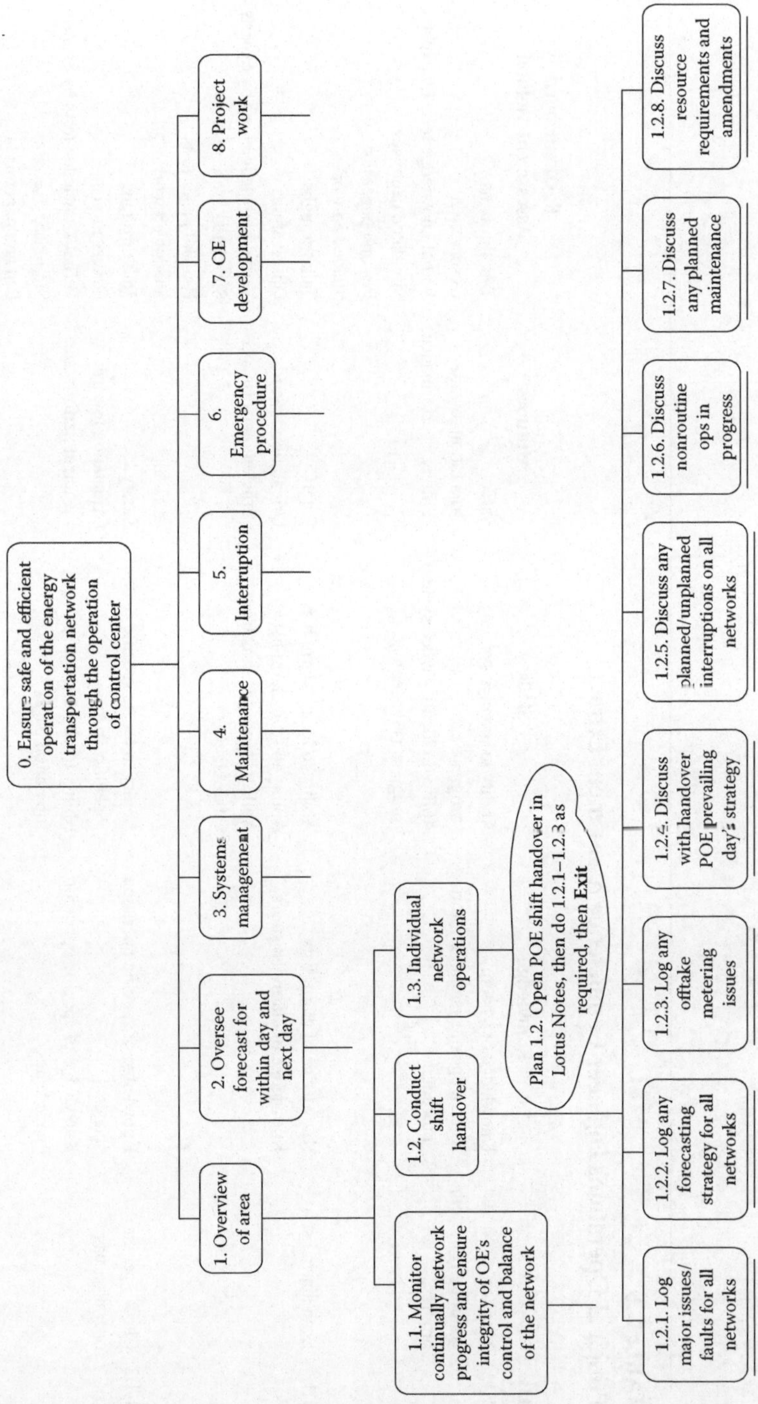

FIGURE 3.7 Principal Operations Engineer HTA extract with example decomposition.

TABLE 3.9
Principal Operations Engineer Competence Assessment Extract

Task		Knowledge	Skills	Attitudes	Recommended Assessment Method
1.1	Monitor continually system progress.	Knowledge of system Knowledge of system monitoring procedures	Ability to monitor system progress Ability to identify when system progress is inadequate	Safety Orientation toward efficient performance	Talk throughs Observation Verbal knowledge tests by experts Simulated exercises Pen and paper tests Shadowed work
1.1.1	Scan alarm lists.	Knowledge of alarm lists Knowledge of alarm list states	Ability to locate alarm lists Ability to scan alarm lists Ability identify alarms from alarm lists	Safety Orientation toward efficient performance	Talk throughs Observation Verbal knowledge tests by experts Simulated exercises Pen and paper tests Shadowed work
1.1.2	Check specific operation pages.	Knowledge of specific operation pages Knowledge of specific operation page states	Ability to locate specific operation pages Ability to check specific operation pages	Safety Orientation toward efficient performance	Talk throughs Observation Verbal knowledge tests by experts Simulated exercises Pen and paper tests Shadowed work

1.1.3	Scan sup dem pages to monitor intake rate and demands.	Knowledge of sup dem pages Knowledge of intake rate/demand monitoring procedure	Ability to locate sup dem pages Ability to monitor intake rate and demand Ability to scan sup dem pages	Safety Orientation toward efficient performance	Talk throughs Observation Verbal knowledge tests by experts Simulated exercises Pen and paper tests Shadowed work
1.1.4	Consult with engineer as required	Knowledge of NCE contact details	Ability to determine NCE contact details Ability to liaise with NCE Ability to articulate progress	Safety Team cohesion Mutual trust Collective orientation Teamwork	Talk throughs Observation Verbal knowledge tests by experts Simulated exercises Pen and paper tests Shadowed work
1.2	Shift handover	Knowledge of shift handover procedure and contents Knowledge of current system state	Ability to liaise with oncoming manager Ability to articulate current system state (i.e., each shift handover element)	Safety Team cohesion Mutual trust Collective orientation Teamwork	Talk throughs Observation Verbal knowledge tests by experts Simulated exercises Pen and paper tests Shadowed work

TABLE 3.10
Operations Engineer/Principal Operations Engineer Gap Analysis Extract

Principal Operations Engineer Tasks		Principal Operations Engineer KSAs	Operations Engineer Gaps	Training Required
1	Overview of area			
1.1	Continually monitor network progress and ensure integrity of Operations Engineer's control and balance of network.	Knowledge of system Knowledge of system monitoring procedures Ability to monitor system progress Ability to identify system issues Safety Orientation toward efficient performance	Knowledge of system Knowledge of system monitoring procedures Ability to monitor system progress Ability to identify system issues	Training in system Training in system monitoring procedures Training in system issues
1.1.1	Scan alarm lists for all 6 to 13 networks to check if alarms are handled by Operations Engineers as required.	Knowledge of alarm lists Knowledge of alarm handling procedures and states Ability to locate and open alarm lists Ability to scan and monitor alarm lists Ability to discriminate between alarms being handled and alarms not being handled	Ability to locate and open alarm lists Ability to scan and monitor alarm lists Ability to discriminate between alarms being handled and alarms not being handled	Training in alarm monitoring for all 6 to 13 networks
1.1.2	Check specific operation pages	Knowledge of specific operations Knowledge of specific operation pages Knowledge of system Ability to operate system Ability to check operation pages	Not applicable	Not applicable

BIBLIOGRAPHY

Annett, J. (1968). A systems approach to training. In E. Evans et al. (Eds.), *Planning Industrial Training*. London: National Institute of Adult Education.

Annett, J. (2005). Hierarchical task analysis. In N.A. Stanton, A. Hedge, K. Brookhuis, E. Salas, and H. Hendrick (Eds.), *Handbook of Human Factors and Ergonomics Methods* (pp. 33.1–33.7). Boca Raton, FL: CRC Press.

Baker, D.P., Prince, C., Shrestha, L., Oser, R.L., and Salas, E. (1993). Aviation computer games for crew resource management training. *International Journal of Aviation Psychology*, 3, 143–156.

Baldwin, T.T. and Ford, J.K. (1988). Transfer of training: a review and directions for future research. *Personnel Psychology*, 41, 63–105.

Barnett, M., Gatfield, D., and Pekcan, C. (2004). A Research Agenda in Maritime Crew Resource Management. Paper presented at the National Maritime Museum Conference on Safer Ships, Safer Lives, London, March 2004 (http://www.he-alert.com/documents/published/HE00240.pdf).

Beard, R.L., Salas, E., and Prince, C. (1995). Enhancing transfer of training: using role-play to foster teamwork in the cockpit. *International Journal of Aviation Psychology*, 5, 131–143.

Brinkerhoff, R. and Montesino, M. (1995). Partnerships for training transfer: lessons from a corporate study. *Human Resources Development Quarterly*, 6(2), 263–274.

Broad, M. and Newstrom, J. (1992). *Transfer of Training*. Cambridge, MA: Perseus.

Burke, L.A. and Hutchins, H.M. (2007). Training transfer: an integrative literature review and implications for future research. *Human Resource Development Review*, 6(3), 263–296.

Byrdorf, P. (1998). Human Factors and Crew Resource Management: An Example of Successfully Applying the Experience from CRM Programmes in the Aviation World to the Maritime World. Paper presented at the 23rd Conference of the European Association for Aviation Psychology, Vienna, September 1998.

Cannon-Bowers, J.A., Tannenbaum, S.I., Salas, E., and Volpe, C.E. (1995). Defining competencies and establishing team training requirements. In R.A. Guzzo and E. Salas (Eds.), *Team Effectiveness and Decision Making in Organizations* (pp. 333–380). San Francisco, CA: Jossey-Bass.

Department of Employment. (1971). *Glossary of Training Terms*, 2nd ed. London: HMSO.

FAA. (2001). *Advisory Circular. Crew Resource Management Training*, AC No. 120-51D. Washington, D.C.: Federal Aviation Administration.

Fiore, S.M., Salas, E., Cuevas, H.M., and Bowers, C.A. (2003). Distributed coordination space: toward a theory of distributed team process and performance. *Theoretical Issues in Ergonomics Science*, 4, 340–364.

Flin, R. and O'Connor, P. (2001). Applying crew resource management on offshore oil platforms. In E. Salas, C.A. Bowers, and E. Edens (Eds.), *Improving Teamwork in Organizations: Applications of Resource Management Training* (pp. 217–233). Mahwah, NJ: Lawrence Erlbaum.

Ford, J.K. and Sego, D.J. (1990). *Linking Training Evaluation to Training Needs Assessment: A Conceptual Model*, AFHRL-TP-90-60. Brooks AFB, TX: Training Systems Division, Air Force Human Resources Laboratory.

Ford, J.K., Quinones, M.A., Sego, D.J., and Sorra, J.S. (1992). Factors affecting the opportunity to perform trained tasks on the job. *Personnel Psychology*, 45, 511–527.

Gaba, D.M. (2000). Anaesthesiology as a model for patient safety in health care. *British Medical Journal*, 320, 785–788.

Gaba, D.M., Howard, S.K., Flanagan, B., Smith, B.E., Fish, K.J., and Botney, R. (1998). Assessment of clinical performance during simulated crises using both technical and behavioural ratings. *Anaesthesiology*, 89, 8–18.

Goldstein, I.L. (1980). Training in work organisations. *Annual Review of Psychology*, 31, 229–272.

Goldstein, I.L. (1986), *Training in Organizations: Needs Assessment Development and Evaluation*. Monterey, CA: Brooks/Cole Publishing.

Goldstein I.L. (1993). *Training in Organizations*. Belmont, CA: Wadsworth.

Health and Safety Executive (HSE). (2002). *Developing and Maintaining Staff Competence*. London: HSE Books.

Health and Safety Executive (HSE). (2003). *Competence Assessment for the Hazardous Industries*, RR086. London: HSE Books.

Health and Safety Executive (HSE). (2007a). *Managing Competence for Safety-Related Systems*. Part 1. *Key Guidance*. London: Health and Safety Executive Report.

Health and Safety Executive (HSE). (2007b). *Developing and Maintaining Staff Competence*, 2nd ed., Railway Safety Publication 1. London: HSE Books.

Helmreich, R.L. and Foushee, H.C. (1993). Why Crew Resource Management? Empirical and theoretical bases of human factors training in aviation. In E.L. Wiener, B.G. Kanki, and R.L. Helmreich (Eds.), *Cockpit Resource Management* (pp. 173–198). San Diego, CA: Academic Press.

Helmreich, R.L., Klinect, J.R., and Wilhelm, J.A. (1999). Models of threat, error, and CRM in flight operations. In *Proceedings of the Tenth International Symposium on Aviation Psychology* (pp. 677–682). Columbus, OH: Ohio State University Department of Aerospace Engineering, Applied Mechanics, and Aviation.

Helmreich, R.L., Wiener, E.L., and Kanki, B.G. (1993). The future of crew resource management in the cockpit and elsewhere. In E.L. Wiener, B.G. Kanki, and R.L. Helmreich (Eds.), *Cockpit Resource Management* (pp. 479–501). San Diego, CA: Academic Press.

Howard, S.K., Gaba, D.M., Fish, K.J., Yang, G., and Sarnquist, F.H. (1992). Anaesthesia crisis resource management training: teaching anaesthesiologists to handle critical incidents. *Aviation, Space, and Environmental Medicine*, 63, 763–770.

Jenkins, D.P., Stanton, N.A., Salmon, P.M., and Walker, G.H. (2009). *Cognitive Work Analysis: Coping with Complexity*. Aldershot, U.K.: Ashgate.

Kirkpatrick, D.L. (1976). Evaluation of training. In R.L. Craig (Ed.), *Training and Development Handbook: A Guide to Human Resource Development*. New York: McGraw-Hill.

Kirwan, B. and Ainsworth, L.K. (1992). *A Guide to Task Analysis*. London: Taylor & Francis.

Klein, G., Calderwood, R., and MacGregor, D. (1989). Critical decision method for eliciting knowledge. *IEEE Transactions on Systems, Man and Cybernetics*, 19(3), 462–472.

Klinect, J.R., Wilhelm, J.A., and Helmreich, R.L. (1999). Threat and error management: data from line operations safety audits. In *Proceedings of the Tenth International Symposium on Aviation Psychology* (pp. 683–688). Columbus, OH: Ohio State University Department of Aerospace Engineering, Applied Mechanics, and Aviation.

Koubek, R.J. and Liang, S. (1999). Training issues in industrial ergonomics. In W. Karwowski and W.S. Marras (Eds.), *The Occupational Ergonomics Handbook* (pp. 1641–1657). Boca Raton, FL: CRC Press.

Kozlowski, S.W.J. and Salas, E. (1997). A multi-level organisational systems approach for the implementation and transfer of training. In J.K. Ford, S.W.J. Kozlowski, K. Kraiger, E. Salas, and M. Teachout (Eds.), *Improving Training Effectiveness in Work Organizations* (pp. 247–287). Mahwah, NJ: Lawrence Erlbaum.

Lauber, J.K. (1984). Resource management in the cockpit. *Air Line Pilot*, 53, 20–23.

Machin, M.A. (2002). Planning, managing, and optimizing transfer of training. In K. Kraiger (Ed.), *Creating, Implementing, and Managing Effective Training and Development* (pp. 263–301). San Francisco, CA: Jossey-Bass.

McGhee, W. and Thayer, P.W. (1961), *Training in Business and Industry*. New York: John Wiley & Sons.

Militello, L.G. and Hutton, J.B. (2000). Applied cognitive task analysis (ACTA): a practitioner's toolkit for understanding cognitive task demands. In J. Annett and N.S Stanton (Eds.), *Task Analysis* (pp. 90–113). London: Taylor & Francis.

Naikar, N. and Sanderson, P.M. (1999). Work domain analysis for training-system definition. *International Journal of Aviation Psychology*, 9, 271–290.

Naikar, N., Hopcroft, R., and Moylan, A. (2005). *Work Domain Analysis: Theoretical Concepts and Methodology*, Tech. Report No. DSTO-TR-1665. Canberra: Australian Government, Department of Defence, Defence Science and Technology Organisation.

Newstrom, J. (1986) Leveraging management development through the management of transfer. *Journal of Management Development*, 5(5), 33–45.

O'Connor, P. and Flin, R. (2003). Crew resource management training for offshore oil production teams. *Safety Science*, 41, 591–609.

O'Hare, D., Wiggins, M., Williams, A., and Wong, W. (2000). Cognitive task analyses for decision centred design and training. In J. Annett and N. Stanton (Eds.), *Task Analysis* (pp. 170–190). London: Taylor & Francis.

Orasanu, J. and Fischer, U. (1997). Finding decisions in natural environments: the view from the cockpit. In C.E. Zsambok and G. Klein (Eds.), *Naturalistic Decision Making* (pp. 343–357). Mahwah, NJ: Lawrence Erlbaum.

Paris, C.R., Salas, E., and Cannon-Bowers, J.A. (2000). Teamwork in multi-person systems: a review and analysis. *Ergonomics*, 43(8), 1052–1075.

Patrick, J. (1992). *Training: Research and Practice*. San Diego, CA: Academic Press.

Roth, E.M., Patterson, E.S., and Mumaw, R.J. (2002). Cognitive engineering: issues in user-centred system design. In J.J Marciniak (Ed.), *Encyclopaedia of Software Engineering*, 2nd ed. (pp. 163–179). New York: John Wiley & Sons.

Rouiller, J.Z. and Goldstein, I.L. (1993). The relationship between organizational transfer climate and positive transfer of training. *Human Resource Development Quarterly*, 4(4), 377–399.

Saari, L.M., Johnson, T.R., McLaughlin, S.D., and Zimmerle, D.M. (1988). A survey of management training and education practices in U.S. companies. *Personnel Psychology*, 41, 731–743.

Salas, E. (2005). Team methods. In N.A. Stanton, A. Hedge, K. Brookhuis, E. Salas, and H. Hendrick (Eds.), *Handbook of Human Factors and Ergonomics Methods* (pp. 43.1–43.4). Boca Raton, FL: CRC Press.

Salas, E. and Burke, C.S. (2002). Simulation for training is effective when.... *Quality and Safety in Healthcare*, 11, 119–120.

Salas, E. and Cannon-Bowers, J.A. (1997). Methods, tools, and strategies for team training. In M.A. Quinones and A. Ehrensstein (Eds.), *Training for a Rapidly Changing Workplace: Applications of Psychological Research* (pp. 249–279). Washington, D.C.: American Psychological Association.

Salas, E. and Cannon-Bowers, J.A. (2000). The anatomy of team training. In S. Tobias and J.D. Fletcher (Eds.), *Training and Retraining: A Handbook for Business, Industry, Government, and the Military* (pp. 312–335). New York: Macmillan.

Salas, E. and Cannon-Bowers, J.A. (2001). The science of training: a decade of progress. *Annual Review of Psychology*, 52, 471–499.

Salas, E., Cooke, N.J., and Rosen, M.A. (2008). On teams, teamwork, and team performance: discoveries and developments. *Human Factors*, 50(3), 540–547.

Salas, E., Fowlkes, J.E., Stout, R.J., Milanovich, D.M., and Prince, C. (1999a). Does CRM training improve teamwork skills in the cockpit? Two evaluation studies. *Human Factors*, 41, 326–343.

Salas, E., Prince, C., Baker, D.P., and Shrestha, L. (1995). Situation awareness in team performance. *Human Factors*, 37, 123–136.

Salas, E., Prince, C., Bowers, C.A., Stout, R.J., Oser, R.L., and Cannon-Bowers, J.A. (1999b). A methodology for enhancing crew resource management training. *Human Factors*, 41, 161–172.

Salas, E., Wilson, K.A., Priest, H.A., and Guthrie, J.W. (2006). Design, delivery, and evaluation of training systems. In G. Salvendy (Ed.), *Handbook of Human Factors and Ergonomics*, 3rd ed. (pp. 475–512). New York: John Wiley & Sons.

Satish, U. and Streufert, S. (2002). Value of a cognitive simulation in medicine: towards optimizing decision making performance of healthcare personnel. *Quality and Safety in Health Care*, 11, 163–167.

Schmeiser, G., Bömmel, T.V., and Bühren, V. (2000). Crew resource management (CRM): cooperation of the police helicopter crew at mountain missions. *Air Medical Journal*, 19, 118.

Schraagen, J.M., Chipman, S.F., and Shalin, V.L. (2000). *Cognitive Task Analysis*. Mahwah, NJ: Lawrence Erlbaum.

Seamster, T. and Kaempf, G. (2001). Identifying resource management skills for pilots. In E. Salas, C. Bowers, and E. Edens (Eds.), *Improving Teamwork in Organizations: Applications of Resource Management Training*. Mahwah, NJ: Lawrence Erlbaum.

Smith-Jentsch, K.A., Baker, D.P., Salas, E., and Cannon-Bowers, J.A. (2001). Uncovering differences in team competency requirements: the case of air traffic control teams. In E. Salas, C.A. Bowers, and E. Edens (Eds.), *Improving Teamwork in Organizations: Applications of Resource Management Training* (pp. 31–54). Mahwah, NJ: Lawrence Erlbaum.

Snow, E.R. and Swanson, J. (1992). Instructional psychology: aptitude, adaptation, and assessment. *Annual Review of Psychology*, 43, 583–626.

Stanton, N.A. (2006). Hierarchical task analysis: developments, applications and extensions. *Applied Ergonomics*, 37, 55–79.

Tannenbaum, S.I. and Yukl, G. (1992). Training and development in work organisations. *Annual Review of Psychology*, 43, 399–441.

Tannenbaum, S.I., Cannon-Bowers, J.A., Salas, E., and Mathieu, J.E. (1993). *Factors That Influence Training Effectiveness: A Conceptual Model and Longitudinal Analysis*, NAWCTSD Tech. Rep. No. 93-011. Orlando, FL: Naval Air Warfare Center.

Tracy, J.B., Tannenbaum, S.I., and Kavanagh, M.J. (1995). Applying trained skills on the job: the importance of the work environment. *Journal of Applied Psychology*, 80(2), 239–252.

Vicente, K.J. (1999). *Cognitive Work Analysis: Towards Safe, Productive, and Healthy Computer-Based Work*. Mahwah, NJ: Lawrence Erlbaum.

Weiner, I.B., Freedheim, D.K., Schinka, J.A., Gallagher, M., Velicer, W.F., Goldstein, A.M., Nelson, R.J., and Lerner, R.M. (2003). *Handbook of Psychology*. New York: John Wiley & Sons.

Wiener, E.L., Kanki, B.G., and Helmreich, R.L. (1993). *Cockpit Resource Management*. San Diego, CA: Academic Press.

Wilson, K.A., Salas, E., Priest, H.A., and Andrews, D. (2007). Errors in the heat of battle: taking a closer look at shared cognition breakdowns through teamwork. *Human Factors*, 49, 243–256.

Wright, M., Turner, D., and Horbury, C. (2003). *Competence Assessment for the Hazardous Industries*, Research Report 086. London: HSE Books.

4 Procedures

4.1 OUTLINE

Operating procedures provide a description of the actions or steps required to perform a task safely and efficiently. A *procedure* has been defined as "a course or mode of action; a step taken" (Ockerman and Pritchett, 2001). The Health and Safety Executive (HSE, 2005a) defined a *procedure* within a work context as "the written description of the steps you need to take or follow to perform a task" (p. 1) and points out that procedures should not be confused with manuals, reference sets, or site and corporate standards. Ockerman and Pritchett (2001) suggested that a procedure "takes a large, complex task, capable of being approached in many different ways, and presents it as a sequence of manageable, communicable steps" (p. 193). The provision of appropriate, efficient procedures is essential for safe and efficient operator performance in safety-critical domains, particularly when faced with nonroutine or emergency situations. Problems associated with operating procedures, such as inappropriate or incorrect procedures or lack of procedures, have been implicated in a number of high-profile catastrophic accidents and incidents, including the Bhopal, Piper Alpha, and Clapham junction disasters (HSE, 2008).

4.2 KEY ISSUES

Although the specification of how to perform a task or job seems, on the face of it at least, to be a relatively simple endeavor, it is actually a complex process, and the end product can often be inappropriate, inaccurate, or unusable. A number of key issues surround the development and application of procedures, including issues related to their development and maintenance, their efficacy and usability, and worker compliance with them. An overview of these key issues is presented below.

4.2.1 Identification of Procedure Requirements

Not all tasks require procedures; also, different tasks require different kinds of procedures in terms of detail and format. The first key issue faced by procedure designers is the identification of those work tasks that require procedural aid. Organizations should have a formal process in place for identifying those safety-critical tasks that require procedures. Approaches presented in the literature (e.g., Embrey, 1989; Marsden, 1996) advocate task analysis, such as hierarchical task analysis (HTA) (Stanton, 2006), followed by error analysis to identify the likelihood of operators failing a task and the level of criticality associated with failing to perform the task successfully.

4.2.2 Development of Procedures

The process used to develop procedures is also critical and includes writing procedures, reviewing and refining them, testing and validating them, and approving them. The process should be consistent across the organization and should involve experienced operators where possible. In order to develop appropriate and efficient procedures, a formal procedural development process is required; procedures developed using informal methods are typically found to be incomplete, incorrect, or generally unrealistic (Marsden, 1996).

4.2.3 Training Staff in Procedures

Once a procedure is approved, the relevant staff must be trained in its use. Any updates in procedures should also lead to additional training. Also, retraining in procedures should be undertaken regularly, particularly for emergency procedures that are rarely used on a day-to-day basis. The HSE (2007) has suggested that organizations should have a formal mechanism in place to train staff in new or updated procedures.

4.2.4 Review, Maintenance, and Update of Procedures

Procedures typically have a finite lifetime. To remain valid, up to date, and usable, procedures must be continually reviewed, maintained, and updated as necessary. Procedures are likely to require updating in the event of new technologies, software, processes, or interfaces being introduced. Organizations should have a formal process in place for reviewing, maintaining, and updating procedures.

4.2.5 Format of Procedures

The format of a procedure is critical to its appropriate usage and efficacy and is an important consideration when developing procedures. Several considerations should be addressed when determining the appropriate format for a procedure, including the task in question, its complexity, the context in which the procedure is likely to be invoked, users that the procedure is intended for, the users' familiarity with the task, and the level of instruction and assistance required. Procedures can be presented in various formats, including checklists, step-by-step instructions, decision trees, standard operating instructions, and flowcharts.

4.2.6 Compliance with Procedures

Operator noncompliance with procedures is a common problem in safety-critical systems, and a large proportion of accidents and incidents occurring in these domains involve operators ignoring procedures (Cox and Cox, 1996; Dekker, 2003; Embrey, 1989; Marsden, 1996). Procedural noncompliance can either take the form of unintentional noncompliance, when the operator unknowingly deviates from a procedure, or intentional compliance, when the operator knowingly ignores a procedure. Organizations need to implement strategies designed to enhance operator compliance with procedures.

4.2.7 Efficacy of Procedures

The efficacy of procedures is obviously a critical factor in operator performance and system safety. The overall efficacy of a procedure is a function of various factors, including its accuracy, validity, usability, presentation, format, training, maintenance, and updating. Organizations should continually assess the efficacy of their procedures and update them where necessary.

4.2.8 Procedural Flaws

Various problems have been found with procedures in safety-critical domains (e.g., Cox and Cox, 1996, Dekker, 2003; Marsden, 1996; Ockerman and Pritchett, 2001). They have been found to be inefficient, burdensome, overly difficult, or incorrect (Ockerman and Pritchett, 2001), and they often are inaccessible, ambiguous, overly demanding on memory, and inadequate when used to diagnose plant problems (Cox and Cox, 1996).

4.3 STANDARDS AND GUIDELINES

4.3.1 *Revitalizing Procedures for Major Hazards*

The HSE's information sheet (HSE, 2007) on revitalizing procedures for major hazards provides an introduction to procedures and guidelines on how to develop and implement procedures. Guidance offered includes tips on how to encourage operator compliance with procedures, how to determine the format in which procedures should be presented, how to develop procedures involving end users, and three steps that organizations wishing to improve their procedures should take.

4.3.2 *Inspectors Toolkit: Human Factors in the Management of Major Accident Hazards*

The HSE's human factors toolkit (HSE, 2005b) contains a section on the reliability and usability of procedures which presents guidance on the development and implementation of procedures in safety-critical systems. The guidance offered includes advice on the preparation, format and layout, and enforcement of procedures, as well as questions designed to assess an organization's current set of procedures and procedure development, training, and maintenance processes.

4.3.3 1999 Control of Major Accident Hazard Regulations

The Control of Major Accident Hazard (COMAH) regulations passed in 1999 were designed to prevent and mitigate major accidents involving dangerous substances (e.g., chlorine, petroleum gas, arsenic pentoxide) that can cause harm to people or the environment (HSE, 2007). The HSE recommends that all COMAH sites should have a "procedure for managing procedures" that includes processes to identify which tasks require procedures, how detailed the procedures should be, how to keep procedures up to date, and how to make sure personnel comply with procedures.

4.3.4 Successful Health and Safety Management

Appendix 5 of *Successful Health and Safety Management* (HSE, 1997) provides guidance on the contribution of procedural failures on the causes of incidents and methods to mitigate them.

4.4 HUMAN FACTORS LITERATURE

4.4.1 Defining Procedures

Within a work context, procedures represent a specification of how to undertake a task or series of tasks. Lind (1979; cited in Dien, 1998) defined a *procedure* as:

> A set of rules (an algorithm) which is used to control operator activity in a certain task. Thus, an operating procedure describes how actions on the plant (manipulation of control inputs) should be made if a certain system goal should be accomplished. The sequencing of actions, i.e., their ordering in time, depends on plans structure and properties, the nature of the control task considered (goal), and operating constraints.

The HSE (2005a) defined procedures within a work context as "the written description of the steps you need to take to follow to perform a task" (p. 1). Ockerman and Pritchett (2001) suggested that a procedure "takes a large, complex task, capable of being approached in many different ways, and presents it as a sequence of manageable, communicable steps" (p. 193). Procedures therefore dictate how a task should be undertaken and are used to both guide and constrain operator activity so tasks can be completed safely and efficiently.

The provision of appropriate, valid, and usable procedures is essential for safe and efficient operator performance in the safety-critical domains, particularly when operators are faced with nonroutine or emergency situations. According to the HSE (2007), procedures are required to minimize operator errors and failures, to protect against loss of operating knowledge (i.e., when experienced personnel leave an organization), to standardize working practices, to provide a basis for training, and to meet statutory requirements. Procedural failures or noncompliance with procedures can have catastrophic consequences, and problems associated with operating procedures have been implicated in a number of catastrophic accidents and incidents, including the Bhopal, Piper Alpha, and Clapham junction disasters (HSE, 2008).

4.4.2 Models of Procedures

The literature on procedures (from a human factors perspective) is sparse, to say the least. Dekker (2003) outlined two different models of procedures that have been adopted within the safety-critical domains. The first model is based on the notion that procedures represent rules that should be followed:

- Procedures represent the best thought out and thus safest way to carry out a job.
- Procedure following is mostly characterized by simple if–then rule-based cognitive activities (if this situation occurs, then this algorithm applies).

- Safety results from people following procedures.
- For progress on safety, organizations must invest in people's knowledge of procedures and ensure that procedures are followed.

The second model describes procedure application as substantive cognitive activity and suggests that procedures do not represent the work itself; rather, workers must interpret procedures with regard to a collection of actions and circumstances that the procedures themselves can never specify. In Dekker's second model:

- Procedures are resources for action. They do not specify all of the circumstances to which they apply and cannot dictate their own application.
- Procedures alone cannot guarantee safety.
- The successful application of procedures across situations can be a substantive and skillful cognitive activity.
- Safety results from being skillful at judging when and how to adapt procedures to local circumstances.
- To achieve safety progress, organizations must monitor and understand the reasons underlying the gap between procedures and actual practice. Organizations must also develop ways that support workers' skill at judging when and how to adapt.

4.4.3 Developing Procedures

The procedure development process is a complex one involving the identification of the tasks that require procedures, a judgment on the level of assistance required, identification of the type or format of procedure required, writing and reviewing the procedure, and then obtaining approval for the procedure. Following this, official publication is required; staff should be trained in the procedure, and the procedure should be regularly reviewed and updated when required. It is essential, then, that organizations have a formal process in place for developing their procedures; however, Marsden (1996) observed that procedures are often developed using informal methods that typically rely on the knowledge of individuals familiar with the system but not with the actual behaviors required to operate it. Procedures developed via such means are often found to be incomplete, incorrect, or generally unrealistic. It is recommended that, to develop appropriate and efficient procedures, a formal procedural development process must be implemented. One such process is specified by Embrey (1989), who argued that the development of efficient procedures should involve six stages (at least) of preparation. Embrey's procedure development process is presented in Figure 4.1.

The first step entails analyzing the task in question using formal methods such as hierarchical task analysis (Stanton, 2006); see Chapter 2 for a full description of the HTA methodology. HTA works by decomposing activities into a hierarchy of goals, subordinate goals, operations, and plans that allows systems and goals, subgoals, and activities to be described exhaustively. Following task analysis, the next step involves undertaking human error analysis of the task in which potential errors are identified; according to Marsden (1996), this serves to highlight areas where the job is vulnerable

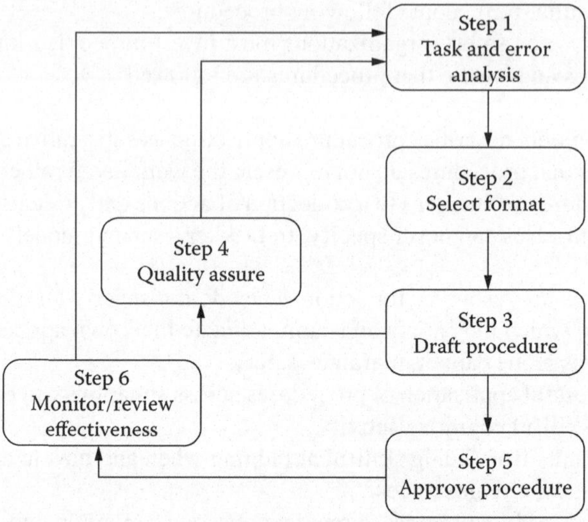

FIGURE 4.1 Procedural development process. (Adapted from Marsden, P., in *Human Factors in Nuclear Safety*, N. Stanton, Ed., Taylor & Francis, London, 1996, pp. 99–116.)

to the actions of the human operator. For error analysis purposes, approaches such as the Systematic Human Error Reduction and Prediction Approach (SHERPA) (Embrey, 1986) can be used; see Chapter 14 for a full description of SHERPA.

Following analysis of the task in question, the next step involves selecting an appropriate format for the procedure. Various forms of procedures exist, including step-by-step task descriptions, checklists, flowcharts, decision trees, and columnar procedures; more specific forms of procedures designed for certain situations also exist, such as accident management procedures, function-based procedures, and symptom-based procedures (Kirwan, 1994). According to Marsden (1996), several factors should be considered, the most important being task complexity and user familiarity. Low-complexity tasks with high operator familiarity require checklist-type procedures, whereas high-complexity tasks with low operator familiarity require step-by-step procedures. A matrix for determining the appropriate procedure format is shown in Figure 4.2. The HSE (2005a) also developed a flowchart for determining what type of procedure should be used which is presented in Figure 4.3.

Marsden (1996) suggested that, once an appropriate format for the procedure is selected, a draft document should be prepared with careful consideration given to the layout of the procedure. Following this, the draft procedure should be formally tested by suitable qualified inspectors within the operating environment where possible. Following appropriate testing, documentation, and refining of the procedure, it can then be approved for general use. The final step in the process involves maintenance and updating of the procedure to ensure its long-term usability.

With regard to the format and presentation of the procedure, the HSE guide recommends that overly long procedures should be divided into shorter chunks, that the use of all capital letters and small fonts should be avoided (as they are more difficult

Operator Familiarity with Task		*Task Complexity*	
		High	**Low**
	High	Checklist format plus supplementary aids (e.g., decision trees, visual aids, etc.)	Checklist procedures
	Low	Step-by-step procedures	Narrative procedures

FIGURE 4.2 Procedure format matrix. (Adapted from Marsden, P., in *Human Factors in Nuclear Safety*, N. Stanton, Ed., Taylor & Francis, London, 1996, pp. 99–116.)

to read), and that a consistent format should be used for all procedures. The HSE (2008) also outlined the following key principles of procedure design:

- Have an appropriate procedure for managing procedures that outlines how to identify which tasks require procedures and how procedures should be developed, complied with, and reviewed and updated.
- Use a format, style, and level of detail appropriate to the user, task, and consequences of failure.
- Use task analysis methods to inform the content of procedures.
- Encourage procedural compliance via user involvement and job design.
- Design out procedural noncompliance (violations).
- Consider links between procedures and competency.

4.4.4 When Procedures Fail

The majority of the literature on procedural failures comes from within the nuclear power domain (e.g., Marsden, 1996), where research suggests that procedural failures have been involved in around 70% of accidents and incidents within the nuclear power domain (Goodman and DiPalo, 1991; INPO, 1986; Marsden, 1996). Within the petrochemical industry, 27% of incidents involving operators are caused by situations for which there are inadequate or no procedures available (Bullemer and Nimmo, 1994).

Various researchers have attempted to identify the different problems involved when procedures have been found to fail. For example, in an analysis of procedural failures within the nuclear power domain, Morgenstern, Barnes, McGuire, Radford,

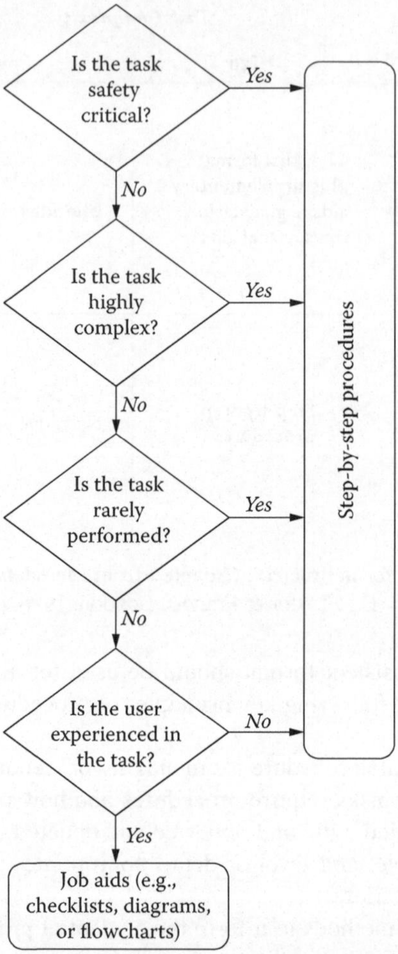

FIGURE 4.3 Procedure format guidance. (Adapted from Health and Safety Executive, *Procedures*, HSE Human Factors Briefing Note No. 4, Health and Safety Executive, London, 2005. Reproduced under the terms of the Click-Use Licence.)

and Wheeler (1987) concluded that procedural failures can be classified into the following four failure types: use of an incorrect procedure, use of the correct procedure at the wrong time, implementing a procedure too late, and failing to undertake a procedure in accordance with a written instruction. Also within the nuclear power domain, Green and Livingston (1992) identified the following eight procedural failure types:

1. Technical accuracy or completeness of the procedure at fault
2. Poor formatting of the procedural document
3. Language or syntax problems associated with the procedural document
4. Inadequate procedure development process

TABLE 4.1
Organizational Weaknesses Involved in Procedural Failures

Procedure Preparation	User Compliance	Procedure System Infrastructure
Incorrect procedure prepared	Intentional noncompliance	Procedure or training failure
Inaccurate procedure prepared	Bad habits	Failure to revise procedure after changes to work practice after plant modifications
No procedure prepared	Routine violations	
	Unintentional noncompliance	
	Slips	
	Lapses	

Source: Adapted from Marsden, P., in *Human Factors in Nuclear Safety*, N. Stanton, Ed., Taylor & Francis, London, 1996, pp. 99–116.

5. Procedure poorly located or cross-referenced
6. Insufficient verification or validation of the procedural document
7. Failure to revise the procedure
8. Training and procedures poorly interrelated

As opposed to the operator-centric view of the two studies described above, Marsden (1996) instead viewed procedural failures from an organizational (rather than operator-focused) perspective, suggesting that the factors involved are more likely to be found within the infrastructure of the organization as opposed to the fallibility of the human operators involved. Three basic failure types involved in procedural failures include weaknesses in the procedure preparation process, problems emerging from the failure of employees to follow procedures, and breakdowns within the infrastructure of the procedure system. See Table 4.1 for Marsden's taxonomy of organizational weaknesses involved in procedural failures.

One of the main problems associated with procedures is the operator's perceptions of them. According to Cox and Cox (1996), procedures are often perceived by workers to be inaccessible, ambiguous, overly demanding on memory, and inadequate when used to diagnose plant problems; further, Ockerman and Pritchett (2001) observed that procedures are often seen to be inefficient, burdensome, overly difficult, or just plain wrong. They also suggested that procedures may often be intentionally ignored because workers perceive more drawbacks to following the procedure than benefits.

Dekker (2003) presented a summary of the some of the more prominent problems associated with the use of procedures in safety-critical systems. First, there is often a mismatch between procedures and actual practice, and this mismatch is not unique to accident sequences. Second, work takes place within a context of limited resources and multiple goals and pressures; the corollary of this is that it can often be impossible to strictly follow procedures and get the job done satisfactorily, so informal work systems are used instead. Third, procedure following can be antithetical to safety. By way of example, Dekker described the Mann Gulch fire that occurred in 1949 in Montana. Firefighters who followed procedures stating that they should carry their tools everywhere perished; ultimately, the firefighters in this case faced a choice between following procedures or surviving. In his discussion of these and other problems, Dekker (2003, p. 234) mused:

> Procedures are in investment in safety—but not always. Procedures are thought to be required to achieve safe practice—yet they are not always necessary, nor likely ever sufficient for creating safety. Procedures spell out how to do the job safely—yet following all the procedures can lead to an inability to get the job done.

Marsden (1996) identified three common organizational weaknesses that are involved in procedure system failures. First, the organization involved fails to specify a formal procedure development process, which may lead to incomplete, inaccurate, or even missing procedures. Second, workers fail to follow procedural instructions. Finally, a breakdown occurs in the interrelationship between the procedure system and other operator support aspects such as training and procedure revision.

4.4.5 Procedural Noncompliance

Workers do not always follow procedures (Dekker, 2003). One of the most prominent problems associated with the use of procedures is operator noncompliance. Violations of procedures are a significant cause of industrial accidents and are a frequent cause of production losses, poor quality, and unreliable maintenance (HSE, 1995). According to Dekker (2003), there is a history of industrial disasters where a lack of procedural compliance played a role. Procedural noncompliance can take the form of unintentional noncompliance, where the operator unknowingly deviates from a procedure due to factors such as memory lapses or misinterpretation, or intentional noncompliance, where the operator knowingly ignores a procedure, often due to a perception that following the procedure offers more drawbacks than benefits (Ockerman and Pritchett, 2001).

Procedural noncompliance is related to the complex error category that is known as *violations* (Reason, 1990), which are categorized as any behavior that deviates from accepted procedures, standards, and rules. Violations can be either deliberate or erroneous (Reason 1997). Deliberate violations occur when an actor deliberately deviates from set rules or procedures; for example, a driver who is consciously exceeding the speed limit is committing a deliberate violation. Erroneous or unintentional violations, however, occur when an actor unintentionally deviates from a set of rules or procedures; for example, a driver who unintentionally exceeds the speed limit, due to either not comprehending the speed of the vehicle or not being aware of the current speed limit, is committing an erroneous or unintentional violation.

Reason (1997) further identified three types of deliberate violations: routine, optimizing, and necessary. Routine violations involve taking short-cuts through procedures to achieve a particular task. Optimizing violations involve optimization of nonfunctional goals during task performance. Reason (1997) cited the example of a driver wishing to get from A to B who optimizes his speed and indulges in aggressive instincts during the journey from A to B. Necessary violations involve essential deviation from the rules to achieve a particular task; for example, when a routine procedure in the workplace is not working, the use of a nonroutine procedure to get the job done would be considered a necessary violation.

Embrey (1989) suggested that a large portion of accidents involve operators intentionally ignoring known procedures (Cox and Cox, 1996). Embrey (1989) outlined the following four reasons why operators may opt not to follow procedures:

1. The procedures in question do not correspond to the way in which the task is actually undertaken.
2. The information within the procedures, despite being correct, is not presented in a usable form.
3. The procedures are not updated on the basis of operational experience.
4. Rules and procedures are not seen to be applicable to the operators or situation in question.

The HSE (2007) suggested that procedural noncompliance occurs for a variety of reasons: The procedures may be incorrect or out of date, they are difficult to use or follow, they are not readily available, or there are easier ways to undertake the task in question. Also, noncompliance may be due to peer pressure, a failure to understand the risks involved, or perceived pressure from management to get the job done.

4.5 GUIDANCE AND RECOMMENDATIONS

Based on a review of the literature, a flowchart depicting the key stages involved in the development, application, and maintenance of procedures has been developed. Guidance surrounding each aspect of the flowchart is outlined in Figure 4.4.

4.5.1 Identify Tasks Requiring Procedures

The process begins with identifying those tasks that require procedural aid; not all tasks require the provision of procedures. In order to identify tasks requiring procedures, it is necessary first to fully understand the range of tasks being undertaken within a system and, second, to understand the potential for error associated with each task and the resultant impact on safety. HTA (Stanton, 2006) can be used to exhaustively describe a system in terms of its goals, subgoals, and component tasks. HTA works by decomposing activities into a hierarchy of goals, subordinate goals, operations, and plans that allows systems and goals, subgoals, and activities to be described exhaustively; it focuses on "what an operator … is required to do, in terms of actions and/or cognitive processes to achieve a system goal" (Kirwan and Ainsworth, 1992, p. 1). HTA outputs, therefore, specify the overall goal of a particular system, the subgoals to be undertaken to achieve this goal, the operations required to achieve each of the subgoals specified, and the plans in place to ensure that the goals are achieved.

For error analysis purposes, human error identification (HEI) approaches such as SHERPA (Embrey, 1986) can be used; see Chapter 14 for a full description of SHERPA. SHERPA uses a behavioral taxonomy linked to an error mode taxonomy and is applied to an HTA of the task under analysis. The behavioral and external

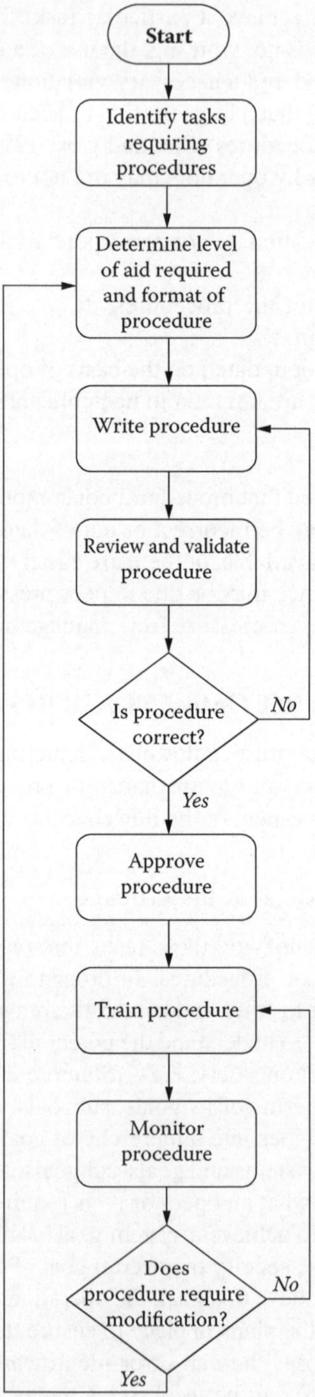

FIGURE 4.4 Procedure development and maintenance flowchart.

error mode (EEM) taxonomies are used to identify credible errors that are likely to occur during each step in the HTA. For each credible error identified, the analyst provides a description of the form that the error would take, such as "pilot dials in wrong airspeed," and identifies any consequences associated with the error and any recovery steps that would have to be taken when the error is made. Finally, ordinal probability (low, medium, or high), criticality (low, medium, or high), and potential design remedies are recorded.

Following the task and error analyses, a judgment should be made on which tasks require procedures. This should be based on task complexity, error potential, and the consequences associated with the errors that are likely to be made.

4.5.2 Determine Level of Procedural Aid Required and Format of Procedure

Different levels of assistance can be provided via procedures. This step involves determining what level of procedural aid is required for the task in question and what format the procedure should take. If the task is highly complex and nonroutine, then detailed, step-by-step procedures are likely to be required. On the other hand, if the task is low in complexity and the operator is likely to be experienced in the task, then checklists, flowcharts, or diagram-based procedures may be sufficient.

4.5.3 Write and Review Procedure

When the level of aid and the format of the procedure have been determined, a draft of the procedure should be produced. Involving the operators who undertake the task in question as part of their job is recommended. Not only does this allow their experience in the task to inform the procedure, but it also gives them a sense of involvement and ownership in the procedure that is likely to enhance their compliance with it. The procedure itself should be clearly and concisely written and easy to read (e.g., avoid the use of all capital letters and small fonts) and should be written in a consistent fashion with other procedures used within the organization in question.

In terms of content, the HSE (1999) suggested that procedures should contain the following information:

- The purpose of the procedure
- Any precautions that must be observed to avoid potential hazards
- Descriptions of any tools or equipment required
- Specification of the initial conditions that must be satisfied before commencing the procedure
- References to any other relevant documentation (e.g., manuals)
- Description of the procedural steps required to perform the task safely and efficiently

When a suitable draft has been developed, it should be formally tested and reviewed by qualified inspectors within the operating environment where possible (Marsden, 1996). Reviewers of the procedure should include operators experienced in the task, supervisors, and managers. Following this, any changes required should then be made to the procedure and it should be tested and reviewed again. This cycle of writing, reviewing, testing, and refining the procedure should continue until the procedure is certified as fit for purpose.

4.5.4 Approve Procedure

The next step involves formally approving the procedure. Organizations should have a formal procedure approval process in place for this purpose. The step should involve establishing a procedure approval panel who meet specifically for the purpose of approving procedures. If the procedure is not ready for approval at this stage (i.e., it requires major corrections), then it should be placed back into the write and review cycle. Procedures requiring minor or no corrections should be approved accordingly.

4.5.5 Train Procedure

Once approved, the relevant workers should be trained in the procedure in question. Workers should also be given further training when a procedure is updated or changed in any way. Also, for emergency or nonroutine procedures, workers should be given regular retraining.

4.5.6 Monitor Procedure

Throughout its lifecycle, the procedure should be continually monitored. This is essential for a number of reasons. First, it is necessary to ensure that the procedure is being used properly. If this is not the case, then steps should be taken to find out why the procedure is not being used correctly so it can be refined accordingly. The procedure may be unclear or may contain mistakes. Second, procedures should be continually monitored to ensure that they are still applicable and appropriate. Changes in the task and technology used may lead to procedures becoming outdated and inappropriate or unfit for their purpose. Finally, procedures should be monitored to ensure that they are being used or complied with. As noted previously, operator noncompliance is a significant problem, so organizations need to ensure that procedures are actually being used. If this is not the case, then the organization should identify why operators are not using a particular procedure; it may be that they have discovered a more appropriate (quicker or easier) approach to undertaking the task in question. Operator feedback is critical when monitoring procedures, and the users of the procedures are likely to be a rich source of information regarding the procedures being monitored, including any problems associated with the procedures and how difficult the procedure is to use. It is recommended that organizations involve end users in the procedure monitoring process and that they have a reporting system in place that allows workers to report any issues associated with the procedures they use.

Any problems identified with the procedure in question would then lead to the procedure being modified accordingly using the process outlined above. The process should continue in an iterative cycle throughout the lifecycle of the procedure.

4.6 ASSESSMENT AND EVALUATION

The assessment and evaluation process of an organization's procedures is a complex one that involves various factors. When assessing procedures it is vital to assess both the procedures themselves and the processes that the organization has in place with regard to procedure development, writing, approval, training, and maintenance. For the procedures themselves, Ockerman and Pritchett (2001) suggested that the efficacy of a procedure can be determined by measuring the following attributes:

- *Comprehensiveness of the procedure*—A procedure's comprehensiveness relates to the range of situations, environments, and operations that the procedure is applicable to.
- *Detail of the procedure*—The level of detail of the procedure relates to the degree to which each step within the procedure specifies the exact actions required by the user.
- *Accuracy of the procedure*—The accuracy of the procedure relates to the extent to which completion of the steps specified will produce the desired results.

It is possible to identify other features of procedures from the literature that should be assessed. Other factors that should be assessed, for example, include the appropriateness of the format used, the consistency of the procedure, the overall ease of use of the procedure, and the clarity of the procedure.

To assess the processes that an organization has in place regarding the development, writing, approval, training, and maintenance of procedures, the HSE's document regarding the reliability and usability of procedures suggests the questions shown in Table 4.2. The process advocated by the HSE focuses not only on the procedures themselves but more so on the processes that the organizational has in place to identify those tasks that require procedures, approving procedures, selecting the format of the procedures, writing procedures, maintaining procedures, removing out-of-date procedures, training staff in the use of procedures, and ensuring staff compliance with procedures. Further, aspects such as staff access to procedures and usage of procedures are considered. The process presented in Table 4.2 involves assessing each factor from the point of view of the site in question and an appropriate independent assessor. Any improvements required are duly noted.

4.7 EXAMPLE

4.7.1 Procedure Development Process

For the purposes of this worked example, the current process for developing, writing, and approving procedures adopted by the example organization was compared to the process depicted in the guidance and recommendations section. Following this, the current procedures for five specific tasks were evaluated using a set of criteria derived from the human factors literature on procedures. The current process adopted by the example organization is presented in Figure 4.5.

TABLE 4.2
Reliability and Usability of Procedures Assessment Questions

Question	Site Response	Inspectors' View	Improvements Needed
1. Is there a formal process in place to determine which safety criteria operations or tasks require procedures (e.g., HAZOPs, risk assessment)?	______	______	______
2. Do the procedures in existence cover the range of areas or operations expected, such as: Maintenance operations? Plant startup and shutdown? Plant operation? Training and competency arrangements?	______	______	______
3. Is there a process in place to consider how the work activities of non-company personnel are managed?	______	______	______
4. Is there an approvals process for operating procedures? Who is responsible for managing this?	______	______	______
5. Are the format, type of procedure, and user considered when writing procedures?	______	______	______
6. Are the types of procedures (e.g., checklists, instructions, flow sheets) appropriate for: Routine operations? Safety-critical operations? Emergency and upset conditions?	______	______	______

7.	Is the procedure consistent across the site or plant where the tasks or operations are identical?			
8.	Are operators involved in the identification and writing of procedures?			
9.	Do operators have ready access to procedures in the normal working environment? How often do they refer them? Why do they refer them?			
10.	Is there a process in place to ensure that procedures remain valid and are up to date? Does this procedure consider both hardcopies and electronically held documents?			
11.	Are formal mechanisms in place to remove all out-of-date procedures? Does this occur in practice, or are older versions still found in control rooms, maintenance offices, etc.?			
12.	Is there a formal mechanism in place to ensure that staff are trained in new or updated procedures?			
13.	Is there an ongoing monitoring system to ensure compliance with procedures? Do the results of this monitoring feed back into the review, revision, validation process?			
14.	Is there a process to ensure that relevant procedural controls are reviewed following an incident or audit noncompliance?			

Source: Health and Safety Executive, *Inspectors Toolkit, Human Factors in the Management of Major Accident Hazards*, Health and Safety Executive, London, 2005. Reproduced under the terms of the Click-Use Licence.

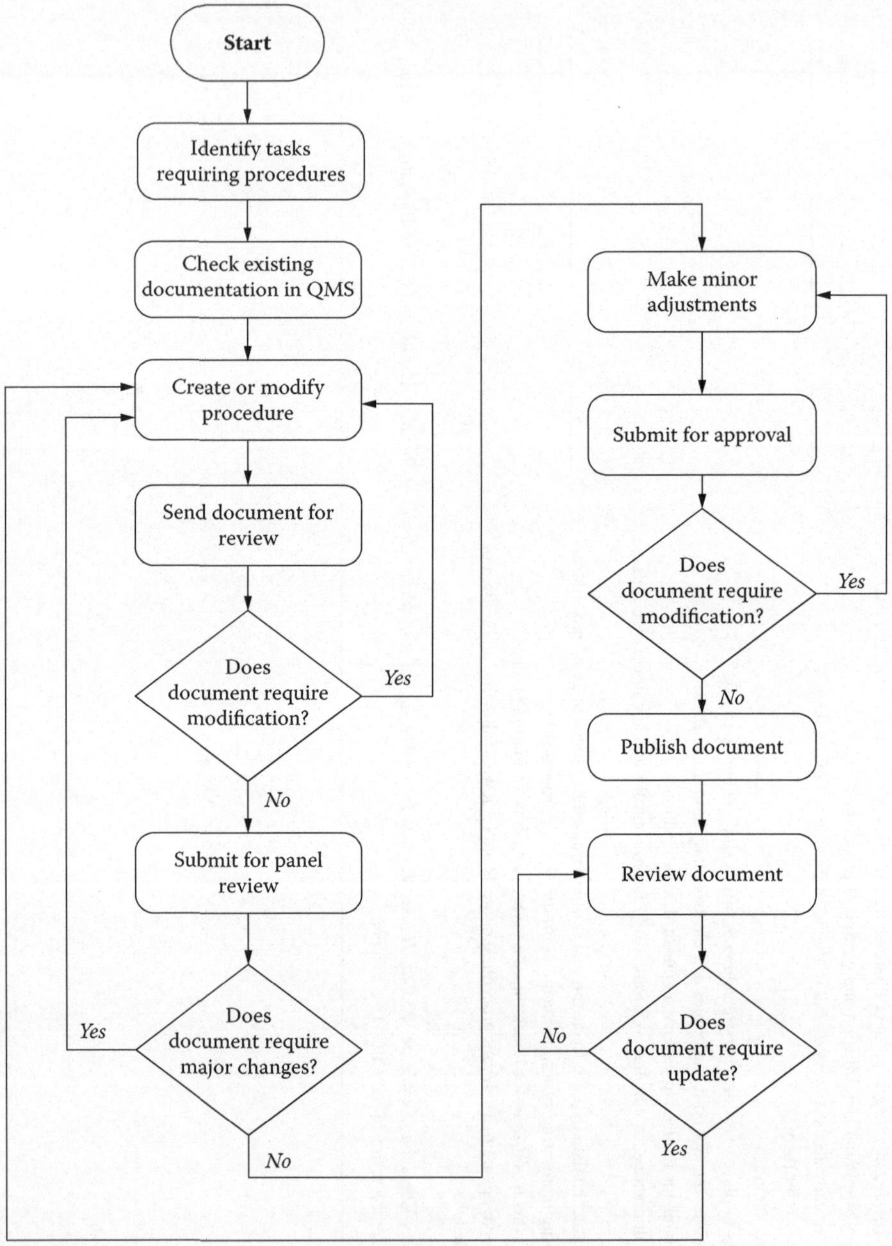

FIGURE 4.5 Quality management system (QMS) procedure process.

The process begins by identifying those tasks that require procedures or those procedures that require modification of some sort (no methods for identifying tasks requiring procedures are specified). Following this, the relevant documentation is reviewed within the quality management system (QMS). If a new document is required, then the "create document" feature is used. This involves initially selecting a document type (e.g., standard instruction, process, STC, training, or other document), for which there is no guidance. Following this, the user enters various details (e.g., category, team name, coauthors, review date) and then selects the procedure type category (e.g., process flow map, process flow diagram, training levels, STC). Following numbering of the document, the user then creates the document as appropriate.

When it has been completed, the document is sent out for review, which involves selecting appropriate reviewers (e.g., line managers, customers) and completing a document overview template, which involves completing a distribution list, detailing the reason for a new document, and stating whether the procedure is permanent or temporary and whether or not training is required for the new procedure. When this is complete, the document is then sent for review, a process that can take up to 3 weeks. Each reviewer is required to review procedural documents and make comments on the technical content and ensure that the document follows a consistent standard. The author of the procedure examines the reviewers' comments and updates the document as required. The document then continues through the review cycle until it requires no further amendments.

Following this, the document is ready for a panel review by the challenge and review panel, which is comprised of line managers. The panel review requires the document administrator to present the document to the panel, who can either accept the document without change or notify the author of any changes required. After being cleared by the panel review, the document is then submitted for approval at management level and upon approval is published by the document administrator. The published document contains a future review date. Once published, documents can be changed via the QMS. The modified document is then subject to the review cycle. Periodic reviews of the QMS documents are also carried out based on the review dates attached at the time of publishing.

Compared to the guidelines presented in this document and in the literature, the process adopted by the organization in question broadly satisfies the majority of the requirements stated. The process involves developing, reviewing, and refining procedures until they are valid, and it has three separate review cycles (i.e., initial review, challenge and review panel, and management approval) which should ostensibly ensure the accuracy, level of detail, and appropriateness of the procedures being developed. The development, review, and approval process is also undertaken for the QMS which ensures consistency levels and involves operators and managers. Following approval and publishing, the procedures are also reviewed and updated accordingly which should ensure that they remain appropriate throughout their lifetime.

The main downside of the QMS process analyzed here appears to be a distinct lack of guidance for procedure developers; for example, there are no guidelines for identifying tasks that require procedures or procedures that require updating, and no guidance is provided on how to determine what type or format of procedures is used. Also, no reference is made to the training of procedures (apart from procedure developers having to denote whether or not training is required for the new procedure being developed).

4.7.2 Evaluation of Existing Procedures

Based on a set of procedures review criteria extracted from the literature (e.g., Embrey, 1989; Marsden, 1996) five existing procedures from the organization in question were evaluated:

1. Control room information recording QMS procedure
2. Telemetry blackout plan QMS procedure
3. Evacuating emergency control center QMS procedure
4. Global alarm response database guidelines procedure
5. Interruption monitoring and review in QMS procedure

The following criteria were used to evaluate the procedures outlined above:

- Format refers to the appropriateness of the format used for the procedures based on Marsden's (1996) procedure format matrix (e.g., high task familiarity and complexity require checklist format plus supplementary aids, whereas high task familiarity and low complexity require checklist procedures).
- Comprehensiveness of the procedures relates to the extent to which it contains all of the information that the literature suggests procedures should include, such as purpose and scope of the procedure, hazards and precautions, related documents, tools and equipment, triggering conditions, and procedural steps.
- Consistency refers to the consistency of the procedure document compared with other procedures used by the organization in question.
- Accuracy of the procedure relates to the extent to which completion of the steps specified will produce the desired results (Ockerman and Pritchett, 2001). In this case, the accuracy of the procedures was evaluated against HTAs of the relevant tasks.
- Detail of the procedure relates to the degree to which each step within the procedure specifies the exact actions required by the user (Ockerman and Pritchett, 2001).
- Clarity of the procedure relates to the level of clarity of the information presented. The information contained within the procedure should be presented in a clear and concise manner at all times.
- Layout of the procedure should be clear and appropriate at all times.
- Ease of use of the procedure relates to how easily users can follow and implement the procedure during operations.
- Review date refers to the specification of a review date. Each procedure should have an appropriate review date specified.

The five procedures outlined above were evaluated using the criteria identified above. The results are presented in Table 4.3.

In general, the procedures analyzed achieved satisfactory ratings for each of the criteria evaluated. Aside from the control room information recording procedure (which should have been in a checklist format), all of the procedures were presented in

TABLE 4.3
Procedure Evaluation Results

Procedure	Format	Comprehensiveness	Consistency	Accuracy	Detail	Clarity	Layout	Ease of Use	Review Date Specified?
Control room information recording procedure	Inappropriate	Medium	High	High	Medium	Medium	Medium	Medium	Yes
Telemetry blackout plan	Appropriate	High	High	High	High	Medium	Medium	Medium	Yes
Emergency control center evacuation procedure	Appropriate	High	High	High	High	Medium	Medium	Medium	Yes
Global alarm response database guidelines procedures	Appropriate	High	High	High	Medium	Medium	Medium	Medium	Yes
Interruption monitoring and review procedure	Appropriate	Medium	High	High	Medium	Medium	Medium	Medium	Yes

the appropriate format based on their complexity and familiarity levels. The majority of the procedures were also comprehensive in that they contained adequate descriptions of the purpose and scope of the procedure, step-by-step task instructions, review dates, related documentation, and conditions. The level of consistency between the procedures evaluated was also high, and all were presented in a standard and consistent manner. The level of accuracy, based on a comparison with HTAs for the tasks in question, was high for all procedures. It was felt that the detail levels of the control room information recording procedure and the global alarm response database procedures were lacking slightly. All of procedures only achieved medium ratings for level clarity, layout, and ease of use. In most cases, the layout of the documents was poor (some of the flowchart diagrams were cut off by the page layout) and large chunks of text were used in some cases which impacted the clarity and ease of use ratings.

To further enhance the efficacy of the procedures evaluated, it is recommended that the content should be expressed more concisely (remove long paragraphs of text), that more clear and usable diagrams (e.g., flowcharts) should be developed, and that appropriate formats be used for the procedures.

BIBLIOGRAPHY

Bullemer, P. and Nimmo, I. (1994). Understanding and supporting abnormal situation management in industrial process control environments: a new approach to training. In *Proceedings of the 1994 International Conference on Systems, Man, and Cybernetics* (pp. 391–396). Piscataway, NJ: IEEE.

Cox, T. and Cox, S. (1996). Work-related stress and control room operations in nuclear power generation. In N. Stanton (Ed.), *Human Factors in Nuclear Safety* (pp. 251–276). London: Taylor & Francis.

Dekker, S.W.A. (2003). Failure to adapt or adaptations that fail: contrasting models on procedures and safety. *Applied Ergonomics*, 34, 233–238.

Dien, Y. (1998). Safety and application of procedures, or "how do 'they' have to use operating procedures in nuclear power plants?" *Safety Science*, 29, 179–187.

Embrey, D.E. (1986). SHERPA: A Systematic Human Error Reduction and Prediction Approach. Paper presented at the International Meeting on Advances in Nuclear Power Systems, Knoxville, TN, April 1986.

Embrey, D.E. (1989). The Management of Risk Arising from Human Error. Paper presented at the Conference on Human Reliability in Nuclear Power, London, October 1989.

Goodman, P.C. and DiPalo, C.A. (1991). Human Factors Information System: A Tool to Assess Error Related to Human Performance in U.S. Nuclear Power Plans. In *Proceedings of the Human Factors Society 35th Annual Meeting* (pp. 662–665). Santa Monica, CA: Human Factors Society.

Green, M.A. and Livingston, A.D. (1992). Procedures development: common deficiencies and possible solutions. In *International Atomic Energy Authority International Working Group on Nuclear Power Plant Control and Instrumentation, Specialist Meeting on Operating Procedures for Nuclear Power Plants and Their Presentation*, Vienna, Austria.

Health and Safety Executive (HSE). (1995). *Improving Compliance with Procedures: Reducing Industrial Violations*. London: HSE Books.

Health and Safety Executive (HSE). (1997). *Successful Health and Safety Management.* London: HSE Books.

Health and Safety Executive (HSE). (1999). *Reducing Error and Influencing Behavior*. London: HSE Books.

Health and Safety Executive (HSE). (2005a). *Procedures*, HSE Human Factors Briefing Note No. 4. London: Health and Safety Executive (http://www.hse.gov.uk/humanfactors/comah/04procedures.pdf).

Health and Safety Executive (HSE). (2005b). Reliability and usability of procedures. In *Inspectors Toolkit: Human Factors in the Management of Major Accident Hazards*. London: Health and Safety Executive (http://www.hse.gov.uk/humanfactors/comah/toolkit.htm).

Health and Safety Executive (HSE). (2007). *Revitalising Procedures for Major Hazards*. London: Health and Safety Executive (http://www.hse.gov.uk/humanfactors/comah/procinfo.pdf).

Health and Safety Executive (HSE). (2008). *Human Factors: Procedures*. London: Health and Safety Executive (http://www.hse.gov.uk/humanfactors/comah/procedures.htm).

Institute of Nuclear Power Operations (INPO). (1986). *1985 Annual Report of the Institute of Nuclear Power Operations*. Atlanta, GA: Institute of Nuclear Power Organizations.

Kirwan, B. (1994). *A Guide to Practical Human Reliability Assessment*. London: Taylor & Francis.

Kirwan, B. and Ainsworth, L.K. (1992). *A Guide to Task Analysis*. London: Taylor & Francis.

Lind, M. (1979). The Use of Flow Models for Design of Plant Operating Procedures. IWG/NPPCI Specialist Meeting on Procedures and Systems for Assisting an Operator During Normal and Anomalous Nuclear Power Plant Operation Situations, Garching, Germany, December 1979.

Marsden, P. (1996). Procedures in the nuclear industry. In N. Stanton (Ed.), *Human Factors in Nuclear Safety* (pp. 99–116). London: Taylor & Francis.

Morgenstern, M.H., Barnes, V.E., McGuire, M.V., Radford, L.R., and Wheeler, W.A. (1987). *Study of Operating Procedures in Nuclear Power Plants*, NUREG CR-3968. Seattle, WA: Battelle Human Affairs Research Center.

Ockerman, J. and Pritchett, A. (2000). A review and reappraisal of task guidance: aiding workers in procedure following. *International Journal of Cognitive Ergonomics*, 4(3), 191–212.

Reason, J. (1990). *Human Error*. Cambridge, U.K.: Cambridge University Press.

Reason, J. (1997). *Managing the Risks of Organizational Accidents*. Burlington, VT: Ashgate Publishing.

Stanton, N.A. (2006). Hierarchical task analysis: developments, applications and extensions. *Applied Ergonomics*, 37, 55–79.

5 Communication Process

5.1 OUTLINE

Communication refers to "information exchanged" or the presence of a "connection." Communication can occur between people in any medium available (e.g., voice, telephone, notes, e-mail), as well as between people and technology (e.g., via information displays and controls). Poor communications in control rooms can result in "missing information, unnecessary information, inaccurate information, poor or variable quality of information, misunderstandings, failing to carry forward information over successive shifts" (HSE, 2007a, p. 1). It is not difficult to appreciate that ineffective information exchange and the lack of an appropriate connection have led to a number of major incidents.

The topic of control room communications is wide ranging and takes in:

- Design of the Supervisory Control and Data Acquisition (SCADA) system
- Design of the alarm system
- Layout of the control room
- Shift handovers
- Teamworking
- Wider concerns about the arrangements for safety critical communications within the organization

5.2 KEY ISSUES

Given that many specific communications issues are dealt with elsewhere, it falls to this chapter to focus on the more generic but no less important aspects. This involves understanding:

- The control room as a sociotechnical system (an interacting mixture of people and technology)
- The structure of communication (who is communicating to whom)
- The process of communication (when are they communicating)
- The function that communication ultimately serves (in terms of shared awareness and teamworking)

5.3 STANDARDS AND GUIDELINES

BS/EN/ISO 11064-6 (Ergonomic Design of Control Centers. Part 6.Environmental Requirements for Control Centers) touches on communications requirements, in particular in the design of the acoustical environment. It states: "Acoustic design should take account of … verbal communications between operators; telephone conversations; hearing alarms … ." Environmental requirements are dealt with fully in Chapter 13. Issues concerning communications, particularly the structure of communications in a control room, are dealt with in BS/EN/ISO 11064-2 (Part 2. Principles for the Arrangement of Control Suites) and in BS/EN/ISO 11064-3 (Part 3. Control Room Layout). The following guidelines based on these standards can be distilled:

- Task zones of individuals requiring frequent interaction with each other should be located close to each other.
- Control suite equipment should be located such that visual contact between operators is possible where and when needed.
- Communications that are extraneous to the control suite functions should not distract personnel.
- Rooms and locations with different functions should be physically separated to avoid potential sources of disturbance.
- The layout of control rooms, where many of the control room operators work, should facilitate teamworking opportunities and social interaction for operators where this factor is deemed important for the primary tasks to be performed.
- Optimizing key operational links, including sightlines, or direct speech communication should be a goal in control room layouts.
- Layouts should allow, wherever practical, direct verbal communication between the control room operators.

Control room layout is dealt with fully in Chapter 10.

The Health and Safety Executive (HSE) recognizes the role of communication processes: "Effective communication is important in all organizations" (HSE, 2007b, p. 2). In particular, "critical times when good communication must be assured include: at shift changeover, between shift and day workers, between different functions of an organization within a shift (e.g., operations and maintenance), and during process upsets and emergencies." From an HSE perspective, effective communications also relate to teamwork, which is considered a psychosocial risk factor.

5.4 HUMAN FACTORS LITERATURE

5.4.1 The Control Room as a Sociotechnical System

Sociotechnical system is a descriptive term given to any practical instantiation of social (e.g., people, teams) and technical (e.g., SCADA systems, displays, alarms) elements engaged in purposeful, interacting, goal-directed behavior. With an appropriate emphasis on a system of sociotechnical systems, the following key features can be noted:

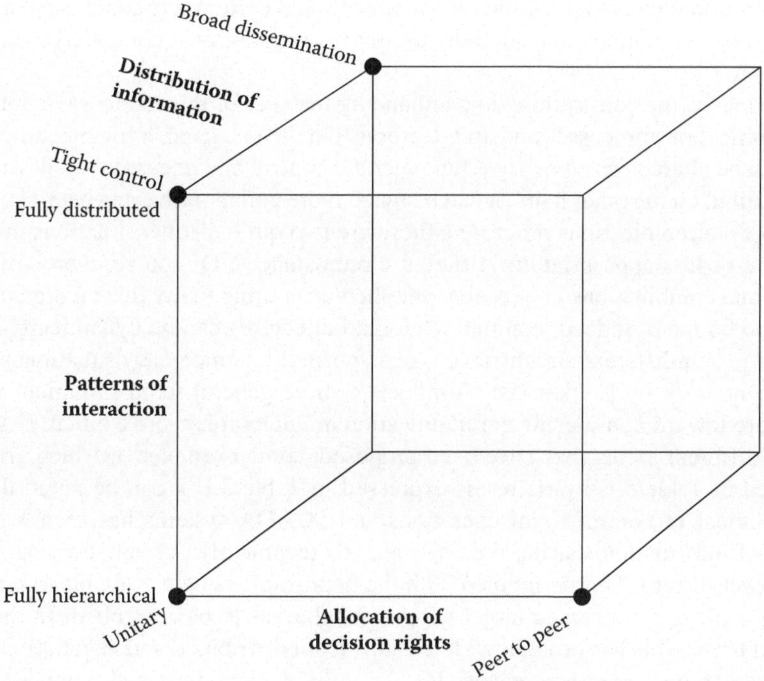

FIGURE 5.1 NATO SAS-050 model of command and control. Different types of teamworking are possible within this three-dimensional space. (From NATO, *SAS-050 Conceptual Model Version 1.0*, North Atlantic Treaty Organization, Brussels, Belgium, 2006.)

- A sociotechnical system is characterized by communication and processing of information.
- A sociotechnical system recognizes that artifacts as well as people can communicate, hold, and process information.
- A sociotechnical system's activities occur over time; thus, it is dynamic.

A particularly useful way of characterizing sociotechnical systems, one with an emphasis on communication processes, is the NATO (2006) SAS-050 conceptual model (Figure 5.1). The model proposes three interlinking factors that characterize the way in which communication and teamworking take place:

- Who is permitted to make decisions (e.g., are decisions made centrally or are they reached through collaboration)
- The nature of the interactions that are permitted or facilitated (e.g., from tightly constrained, in which the optimum means to an end are well specified, to unconstrained, in which interactions are only minimally specified)
- The distribution of information (e.g., from tight bureaucratic control, in which role incumbents only receive the information that the organizational structure deems necessary for that role, through to broad dissemination, in which everyone knows everything)

An appropriate metaphor for the two organizational extremes is comparing an organization or communications architecture that is programmed compared to one that can learn.

Obviously, the constraining and enhancing features of this model say a lot about communications processes and architectures. On the one hand, a sociotechnical system can be characterized by rigid bureaucratic control and one-way, top-down communication; on the other hand, it can be much more collaborative and peer to peer. In terms of communications processes, these are two quite distinct situations that may be more or less appropriate to different circumstances. Of course, there are many points and combinations in between, and there is nothing to say that an organization needs to be fixed. Indeed, communication architectures can be dynamically reconfigured to suit different circumstances (e.g., normal vs. emergency situations).

Recent work by Dekker (2002) reflects a more general trend in human factors literature toward considering communication architectures more explicitly. A number of different archetypes have been proposed, some examples of which are summarized in Table 5.1. In the terms expressed in Table 5.1, it can be noted that the development of control room operations and SCADA systems has seen a greater extent of information sharing (i.e., it is at least technically possible for everyone to know everything). This, combined with the opportunities now available to remotely control a plant, has enabled larger geographical areas to be controlled. In the most general terms, this has brought with it a shift from distributed and negotiation architectures (where teams are responsible for individual regions and plant locations) to forms of centralized and split architectures (where teams are responsible for all regions and plant locations). The purpose in raising this general point is to say that communication architectures continually evolve and that different architectures are suited to different environments. Current human factors best practice in this area is represented by numerous methods that allow communications architectures to be modeled and for critical features to be diagnosed.

5.4.2 The Purpose of Communication: Teamwork and Shared Awareness

In conceptual terms we have defined communications as part of an information processing activity undertaken by the control room staff and the organization as a whole. Implicit in this is the idea of teamwork. A dictionary definition of teamwork is "work done by several associates with each doing a part but all subordinating personal prominence to the efficiency of the whole" (Merriam-Webster, 2007). In a practical sense, this takes the form of "two or more people, a common goal, specific role assignment, and interdependence" (Salas, Bowers, and Cannon-Bowers, 1995, p. 56). In more detail:

1. Teams make decisions in the context of a larger task.
2. Team members have specialized knowledge and skills relevant to the task and decision.
3. Task conditions under which teams operate often include high workload and time pressure.

TABLE 5.1
Dekker's (2002) Communication Architecture Archetypes

Architecture	Distribution of Information	Patterns of Interaction	Decision Rights
Centralized architecture without information sharing	*Tight control*—Individuals possess information relevant to their own roles only.	*Fully hierarchical*—Information flows from top to bottom through a shallow hierarchy.	*Unitary decision rights*—Teamworking and communications are orientated around the enactment of centrally generated instructions.
Split architecture without information sharing	*Tight control*—Individuals possess information relevant to their own roles only.	*More distributed*—Information flows from top to bottom through a deeper hierarchy.	*Some autonomy*—Intermediate layers have some degree of autonomy.
Distributed architecture without information sharing	*Tight control*—Individuals possess information relevant to their own roles only.	*Distributed*—Individuals know what they need to know (but not what other people know).	*More autonomy*—Decisions are made at lower levels without reference to higher levels.
Negotiation architecture without information sharing	*Some dissemination*—Individuals possess information relevant to their own roles but can choose to pass this on.	*Fully distributed*—Individuals can interact with other individuals.	*Peer to peer*—Decisions are taken on a more negotiated basis.
Architectures with information sharing	*Broad dissemination*—Everyone knows everything.	*Fully distributed*—Individuals can interact with other individuals.	*Peer to peer*—Decisions are taken on a more negotiated basis.

In general, teamworking research tends to fall into two distinct camps. On the one hand is the "too many cooks spoil the broth" school of maladaptive team phenomena such as groupthink (e.g., Janis, 1982), group polarization (e.g., Myers and Lamm, 1976), minority influence (e.g., Nemeth, 1986), and so forth. On the other is the "two heads are better than one" view in which the implicit goal of all teams (and teamworking interventions) is to create synergy; in other words, the team becomes more than the sum of its parts, and the role of connections or communication is as much a determinant of performance as the individuals or artifacts that are communicating.

The more fundamental concept this relates to is shared understanding. MacMillan, Paley, Entin, and Entin (2005) noted that "team members need a shared awareness of the situation and of the roles, tasks, and actions of other team members." This

TABLE 5.2
Minimum Requirements

Minimum Requirements	Yes or No (Y/N)
Task zones of individuals requiring frequent interaction with each other are located close to each other.	Y/N
Control suite equipment is located such that visual contact between operators is possible where and when needed.	Y/N
Communications that are extraneous to the control suite functions do not distract personnel.	Y/N
Rooms and locations with different functions are physically separated to avoid potential sources of disturbance.	Y/N
The layout of the control room facilitates teamworking opportunities and social interaction for operators.	Y/N
Key operational links, including sightlines or direct speech communication, have been optimized.	Y/N
Layout allows direct verbal communication between the control room operators.	Y/N

Source: Permission to reproduce table from BS/EN/ISO 11064 is granted by the British Standards Institution.

awareness seems to break down into a form of meta-awareness of the team (and what other team members are doing) and an awareness of a common situation. In terms of human factors best practice, a slight note of caution should be raised. Shared awareness does not necessarily mean "identical awareness" (indeed, it might be surprising if it were identical, given the different tasks and perspectives of the individuals in the team).

Current thinking recognizes this and puts forward the idea of "compatible situational awareness" as a complementary process. It is also important to point out that high degrees of synergy and emergence can arise from teams that have very little shared awareness. Teams that have only a limited local awareness can show extremely high levels of synergy and emergence. The wider point, then, is that the notion of shared awareness, the fundamental concept to which teamworking often applies, while often desirable, is also complex.

5.5 GUIDANCE AND RECOMMENDATIONS

5.5.1 Minimum Requirements

Minimum requirements for control room communications, synthesized from BS/EN/ISO 11064 (Ergonomic Design of Control Centers) and HSE guidance, are shown in Table 5.2. In addition, the Health and Safety Executive has developed a useful checklist that reflects a slightly more wide-ranging concern with not only interpersonal communication but also the communication of safety-critical information. This checklist is shown in Table 5.3.

5.5.2 Human Factors Best Practice

Minimum requirements drawn from legislation provide some highly useful generic insights. Human factors best practice supplements this with more focused methods that enable facets of the structure and process of communication to be diagnosed.

5.5.2.1 Social Network Analysis

Social network analysis (SNA) can be used to characterize, model, and diagnose the structure of communications in a control room. It is used to analyze and represent the relationships between groups of agents or teams in the form of a network. A *social network* is defined as a "set or team of agents that possess relationships with one another" (Driskell and Mullen, 2004). SNA can be used to demonstrate the type, importance, and number of relationships within a specified group. The output typically provides a graphical depiction and a mathematical analysis of the relationships within the group under analysis. Depending on the focus of the analysis, a number of diagnoses can be made—for example, centrality, closeness, and betweenness—all of which provide an indication of agent importance within the network based on communications. A network density figure can also be derived which gives an indication of the overall level of interconnectivity within the network (higher levels of density relate to the extent of information sharing, among other things).

5.5.2.2 Communications Usage Diagram

The communications (or comms) usage diagram (CUD) (Watts and Monk, 2000) focuses more on the process of communication. In particular, it is used to describe collaborative activity between teams of actors dispersed across different geographical locations. A CUD output describes how and why communications between actors occur, which technology is involved in the communications, and the advantages and disadvantages associated with the technology used. It is a useful adjunct to SNA.

5.5.2.3 Team Communications

As well as the objective measures of communications described above, there is a more subjective, psychological perspective that can be measured using self-rating/questionnaire techniques. Among other things, they probe levels of shared and compatible awareness. MacMillan et al. (2005) described a useful set of self-rating questionnaires designed to assess team member mutual awareness. Based on a model of team mutual awareness, the methodology they proposed is comprised of three questionnaires: task mutual awareness questionnaire, team workload awareness questionnaire, and teamwork awareness questionnaire:

- The task mutual awareness questionnaire involves participants recalling salient events that occurred during the task under analysis and then describing the tasks that they were performing during these events as well as the tasks they think the other team members were performing.
- The team workload awareness questionnaire is a subjective workload assessment technique based on an established workload questionnaire (the NASA TLX; see Hart and Staveland, 1988). It is used to elicit subjective ratings

TABLE 5.3
Safety Critical Communications Checklist

Questions	Site Response (Y/N)	Inspectors' View (Y/N)	Improvements Needed? (Y/N)
Is there a process to identify and define the communications requirements for the safe operation of the plant (e.g., shift handover, control room to field or maintenance communications)?	Y/N	Y/N	Y/N
Management Questions			
Has the company considered effective ways to communicate major hazard information to contractors, temporary staff, and visitors?	Y/N	Y/N	Y/N
Is there a process to select the key information to be communicated (from internal and external sources)?	Y/N	Y/N	Y/N
Is there a system to communicate changes in practice and lessons following an incident?	Y/N	Y/N	Y/N
Does a process exist to monitor the effectiveness of the communication of major hazard information?	Y/N	Y/N	Y/N
Is the workforce actively involved in communications (not just passive receivers of information)?	Y/N	Y/N	Y/N
Is there a process to ensure that modifications to processes, procedures, systems, and organization are communicated to all relevant staff?	Y/N	Y/N	Y/N
Operator Questions			
Is there evidence that the key major hazard information has been understood by the target audience (staff, contractors, visitors)?	Y/N	Y/N	Y/N
Is there evidence that changes to practices, as a result of an incident, are understood by staff?	Y/N	Y/N	Y/N
Is there an audit trail documenting the monitoring process?	Y/N	Y/N	Y/N

Have the communication routes been clearly identified?	Y/N	Y/N	Y/N
Management Questions			
Have communications been considered in defining the plant's safety command and control structure?	Y/N	Y/N	Y/N
Are there routes by which operators can raise safety concerns with management, and is the loop then closed adequately?	Y/N	Y/N	Y/N
Is there a process to review the communications routes in light of plant or organizational change?	Y/N	Y/N	Y/N
Are the key communication aspects of critical procedures assured?	Y/N	Y/N	Y/N
Is there a defined structure for shift handover arrangements?	Y/N	Y/N	Y/N
Operator Questions			
Is there evidence that key communication channels are documented?	Y/N	Y/N	Y/N
Do operators know when to report safety concerns?	Y/N	Y/N	Y/N
Do operators understand where they can obtain key safety information?	Y/N	Y/N	Y/N
Can staff describe the safety implications of any recent changes to the plant?	Y/N	Y/N	Y/N

Source: Health and Safety Executive, *Common Topic 3: Safety Critical Communications*, Health and Safety Executive, London, 2007, p. 4. Reproduced under the terms of the Click-Use Licence.

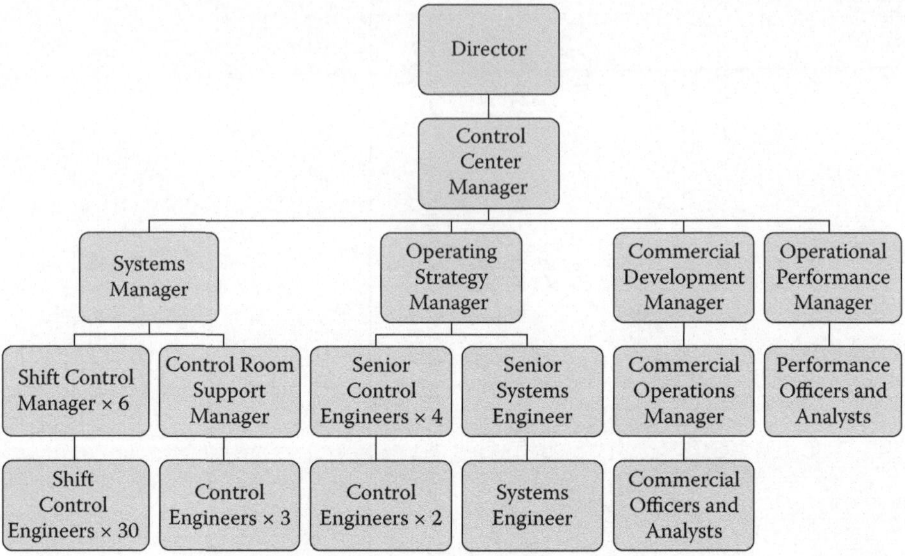

FIGURE 5.2 Formal organization hierarchy for example control room.

of team member workload on the following dimensions: mental demand, temporal demand, performance effort, and frustration. Team members also provide an overall rating of other team members' mental workload (MWL) and a rating on each dimension for the team as a whole.

- The teamwork awareness questionnaire rates the team on four components of teamwork processes. Team members provide subjective ratings of the team's performance on the four team behaviors of communication, backup, coordination and information management, and leadership/team orientation.

Each of the questionnaires is administered post-trial to gain a measure of team mutual awareness.

5.6 ASSESSMENT AND EVALUATION

The organization chart (such as that shown in Figure 5.2) maps out super- and subordinate positions based on responsibility and the division of tasks. Although it hints at communication flows, the situation does become more complex when individuals are placed in their working context, as Figure 5.3 shows. An analysis of the structure of communications in an actual control center helps to diagnose certain issues such as information bottlenecks, who (based on the number of communication links) is the most central individual in the structure, and what methods are currently used to communicate information. Figure 5.3 shows a rough schematic outline of a control room environment, who is present in that environment, and the communication links that exist between them. The result is visually complex, and it is the role of the selected human factors methods to reduce that complexity and arrive at some simpler, more concrete diagnostics.

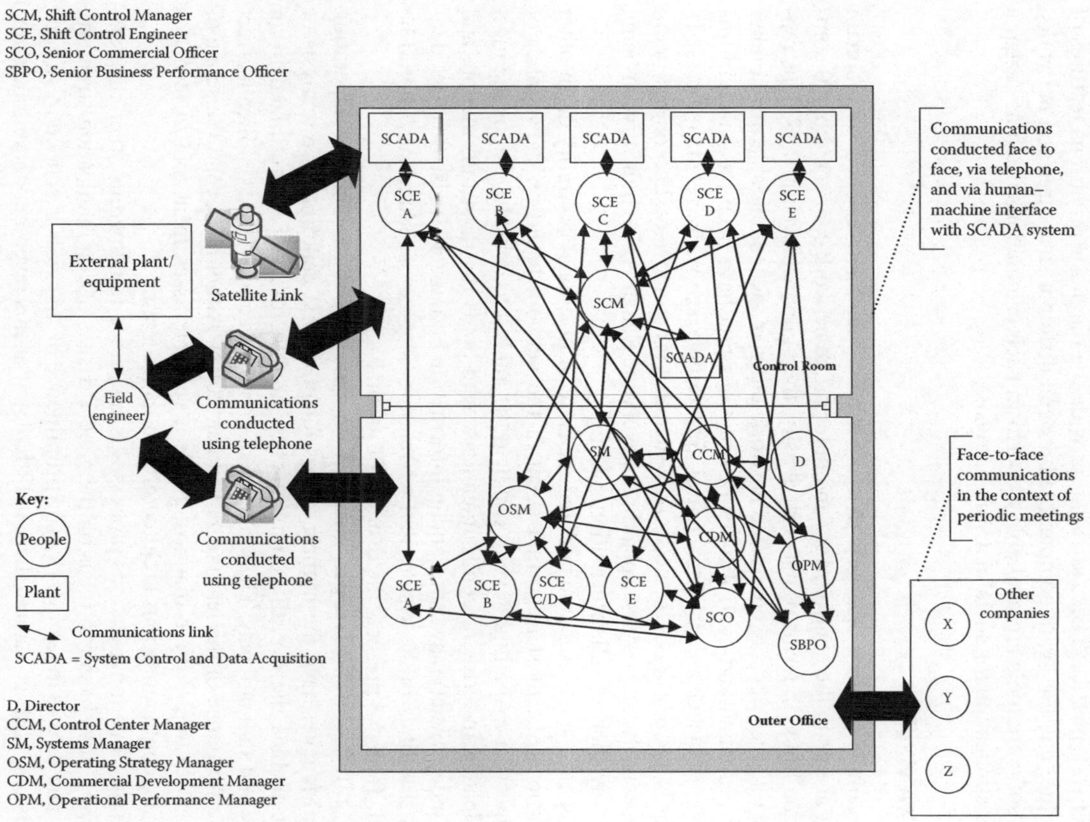

FIGURE 5.3 Diagram showing who communicates with whom and by what means they communicate (the focus of this study was on the communications within and between the control room and outer office).

The visual complexity of Figure 5.3 is made considerably more tractable by using social network analysis. The first output of the analysis is a social network diagram. The social network diagram simplifies matters by ignoring the geographical location of people and things, creating a standardized representation, and thus enabling different structures to be more easily compared. Greater simplification and diagnostic resolution are achieved by using the network diagram to calculate a number of social network metrics. These are what provide the diagnosis of various features of the structure and are again comparable between different structures and networks. Various software tools are available to perform such calculations (such as Agna, a freeware application for social network analysis).

5.7 EXAMPLE

In this example, the first diagnosis to make is purely descriptive. There are 18 main actors or agents in the control room and outer office network connected by 70 communication links. In terms of network density, the figure achieved is 0.27. This represents a relatively loosely connected network and is in fact comparable to a number of other command-and-control types of situations that have been measured in this way previously (such as air-traffic control, for example; see Walker et al., in press). In other words, the density of the network is such that it is some way short of maximum interconnectivity, or everyone speaking to everyone else. It is also indicative of being in some way short of full information sharing.

Sociometric status is a standardized measure of positional centrality or, in other words, who is the most important agent in the network based on the number of interconnections. At a high level it can help the analyst to consider who is the most heavily loaded in terms of communications, who holds the most information, and who (or indeed what) may represent an information bottleneck. It also helps to identify vulnerabilities in the system and where some degree of redundancy should be provided. Table 5.4 presents the results of this analysis.

The Shift Control Manager, Senior Commercial Officer, and the Operating Strategy Manager have the highest scores for positional centrality. An analysis of this sort enables the overall communication architecture to be scrutinized in an objective and standard way. The structure of communications has a significant effect on the performance of the team (e.g., Dekker, 2002). Although beyond the scope of the current example, these diagnostics are a valuable assessment tool in terms of considering communications architectures in more detail.

The issue of communications processes is addressed in numerous other chapters, thus leaving this chapter to focus on the generic issues of communication structure, process, and its role in teamwork. The prominence of this view is echoed in various standards and HSE guidance; there is little doubt at this level that communications play a significant role in safe and effective operations. Achieving effective communications relies on the physical attributes of the control rooms and equipment, but it also relies on various sociotechnical design parameters—for example, who is communicating to whom, the role of technical artifacts in communication, and the synergistic effects of teamwork. Human factors best practice provides greater insight into these effects and an efficient way to model and understand them.

TABLE 5.4
Sociometric Status

Node	Status
Shift Control Engineer A	0.471
Shift Control Engineer B	0.471
Shift Control Engineer C	0.471
Shift Control Engineer D	0.471
Shift Control Engineer E	0.471
Shift Control Manager	*0.941*
Systems Manager	0.588
Control Center Manager	0.588
Director	0.118
Operational Performance Manager	0.412
Senior Business Performance Officer	0.647
Commercial Development Manager	0.471
Senior Commercial Officer	*1.177*
Operating Strategy Manager	*0.941*
Senior Control Engineer A	0.353
Senior Control Engineer B	0.353
Senior Control Engineer C/D	0.471
Senior Control Engineer E	0.353

Note: Based on communications traffic, this table shows who the most important agent in the network is. The top three ranking agents are italicized.

BIBLIOGRAPHY

Dekker, A.H. (2002). *C4ISR Architectures, Social Network Analysis and the FINC Methodology: An Experiment in Military Organisational Structure*. Edinburgh, South Australia: DSTO Electronics and Surveillance Research Laboratory.

Driskell, J.E. and Mullen, B. (2005). Social network analysis. In N. Stanton, A. Hedge, K. Brookhuis, E. Salas, and H. Hendrick (Eds.), *Handbook of Human Factors and Ergonomics Methods*. Boca Raton, FL: CRC Press.

Hart, S.G. and Staveland, L.E. (1988). Development of a multidimensional workload rating scale: results of empirical and theoretical research. In P.A. Hancock and N. Meshkati (Eds.), *Human Mental Workload* (pp. 139–183). Amsterdam: Elsevier.

Health and Safety Executive (HSE). (2007a). *Common Topic 3: Safety Critical Communications*. London: Health and Safety Executive (http://www.hse.gov.uk/humanfactors/comah/common3.pdf).

Health and Safety Executive (HSE). (2007b). *Safety Critical Communications*, HSE Human Factors Briefing Note No. 8. London: Health and Safety Executive (http://www.hse.gov.uk/humanfactors/comah/08communications.pdf).

Janis, I.L. (1982). *Groupthink*. Boston: Houghton Mifflin.

MacMillan, J., Paley, M.J., Entin, E.B., and Entin, E.E. (2005). Questionnaires for distributed assessment of team mutual awareness. In N. Stanton, A. Hedge, K. Brookhuis, E. Salas, and H. Hendrick (Eds.), *Handbook of Human Factors and Ergonomics Methods* (pp. 484–494). Boca Raton, FL: CRC Press.

Merriam-Webster. (2007). *Merriam-Webster OnLine* (www.merriam-webster.com).

Myers, D.G. and Lamm, H. (1976). The group polarization phenomenon. *Psychological Bulletin*, 83, 602–627.

NATO. (2006). *SAS-050 Conceptual Model Version 1.0*. Brussels, Belgium: North Atlantic Treaty Organization (http://www.dodccrp.org/SAS/SAS-050%20Final%20Report.pdf).

Nemeth, C.J. (1986). Differential contributions of majority and minority influence. *Psychological Review*, 93, 23–32.

Salas, E., Bowers, C.A., and Cannon-Bowers, J.A. (1995). Military team research: 10 years of progress. *Military Psychology*, 7(2), 55–75.

Walker, G.H., Stanton, N.A., Wells, L., Gibson, H., Young, M.S., and Jenkins, D.P. (in press). Beyond ethnography: a new approach to "doing" distributed cognition. *Theoretical Issues in Ergonomic Science*.

Watts, L.A. and Monk, A.F. (2000). Reasoning about tasks, activities, and technology to support collaboration. In J. Annett and N. Stanton (Eds.), *Task Analysis* (pp. 55–78). London: Taylor & Francis.

6 Manning and Workload

6.1 OUTLINE

6.1.1 MANNING

Manning refers to the staffing levels used in plant process control rooms (i.e., the number of operators working together at any one time). The level of manning used is of critical importance to process control performance and overall system safety. In recent times, an increase in the levels of automation, teamworking, and multiskilling has led to a general reduction in the manning levels used, and there are concerns that this could adversely impact process control team performance and system safety through increases in operator workload and its associated consequences (Brabazon and Conlin, 2001).

The level of manning present in a process control system is a key factor in the safety and efficiency of the plant. Manning levels are intrinsically related to the level of workload experienced by process control room operators and the quality of task performance. The Control of Major Accident Hazard (COMAH) regulations passed in 1999 require that organizations must be able to demonstrate a structured, effective process for ensuring that plant staffing levels are adequate for normal and abnormal situations (Brazier, Waite, and Gait, 2004). Inappropriate manning levels can have serious consequences for process control room team and operator performance efficiency and thus the overall safety of the plant. For example, the Health and Safety Executive (HSE) has stated that a control room designed for use by four operators is dangerous when staffed by only three operators.

6.1.2 WORKLOAD

Workload is a critical issue in control room design, and contemporary plants are designed with the aim of reducing the level of operator workload through improved interface design and increased levels of automation (Sebok, 2000). According to Hart and Wickens (1990), workload is the cost of accomplishing task requirements for the human element of sociotechnical systems. Inappropriate workload levels (both too high and too low) have a range of adverse consequences, including fatigue, errors, monotony, mental saturation, reduced vigilance, and stress (Spath, Braun, and Hagenmeyer, 2007). For the purposes of this chapter, workload can be divided into two forms: *mental* and *physical*. A brief overview of each is given below.

6.1.2.1 Mental Workload

The level of mental workload experienced by operators is a key element in the safety, reliability, and efficiency of complex sociotechnical systems (Gregoriades and Sutcliffe, 2007). Mental workload is a contentious area, and there has been great debate over how to define the concept. It is generally agreed that human operators possess a finite attentional capacity and that during task performance these attentional resources are allocated to component tasks. Workload is thus a function of the human operator's attentional capacity and the demand for resources imposed by the task. The level of mental workload represents the proportion of resources that are required to meet the task demands (Welford, 1978). Young and Stanton (2001) formally defined *mental workload* as follows: "The mental workload of a task represents the level of attentional resources required to meet both objective and subjective performance criteria, which may be mediated by task demands, external support, and past experience." Mental workload is therefore a multidimensional construct that is characterized by the task (e.g., complexity, demands) and the individual involved (e.g., skill, experience, training).

When operators are faced with excessive task demands and their attentional resources are exceeded, they become *overloaded*. Mental overload therefore occurs when the demands of the task are so great that they are beyond the limited attentional capacity of the operator. Conversely, when operators experience excessively low task demands, they may experience a state of mental *underload*. Both conditions can be detrimental to task performance (Wilson and Rajan, 1995) because operators become less likely to attend to potentially important sources of information (Lehto and Buck, 2008).

Various consequences associated with inappropriate levels of workload can potentially lead to performance decrements. These include inattention, complacency, fatigue, monotony, mental saturation, mental satiation, reduced vigilance, and stress. The Health and Safety Executive stated that "for mental workload, conditions of over and under-arousal should be avoided. The duration of tasks that have an associated low or high level of mental workload should be limited. Both these extremes will increase the likelihood of human error affecting the system" (HSE, 2008).

Ostensibly, then, there is an optimal level of workload for optimal task performance, and system and procedural designers should aim to optimize operator workload to ensure efficient task performance. Many human factors practitioners have cited workload optimization as being critical to maintaining effective task performance in complex systems (e.g., Young and Stanton, 2002; Sebok, 2000). Workload optimization involves achieving a balance between task demands and operator resources.

A diagram depicting the factors impacting workload, the different workload states, and some of the likely consequences of mental *overload* and *underload* is presented in Figure 6.1.

6.1.2.2 Physical Workload

Although advances in technological capability and increased automation have reduced the level of physical workload imposed on control room operators, it is still a pertinent issue to consider during the design and assessment of process control rooms.

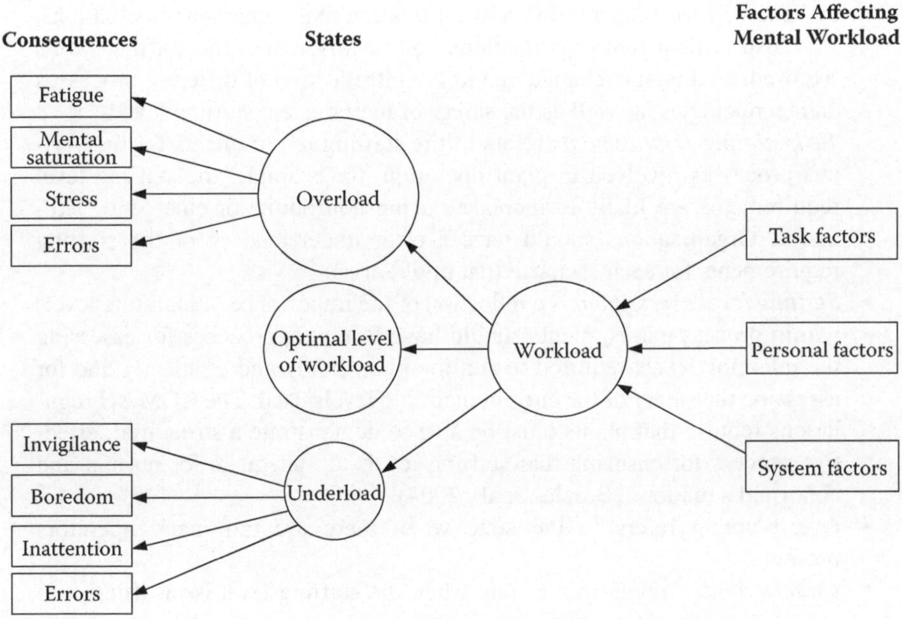

FIGURE 6.1 Factors affecting mental workload and the consequences associated with overload and underload.

Physical workload represents the level of physical exertion or demand imposed on an operator by a task or process. According to Sanders and McCormick (1993), many factors contribute to the level of workload experienced by operators undertaking physical activity, including the nature of the work itself, somatic factors (e.g., age, sex), training, motivation, and environmental conditions (e.g., noise, heat). As with mental workload, various consequences are associated with inappropriate levels of physical workload. Too great a level of physical workload, and operators can become fatigued and overstressed; too low a level of physical workload, and they can become bored or inattentive.

6.2 KEY ISSUES

6.2.1 Manning

The key issues related to the level of manning used in process control include the staffing level required to operate a particular plant safely and efficiently during both routine and nonroutine situations and the issues that arise when a plant is either overmanned (i.e., too many operators) or undermanned (i.e., insufficient number of operators). A brief summary of the key issues is given below:

- *Staffing levels* refers to the number of team members available to work in a control room (Sebok, 2000). The staffing level has a direct impact on performance and safety of process control tasks. Safe staffing levels should be specified for both routine and nonroutine (e.g., emergency) situations. It is also critical that organizations continually assess the staffing levels required, as this may change due to the introduction of different processes and technologies, as well as the safety of their current staffing levels.

- *Task staffing requirements* refers to the staffing requirements for the various processes involved in plant operation; for example, the staffing level requirements are likely to increase during nonroutine or emergency situations. Organizations should have a clear understanding of the staffing requirements for each scenario that might arise.
- *Staffing level assessment* is a reflection of the importance of manning levels within process plants; plants should have formal approaches for assessing the manning levels required to run the plant safely and efficiently and for assessing the safety of the current manning levels used. The COMAH regulations require that plants must be able to demonstrate a structured, effective process for ensuring that staffing levels are adequate for normal and abnormal situations (Brazier et al., 2004).
- *Overmanning* refers to the state when there are too many operators present.
- *Undermanning* refers to the state when the staffing level is insufficient to operate the plant efficiently and effectively.
- *Workload*, both mental and physical, imposed on process control operators is directly related to staffing levels. Inappropriate staffing levels can potentially lead to either overload or underload, both of which can impact task performance adversely.
- *Automation* is impacting staffing levels, as it allows system designers to automate various system control tasks and thus change the role of human operators from frontline controllers to system monitors or supervisors. An important consideration here is appropriate allocation of functions—that is, identifying the tasks that should be automated and which tasks should continue to be the responsibility of human operators.
- *Team organization* is important, as teams rely on the appropriate designation of tasks and roles so they can share the level of workload effectively (Sebok, 2000). It is therefore critical that jobs and roles are appropriately and clearly allocated between process control teams.

6.2.2 Workload

The use of improved interface design and increased levels of automation are expected to reduce the level of workload experienced by process control operators; however, attempting to work with automation and computerized systems and in unfamiliar types of plants could actually increase workload (Sebok, 2000). The key issues surrounding the concept of mental workload and process control are therefore related to the factors that either increase or decrease workload and the consequences arising from those instances in which operator workload is either too high (i.e., overload) or too low (i.e., underload). A brief overview of the main issues is presented below:

- *Interface, display, and system design*—The system, interfaces, and displays used by process control operators impact the level of workload imposed on them. Factors such as information presentation on displays, the signal discriminability of alarms, and the environmental conditions within the control room such as temperature and noise levels all contribute to operator workload. Appropriate consideration should therefore be given to these factors during the system design process.
- *Task complexity*—The complexity of the tasks being performed has a direct bearing on the level of workload that human operators experience. It is critical that process control system design considers the human operator's competencies and capabilities for performing control tasks. Automation is typically used to remove high-complexity tasks from the human operator's responsibility.
- *Overload*—Operators may experience a state of overload when the demands of a task are beyond their limited attentional capacity (Young and Stanton, 2002). Various consequences are associated with overload, including mental saturation, fatigue, stress, and human error.
- *Underload*—An operator may experience a state of underload when the task demands are excessively low. Malleable attentional resources theory (MART) (Young and Stanton, 2002) suggests that mental underload leads to performance degradation due to a shrinkage in attentional resources. Various consequences are associated with operator underload, including inattention and lack of vigilance, fatigue, boredom, and monotony.
- *Workload optimization*—When considering operator workload, most agree that one of the key goals of system and task design is to maximize the match between task demands and human capacity (Young and Stanton, 2002). System and procedure designers should therefore strive to optimize the level of workload experienced by process control operators. When workload falls below or rises above this optimal level, performance decrements will occur.
- *Vigilance*—Operator vigilance plays a vital role in automated human machine systems (Grier et al., 2003). Vigilance refers to an operator's capacity for sustained effective attention when monitoring a display for critical signals or events to which the operator must respond (Donald, 1999). Incorporated into this view of vigilance is the ability to effectively observe, process, and analyze signal information. Vigilance decrements can occur as a result of both overload and underload and can lead to human operators failing to attend to or process incoming stimuli due to boredom or difficulty in maintaining attention (Lehto and Buck, 2008).
- *Automation*—Advances in technology have allowed complex system designers to automate various system control tasks and thus change the role of human operators from frontline controllers to system monitors or supervisors. Despite being originally conceived as a way of reducing operator workload in complex systems, it is apparent that automation can potentially cause both decreases and increases in workload levels (Young and Stanton, 2002).

- *Out-of-the-loop performance problem*—The out-of-the-loop performance problem typically occurs within automated systems when human operators are removed from the control loop and placed in the job of system monitor (Endsley, 1995). Because human operators are generally ill suited to the role of monitoring complex systems, out-of-the-loop performance has been found to lead to problems such as vigilance decrements, complacency, loss of operator situation awareness, poor feedback under automated conditions, and manual skill decay (Endsley and Kaber, 1999). The various consequences associated with operators performing out of the loop have been found to cause inadequate system monitoring and problems in humans assuming control when necessary.
- *Time pressure*—The level of time pressure faced by human operators when undertaking work activities also has an impact on the level of workload imposed on them. As the time available for task performance diminishes, workload can increase. The less time available for a particular task, the greater the level of physical and cognitive workload may be, as more tasks must be performed in a shorter space of time. Alternatively, the level of workload imposed on an individual during task performance can be moderated by increases in the time available to perform the task in question. Increases in the time available to perform a task allow individuals to organize the tasks required optimally so workload can be moderated.
- *Allocation of functions*—Allocation of functions is used during the design process to allocate jobs, tasks, functions, and responsibility to the human or machine in the system in question (Marsden and Kirby, 2004). Allocation of functions analysis is particularly important when considering system automation and operator workload.
- *Human error*—In most cases, errors result from either human or system limitations, such as limited memory and mental processing capacities, failure to cope with stress, tunnel vision, and physiological factors such as fatigue and hunger. Increased levels of workload can increase the probability of errors being made. For example, Helmreich (2000) suggested that a major cause of human error is cognitive overload. Conversely, reduced levels of workload, or underload, can also lead to more errors being made due to factors such as lack of vigilance, inattention, and complacency. According to Wilson and Rajan (1995), mental overload and underload can potentially lead to performance degradation, attentional lapses, and errors.
- *Procedural design*—Operator workload is also impacted by the procedures used during process control. Procedures should be designed so tasks are allocated appropriately between different human operators and between the human operators and the system itself (automation).
- *Training*—Training programs can be used to provide control room operators with the knowledge, skills, and abilities required to reduce the level of workload associated with process control tasks. Factors such as operator experience and expertise can serve to minimize the level of workload imposed by certain tasks.

6.3 STANDARDS AND GUIDELINES

6.3.1 Manning

6.3.1.1 Control of Major Accident Hazards Regulations

The Control of Major Accident Hazard (COMAH) regulations passed in 1999 were proposed in order to prevent and mitigate the effects of major accidents involving dangerous substances, such as chlorine, liquefied petroleum gas, explosives, and arsenic pentoxide, that can cause serious damage or harm to people and the environment. The COMAH regulations require that plants must be able to demonstrate a structured, effective process for ensuring that staffing levels are adequate for normal and abnormal situations.

6.3.2 Workload

6.3.2.1 Mental Workload ISO Standards

- ISO 10075-1:1991 (Ergonomic Principles Related to Mental Workload. Part 1. General Terms and Definitions) is an extension of ISO 6385:1981 (Ergonomic Principles in the Design of Work Systems) which focuses specifically on the concept of mental workload. ISO 10075-1:1991 presents definitions of mental stress and mental strain and briefly discusses a range of consequences associated with mental strain, including facilitating effects (e.g., activation), impairing effects (e.g., fatigue and reduced vigilance), and other types of effects (e.g., practice effects). A model is presented that details the relationship between mental stress and mental strain and the situational (task requirements, physical conditions, and social and organizational factors) and individual characteristics relating to mental stress and strain.
- ISO 10075-2:1991 (Ergonomic Principles Related to Mental Workload. Part 2. Design Principles) is an extension of ISO 6385:1981 (Ergonomic Principles in the Design of Work Systems) that focuses specifically on the concept of mental workload. ISO 10075-2:1991 provides system design guidelines related specifically to the design of work systems, including task, equipment, workplace, and working conditions, to prevent mental overload and underload and the range of effects presented in ISO 10075-1:1991. This document is aimed at those involved in the design and use of work systems, including system and equipment designers, employers, and employee representatives. ISO 10075-2:1991 presents a series of general system design principles followed by system design guidelines related to the issues of fatigue, monotony, reduced vigilance, and satiation. In addition to information and training guidance, ISO 10075-2:1991 also presents examples of design solutions for avoiding the effects of mental overload and underload at the different levels of the design process, including task design, work equipment design, environmental design, organizational design, and temporal organization design.

- ISO 10075-3:2004 (Ergonomic Principles Related to Mental Workload. Part 3. Principles and Requirements Concerning Methods for Measuring and Assessing Mental Workload) presents a series of guidelines related to the development, evaluation, and selection of appropriate methods for assessing mental workload. The document presents definitions of a variety of factors related to mental workload measurement, such as validity, consistency, sensitivity, objectivity, and generalizability, and outlines a series of guidelines relating to the measurement of mental workload, including general principles, procedural requirements, quantitative requirements, and documentation requirements.
- ISO 11064-1:2001 (Ergonomic Design of Control Centers. Part 1. Principles for the Design of Control Centers) provides a generic framework for applying ergonomics and human factors requirements and recommendations to the design and evaluation of control centers.

6.3.2.2 Physical Workload ISO Standards

- ISO 11226:2000 (Ergonomics—Evaluation of Static Working Postures)—Pain, fatigue, and musculoskeletal disorders are all consequences of sustained inadequate working postures and repetitive movements caused by inappropriate working situations (Delleman and Dul, 2006). ISO 11226:2000 provides ergonomic recommendations for the development of acceptable static working postures.
- ISO 11228-1:2003 (Ergonomics—Manual Handling. Part 1. Lifting and Carrying)—This standard presents recommendations on the limits for manual lifting and carrying and provides guidance on the assessment of lifting and carrying to evaluate the health risks associated with the manual and lifting tasks taking place within an organization.
- ISO 11228-2:2007 (Ergonomics—Manual Handling. Part 2. Pushing and Pulling)—Overly forceful or awkward physical pushing or pulling tasks can lead to pain, fatigue, and disorders of the musculoskeletal system. ISO 11228-2:2007 describes two approaches for identifying the potential risks and hazards associated with whole body pushing and pulling and proposes recommendations for reducing the risk of injury or ill health resulting from pushing or pulling.
- ISO 11228-3:2007 (Ergonomics—Manual Handling. Part 3. Handling of Low Loads at High Frequency)—Handling of low loads at high frequency can cause pain and fatigue which in turn can potentially result in musculoskeletal disorders, reduced productivity, deteriorated posture, decreased movement coordination, errors, reduced quality, and hazardous situations. ISO 11228-3:2007 outlines a series of recommendations for repetitive work tasks involving manual handling of low loads at high frequency. The document presents a series of related definitions and outlines a procedure and guidance for the identification and assessment of associated risk factors.

6.4 HUMAN FACTORS LITERATURE

6.4.1 Manning

Safe manning levels are likely to differ depending on the plant and process involved, so little guidance on appropriate manning levels is presented in the literature. The effects of differing levels of manning within process control rooms has also received only limited attention within the human factors literature. Sebok (2000) conducted a study designed to compare the effects of interface design and staffing levels on various aspects of nuclear power plant process control team performance, including workload, situation awareness (SA), team interactions, and overall task performance. The study involved teams of nuclear power plant operators undertaking scenarios in a simulated conventional plant with a hard-control interface or in a simulated advanced plant with a computerized interface. At the conclusion to the study, Sebok (2000) reported that the larger crews (i.e., increased staffing levels) consistently performed better than smaller crews in the conventional plant and that the level of workload was consistently higher for the smaller crews than for the larger crews, irrespective of plant and interface type. Interestingly, it was also found that the smaller crews had better SA than the larger crews. Based on the study findings, Sebok (2000) concluded that, if the interface in question presents data rather than information (i.e., conventional interface) and requires operators to integrate information, then a larger crew is more effective, but additional personnel appeared to be of little benefit when using more advanced, highly computerized interfaces that present already integrated information. In a study on process control operator errors, Jacobsson Kecklund and Svenson (1997) found that one of the strategies used to maintain performance by control room operators was to hand over tasks to other operators.

6.4.2 Workload

6.4.2.1 Mental Workload

The concept of mental workload is a contentious one. Sanders and McCormick (1993) suggested that the concepts of stress and strain are relevant to human workload. *Stress* refers to "some undesirable condition, circumstance, task or other factor that impinges upon the individual," and sources of stress include heavy work, immobilization, heat and cold, noise, sleep loss, danger, information overload, and boredom (Sanders and McCormick, 1993). *Strain*, on the other hand, refers to the effects of the stress on the individual and is measured by observing changes in elements such as blood oxygenation, electrical activity of the muscles, heart rate, body temperature, and work rate (Sanders and McCormick, 1993). It is generally agreed that human operators possess a finite attentional capacity and that operator workload is a function of the human operator's attentional capacity and the demand for resources imposed by the task. Tsang and Vidulich (2006), for example, observed that it is generally accepted within the literature that operator workload is a function of the supply and demand of attentional resources. The level of mental workload therefore

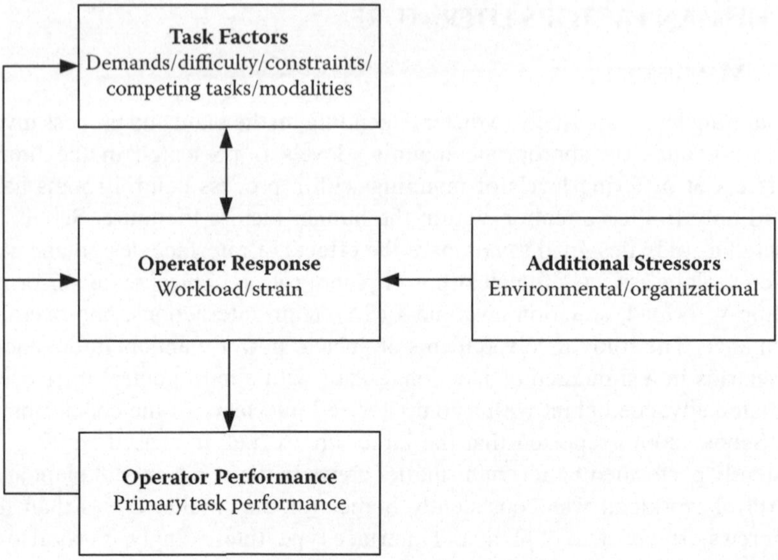

FIGURE 6.2 Framework of interacting stressors impacting workload. (Adapted from Megaw, T., in *Evaluation of Human Work*, J.R. Wilson and N. Corlett, Eds., CRC Press, Boca Raton, FL, 2005, pp. 525–553.)

represents the proportion of resources required to meet the task demands (Welford, 1978). According to Tsang and Vidulich, the two main determinants of workload are the exogenous task demand (as specified by factors such as task difficulty and priority and contextual factors) and the endogenous supply of attentional resources available to support information processing activities.

In an attempt to clarify the concept and the factors impacting it, Megaw (2006) presented a framework of interacting stressors on an individual (see Figure 6.2). Megaw's model suggests that workload is a multidimensional construct and that operator workload is impacted by a variety of factors, including task factors (e.g., task demands, task constraints), environmental factors (e.g., workspace, interfaces), and organizational factors (e.g., staffing levels, team organization, allocation of functions).

In terms of theoretical underpinning, various models of attention and workload are presented in the literature. Kahneman (1973), for example, presented a capacity model of attention that suggests that individuals possess a single pool of attentional resources that has a finite limit. The ability to perform concurrent tasks is therefore dependent on the effective allocation of attention to each (Young and Stanton, 2002). More recently, multiple resource theory models have contended the notion of a single pool of attentional resources; for example, Wickens' (1992) multiple resource view of attentional resources argued that, rather than simply having one single supply of resources, individuals instead have several different capacities with resource properties. This means that tasks will interfere more when resources are shared. Wickens' multiple resource model is presented in Figure 6.3. The mental workload literature is too great to describe adequately here. Regardless of the theoretical standpoint, it

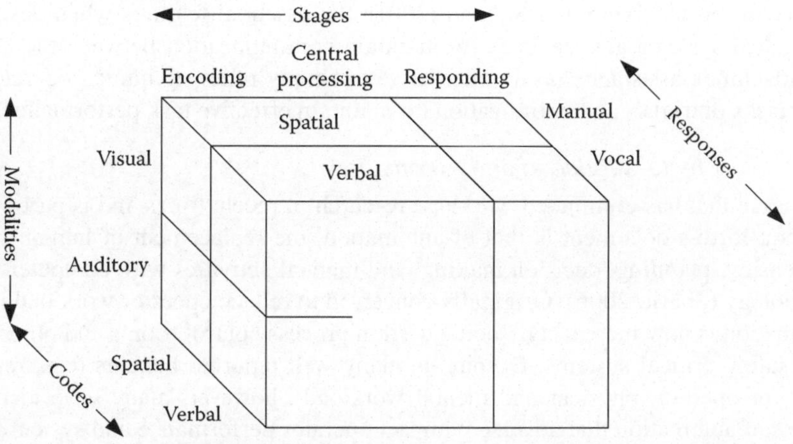

FIGURE 6.3 Multiple resource model. (Adapted from Wickens, C.D., *Engineering Psychology and Human Performance*, 2nd ed., Harper Collins, New York, 1992.)

is now generally agreed that individuals possess a finite pool of attentional resources and that these resources are allocated to tasks during task performance.

The importance of operator workload as a critical consideration in complex system design has been noted by many. Nachreiner (1995), for example, pointed out that the level of workload imposed on operators is one of the key criteria for evaluating system design because it is directly connected to the operator's capabilities and skills and the impairing effects of work on operators. Most in the field agree that when considering operator workload in complex systems, designers should aim to maximize the match between task demands and human capacity (Young and Stanton, 2002). This is known as *workload optimization*. It is hypothesized that when operator workload falls below or rises above this optimal level of workload performance decrements will occur.

System and procedure designers should therefore strive to optimize the level of workload experienced by process control operators. Young and Stanton (2001) outlined the following ways in which system designers can optimize operator workload:

- Designers should recognize that operators are not inherently flawed and subsequently should not simply remove or automate as many tasks as possible. This approach is likely to lead to mental underload and could leave operators with incoherent sets of tasks. The goal instead should be to design the system and procedures to exploit the skills of human operators.
- Physical workload should be kept to a minimum, whereas mental workload should be optimized. Ways to optimize mental workload include the use of adaptive interfaces (systems that can detect periods of elevated workload and then assist the operator accordingly) and dynamic allocation of function (systems that relieve the operator of task elements during high workload portions of the task).
- Mental workload should be reduced without affecting task performance.

According to Young and Stanton (2001), the essential features when designing automated systems are feedback (to maintain communication between the human and machine), assistance (as an alternative to simply relieving human operators of their tasks outright), and optimization (to maintain effective task performance).

6.4.2.1.1 Mental Workload and Automation

One issue that has dominated workload research in recent times and is particularly relevant to this document is that of automation, the replacement of human operator sensing, planning, decision making, and manual activities with computer-based technology (Sharit, 2006). Originally conceived to reduce operator workload levels, automation is now increasingly being used in process control rooms and other complex safety critical systems. Despite its many well-reported benefits (e.g., reduced levels of operator physical and mental workload), however, many have identified effects of automation that adversely impact operator performance and system safety (e.g., Bainbridge, 1987; Lee, 2006; Young and Stanton, 2002). These include manual and cognitive skill degradation, operator underload, out-of-the-loop performance decrements (Endsley and Kiris, 1995) and increased levels of workload due to system monitoring requirements. Lee (2006), for example, identified the following automation pitfalls:

- Out-of-loop unfamiliarity
- Clumsy automation
- Automation-induced errors
- Inappropriate trust (misuse, disuse, and complacency)
- Behavioral adaptation
- Inadequate training and skill loss
- Job satisfaction and health
- Eutectic behavior

It is apparent that, despite the impetus for most automated systems being a requirement to reduce operator workload, automation can potentially cause both decreases and increases in workload levels (Young and Stanton, 2002). It has been postulated, for example, that automation can lead to overload through a lack of feedback, increased vigilance demands (Hancock and Verwey, 1997), and increased decision options in certain situations (Hilburn, 1997). Various studies have also identified poorer performance in automated control conditions (e.g., Stanton et al., 1997).

The general consensus is currently that great caution should be taken when using automation in an attempt to enhance performance efficiency and safety in complex systems. Lee (2006) proposed a series of strategies for designing effective automation, including effective allocation of functions, the use of dynamic function allocation (adaptive and dynamic automation), matching automation to human performance characteristics, representation aiding and multimodal feedback, matching automation to user mental models, and the use of formal automation analysis techniques.

One theory that has been posited to explain why automated systems may lead to performance decrements is the malleable attentional resources theory (MART) (Young and Stanton, 2002). MART introduces the concept of malleable (as opposed to static) attentional resource pools and suggests that during low demand tasks an operator's resources shrink to accommodate the reduced level of demand. According to the theory, this reduced pool of attentional resources explains why operator performance suffers during low workload tasks. Further, the model suggests that operator performance suffers when the task demand increases quickly (e.g., emergency or nonroutine scenarios) because the diminished resources cannot keep up.

6.4.2.1.2 *Mental Workload and Situation Awareness*

Mental workload is intrinsically linked to the concept of situation awareness (SA), which refers to the level of awareness that an individual has of a situation, an operator's dynamic understanding of what is going on, and has been formally defined as "the perception of the elements in the environment within a volume of time and space, the comprehension of their meaning, and the projection of their status in the near future" (Endsley, 1995, p. 36). SA is a critical commodity for process control room operators and teams.

Many researchers have hypothesized over the link between SA and mental workload (e.g., Endsley 1993; Wickens, 2002). The first and most obvious relationship between the two is that the acquisition and maintenance of SA exploit an individual's attentional resources and thus impacts the level of mental workload imposed on them—that is, the greater the SA requirements, the greater the level of workload invested in acquiring SA. According to Endsley et al. (2003), when workload demands exceed human capacity, SA may suffer. Likewise, a reduced level of mental workload could potentially lead to higher levels of SA, because the operator has more resources to devote to the acquisition and maintenance of SA. Whereas these relationships may be apparent in many situations, it is clear that there are many more factors that impact the relationship between the two constructs. Endsley (1993) pointed out that, although SA and workload are related, they are essentially independent constructs, and high SA is not necessarily the result in low workload conditions. Conversely, Endsley stressed that poor SA is not always guaranteed in high workload situations. Endsley described the following four possible relationships between the SA and mental workload:

- *Low SA and low workload*—Incomplete or inaccurate SA can still be evident when the operator is experiencing a very low level of workload (underload). This may be due to problems such as lack of vigilance and inattentiveness.
- *Low SA and high workload*—An operator who is experiencing a high level of workload (overload) may fail to achieve an adequate level of SA due to a diminished level of resources.
- *High SA and high workload*—An operator who is experiencing a high level of workload (overload) can still achieve and maintain higher levels of SA. Experienced operators can still achieve higher levels of SA in these conditions through the use of schemata or mental models that facilitate directing one's attention to relevant elements in the environment, as well as the comprehension of these elements and the projection of their future status.
- *High SA and low workload*—Operators working under low workload conditions should achieve high levels of SA. This is the design optimum.

In studies of the operational errors made by air-traffic controllers, Endsley and Smolensky (1998) and Rodgers and Nye (1993) found evidence of all of the above relationships between workload and SA. A number of studies have investigated the relationship between SA and workload. Endsley (1993) assessed the workload and SA of six former military pilots during the performance of a set of simulated air-to-air engagements. The Situation Awareness Global Assessment Technique (SAGAT) (Endsley, 1995) was used to measure participant SA, and the Subjective Workload Assessment Technique (SWAT) (Reid and Nygren, 1987) was used to measure participant mental workload (MWL). It was found that overall there was no significant relationship between participant SA and MWL; however, two of the participants demonstrated a significant relationship between SA and workload for the low SA/high workload and high SA/low workload conditions. Endsley and Rodgers (1998) found that air-traffic controllers reported significantly fewer of the aircraft present and identified aircraft heading correctly significantly less often for those aircraft as subjective workload increased. Further, as workload increased, participants were significantly less likely to identify aircraft completion of assigned clearance and whether an aircraft had correctly received its assigned clearance or not.

The effect of cognitive underload on SA has also been investigated. Endsley and Kiris (1995) concluded that reduced SA can be caused by a loss of vigilance and an increase in complacency brought about by automated systems after they investigated the out-of-the-loop performance problem, whereby operators working with automated systems are removed from the control loop which leads to a diminished ability to detect system problems and perform tasks manually in the event of an automation breakdown. Endsley and Kiris used varying levels of automation on an automobile navigation task to assess the effect of automation on participant workload, SA, and decision time. SA and decision time were negatively affected following loss of automation, but no reduction in workload was observed; consequently, no significant relationship between SA and workload could be derived. In the conclusion to their investigations into the relationship between air traffic controller workload and SA, Endsley and Smolensky (1998) noted that numerous studies have observed an increase in the percentage of operational errors under light and moderate levels of workload as opposed to high workload conditions (e.g., Schroeder, 1982; Schroeder and Nye, 1993; Stager and Hameluck, 1990). Endsley (1993) provided a graph depicting the relationship between SA and workload (Figure 6.4). According to Endsley and Smolensky (1998), disconnects between workload and SA can occur for many reasons related to system design, the tasks being performed, and the individual involved. Presumably, other factors such as experience, training, the tools being used, and context can also dissociate the two.

In summary, the literature indicates that the relationship between workload and SA is a complex one with many dimensions. Increased levels of workload can potentially adversely impact SA, as resources become diminished. If mental capacity is exceeded, SA may become degraded; however, individuals experiencing very high levels of workload can still achieve maximum levels of SA due to factors such as experience and the use of mental models (Endsley, 1995), which direct attention to SA elements and facilitate their comprehension. The effect of low levels of workload on SA is equally divergent. Individuals working under low workload conditions

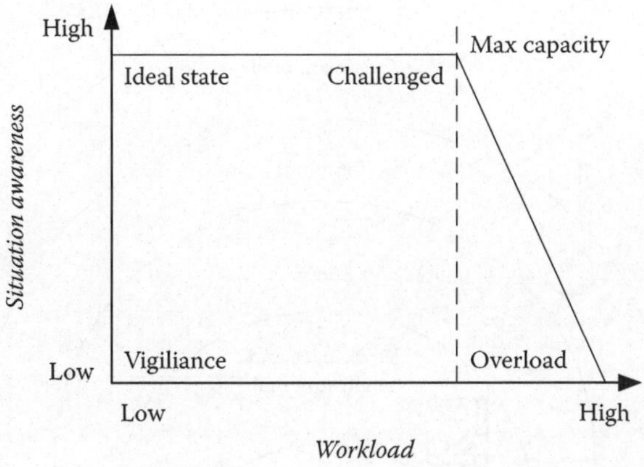

FIGURE 6.4 The relationship between workload and SA. (Adapted from Endsley, M.R., *International Journal of Aviation Psychology*, 3, 157–168, 1993.)

should possess high levels of SA, because they have more resources available to devote to the acquisition and maintenance of SA; however, low workload conditions can often lead to degraded SA due to problems of complacency and reduced vigilance (Endsley and Smolensky, 1998).

6.4.2.2 Physical Workload

An abundance of human factors literature surrounds the area of physical workload; however, for obvious reasons, very little research has focused on the level of physical workload imposed on process control room operators. The type of manual tasks investigated within the context of physical workload include lifting, pushing, and pulling types of tasks, most of which are not required in contemporary control rooms.

6.5 GUIDANCE AND RECOMMENDATIONS

6.5.1 Manning

As pointed out previously, the literature on safe staffing levels in process control is sparse. Nevertheless, the use of appropriate staffing levels by organizations is of critical importance to the safety and efficiency of process control performance. It is therefore recommended that organizations take steps to identify exactly what the safe staffing levels are for each process undertaken within their plant. Steps should then be taken to ensure that staffing levels are appropriate.

6.5.2 Mental Workload

It is clear from the literature that operator workload is a complex area. We have seen that mental workload is a multidimensional concept impacted by numerous factors, so much so that designing systems with the aim of optimizing the workload levels is

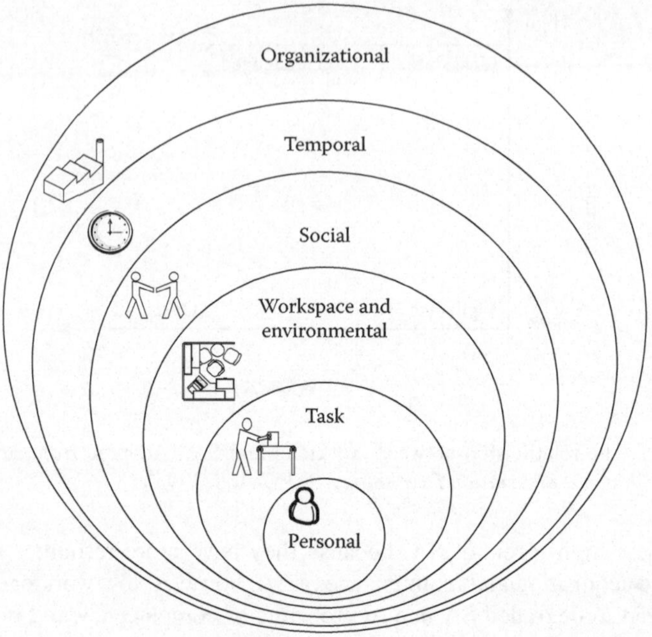

FIGURE 6.5 System design guidelines and recommendations with regard to operator workload.

likely to be a difficult proposition. Despite this caution, however, sufficient knowledge is contained within the literature to formulate a series of general guidelines relating to the design of systems for workload optimization. An outline of these guidelines is presented below.

The guidelines and recommendations for system design regarding operator workload can be categorized under the following headings (as represented in Figure 6.5):

- Personal factors include system design recommendations relating to the personal factors that impact operator workload levels, such as training, experience, and mental models.
- Task factors include system design recommendations relating to the tasks being performed within the system and the task characteristics that impact operator workload, such as task complexity, task allocation, and task compatibility.
- Workspace and environmental factors include system design recommendations relating to the aspects of the physical workspace and work environment that impact operator workload, such as interface and display design and noise, temperature, and ambience.
- Social factors include system design recommendations relating to the social factors impacting operator workload, such as task allocation, team organization, and social interaction.

- Temporal factors include system design recommendations relating to the temporal factors that impact workload, including time pressure, shiftwork, and the temporal distribution of workload.
- Organizational factors include system design recommendations relating to the organizational factors that impact operator workload, such as procedures and training.

6.5.2.1 Personnel Guidelines and Recommendations (ISO 9001:2008)

- *Processing*—Serial processing is preferable to parallel processing due to the demands imposed on processing resources; however, if comparisons are to be made between different sources of information, then the information in question should be presented in a parallel manner.
- *Mental models*—Mental models are defined as "mechanisms whereby humans generate descriptions of system purpose and form, explanations of system functioning and observed system states, and predictions of future system states" (Rouse and Morris, 1986, p. 360). System users possess mental models of how the system or process works, and the key to user-centered design is to design the new system or process to fit in with, or reflect, user mental models of that system or process. A design that is incongruent with the user's mental model may increase the level of workload required to operate the system in question.
- *Absolute vs. relative judgments*—Relative judgments are easier to perform and so are preferable to absolute judgments. It is recommended that this be achieved by providing displays that allow for the presentation of reference standards to which critical information can be compared.
- *Working memory*—Overloading of working memory should be avoided by providing adequate time for operators to select and memorize serially presented information.
- *Long-term memory*—Adequate information retrieval facilities (e.g., help functions) should be provided to avoid long-term memory overload.
- *Recognition vs. recall from memory*—Recognition of memorized items is easier and more effective than recall; therefore, it is recommended that alternatives be displayed to operators.

6.5.2.2 Task Guidelines and Recommendations (ISO 9001:2008)

- *Optimization of workload*—Tasks and procedures should be designed so operator workload is optimized.
- *Task complexity*—The level of task complexity has a direct impact on operator workload. Operator tasks should be designed so they are neither overly complex nor of insufficient complexity levels. Overly complex tasks should be considered for automation.
- *Task goals*—Task goals should be clear and made explicit to the operator. Clear task goals should be provided, and the priorities of different goals should be specified.

- *Task duration*—The duration of tasks should be designed so workload is not increased via long duration tasks. Lengthy sustained attention tasks should also be avoided.
- *Task allocation*—Tasks should be allocated between human operators and system automation appropriately; for example, to avoid mental satiation, simple repetitive tasks should be automated.
- *Mental and physical activities*—An appropriate balance between mental and physical activities should be used; for example, to reduce monotony, appropriate opportunities for physical activities should be provided to operators.

6.5.2.3 Workspace and Environment (ISO 9001:2008)

- *Interface and display design*—The displays and interfaces within the system should be designed considering the following issues:
  - *Adequacy of information*—The information required for efficient and effective task performance should be explicitly provided to the operators as and when required. Missing or unnecessary information increases operator workload as they have to make decisions on the basis of incomplete information or filter the appropriate information from the total pool of information supplied to them.
  - *Ambiguity of information*—The presentation of ambiguous information should be avoided.
  - *Signal discriminability*—Adequate signal discriminability should be provided. Signal discriminability can be enhanced by manipulating the intensity of signals; by coding signals using shape, color, duration, or time; by reducing background noise; and through masking and filtering via technical systems.
  - *Redundancy*—Redundancy should be planned according to operational requirements. On the one hand, some level of redundancy of information is required so operators can cross check displayed information; however, on the other hand, too much redundant information can increase operator workload through distraction.
- *Noise*—Appropriate noise levels should be used within the system so perception and information processing conditions are optimal.
- *Lighting*—Appropriate lighting levels should be used within the system so perception and information processing conditions are optimal.

6.5.2.4 Social Factors (ISO 9001:2008)

- *Social interaction*—Opportunities for social interaction can help by providing social support in critical decision making. Task, system, and procedural design should therefore be used to encourage appropriate levels of social interaction between operators.
- *Team organization*—The team should be organized so workload is appropriately distributed among operators.
- *Codependency*—Dependency on others increases strain and should be avoided.

6.5.2.5 Time Factors (ISO 9001:2008)

- *Time pressure*—Increased levels of time pressure increase operator workload and the likelihood of errors. Tasks and procedures should be designed without the presence of significant time pressure on operators where possible.
- *Time of day*—Operator performance is impacted by the time of day and variations in their circadian rhythm. As operator performance during night hours is generally inferior to performance during day hours, the performance demands for night work should be reduced relative to day work.
- *Shiftwork*—Shiftwork should be avoided wherever possible; however, where shiftwork is necessary, ergonomically designed shiftwork schedules should be used.
- *Shift duration*—The duration of operator shifts is an important factor in workload optimization and should be considered accordingly. Working hours should be limited to a point where fatigue effects do not manifest themselves.
- *Time off*—Adequate levels of time off should be provided so operators working shifts can recover appropriately.

6.5.2.6 Organizational Factors (Tsang and Vidulich, 2006)

- *Training design*—There is great potential for tasks to be accomplished with fewer resources at a lower level of workload through the provision of appropriate training. When determining what should be trained to enhance operator workload and SA, a thorough understanding of the system and its workings (including in nonroutine situations) is helpful, along with training in strategic task management and reinforcing the knowledge structure of experts.
- *Procedural design*—The design of procedures should consider workload and use task allocation and procedural factors to distribute workload accordingly.
- *Allocation of functions*—Functions should be allocated appropriately between the operators and the system (automation).

6.6 ASSESSMENT AND EVALUATION

6.6.1 Manning

There is a general lack of formalized procedures for assessing the manning levels in process plants. In response to this, Entic UK, Ltd., and the HSE developed a staffing level assessment methodology that can be used by companies to assess the safety of their current staffing arrangements. The CRR348/2001 methodology is designed to flag when too few staff are being used to control a process and focuses on staffing level requirements during hazardous incident responses, in particular how staffing levels and organization impact the reliability and timeliness of incident detection, diagnosis, and recovery (Brabazon and Conlin, 2001). The methodology has two components: (1) a physical assessment component that is used to assess the ability of staff to detect, diagnose, and recover hazardous situations, and (2) a ladder assessment component that is used to benchmark organizational factors in relation to industry best practice. The method specifies six principles that should be used to assess an organization's staffing arrangements:

- Skilled operators should provide continuous supervision of the process.
- Distractions, such as answering phones, talking to people in the control room, administration tasks, and nuisance alarms, should be minimized to reduce the possibility of missing alarms.
- Additional information required for diagnosis and recovery should be accessible, correct, and intelligible.
- Communication links between the control room and field should be reliable.
- Staff required to assist in diagnosis and recovery should be available with sufficient time to attend when required.
- Operating staff should be allowed to concentrate on recovering the plant to a safe state.

The assessment itself involves administering a series of simple yes or no questions. The questions are arranged in eight trees; the first three are designed to detect potential problems, and the remaining five are designed to test the ability to diagnose and recover from the problems in time. An example question tree is presented in Figure 6.6. A flowchart depicting the CRR348/2001 methodology procedure is presented in Figure 6.7. Further information related to the background and development of the approach can be found in Brabazon and Conlin (2001). A user guide providing practical guidance on how to apply the CRR348/2001 methodology and example applications of the approach can be found in Brazier et al. (2004).

6.6.2 Workload

6.6.2.1 Mental Workload

Various mental workload assessment approaches are available, including primary and secondary task performance measures, physiological measures, and subjective-rating techniques. Typically, a combination of all four is used during mental workload assessment efforts (Young and Stanton, 2001). A brief summary of each method type is presented below.

Primary task performance measures are used to assess an individual's ability to perform a particular task under varying levels of workload, based on the supposition that an individual's performance will diminish as his or her level of workload increases. When assessing the level of workload imposed by driver support systems, Young and Stanton (2004) measured speed, lateral position, and headway as indicators of performance on a driving task. Poor performance on the primary task is taken to be indicative of an increased level of workload, while efficient task performance indicates that the level of workload is manageable. According to Wierwille and Eggemeier (1993), primary tasks measures should be included in any assessment of operator mental workload. The main advantages associated with the use of primary

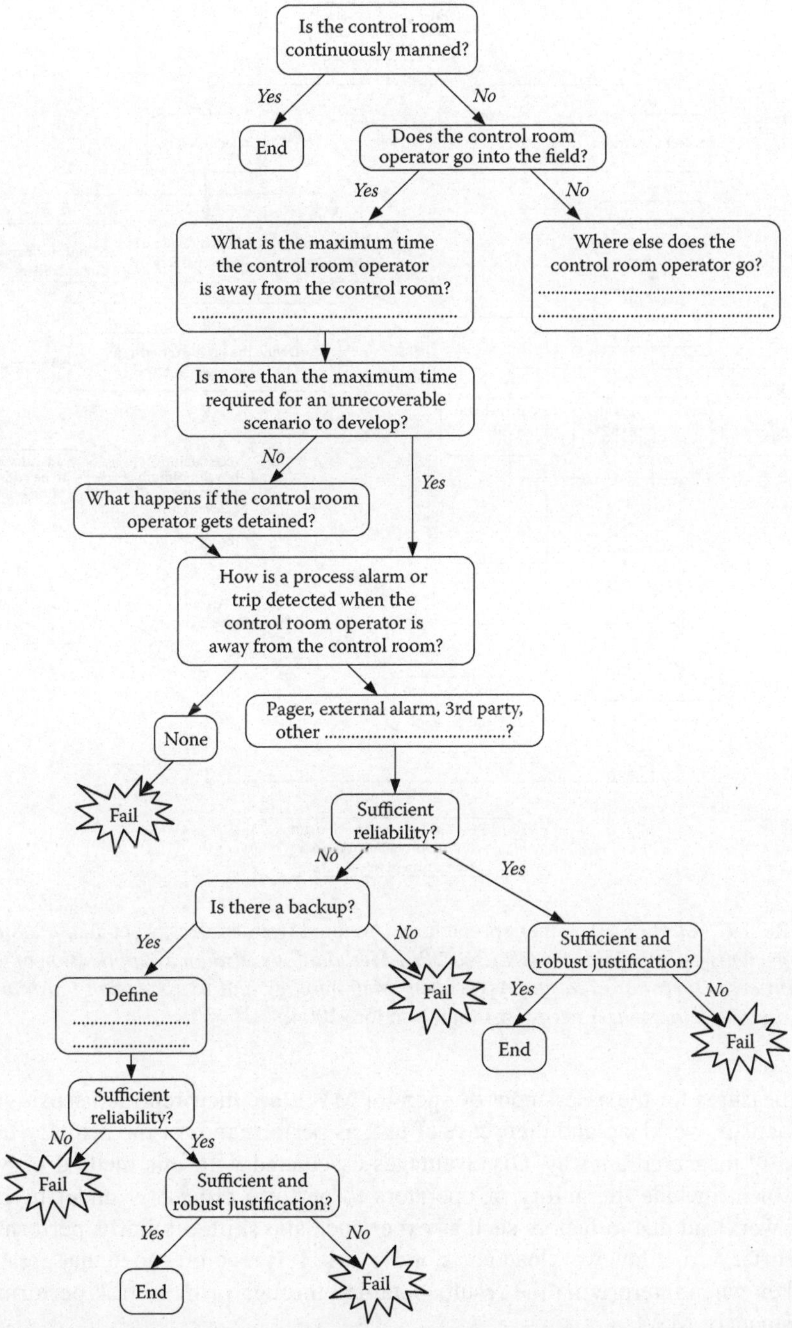

FIGURE 6.6 Example question tree. (From Brabazon, P. and Conlin, H., *Assessing the Safety of Staffing Arrangements for Process Operations in the Chemical and Allied Industries*, Contract Research Report 348/2001, HSE Books, London, 2001. Reproduced under the terms of the Click-Use Licence.)

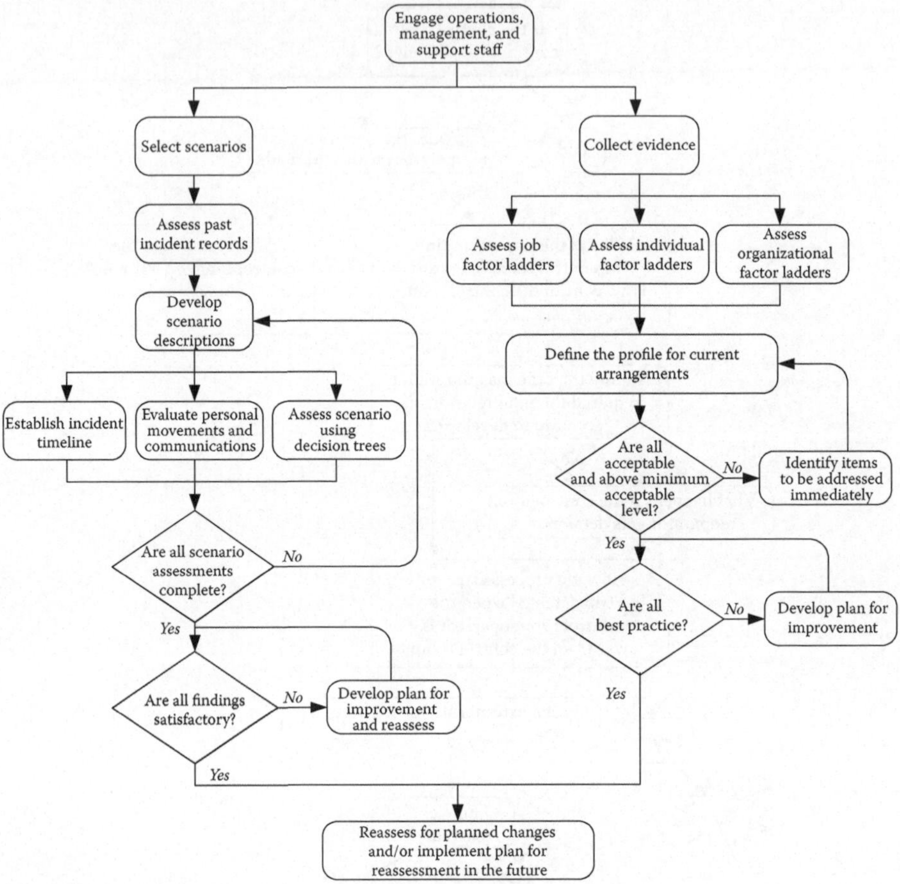

FIGURE 6.7 Staffing assessment procedure. (Adapted from Brazier, A. et al., *Safe Staffing Arrangements—User Guide for CRR348/2001 Methodology: Practical Application of Entec/HSE Process Operations Staffing Assessment Methodology and Its Extension to Automated Plant and/or Equipment*, Energy Institute, London, 2004.)

task measures for the assessment of operator MWL are their reported sensitivity to variations in workload and their ease of use, as performance of the primary task is normally measured anyway. Disadvantages associated with this method of MWL assessment include the ability of operators to perform efficiently under high levels of workload due to factors such as experience and skill. Similarly, performance may suffer during low workload parts of the task. It is recommended that great care be taken when interpreting the results obtained through primary task performance assessment of MWL.

Secondary task performance measures are used to measure an individual's ability to perform a secondary task (in addition to the primary task), the hypothesis being that as the level of workload imposed by the primary task increases performance on the secondary task will diminish, as the resources available for the secondary task are reduced. A variety of secondary task performance measures is available, including memory recall tasks, mental arithmetic tasks, reaction time measurement, and tracking tasks. The main disadvantages associated with secondary task performance assessment techniques are a reported lack of sensitivity to minor workload variations (Young and Stanton, 2004) and their intrusion on primary task performance. One way around this is the use of embedded secondary task measures, whereby the operator is required to perform a secondary task with the system under analysis. Because the secondary task is no longer external to operating the system, the level of intrusion is reduced. According to Young and Stanton (2004), researchers adopting a secondary task measurement approach to the assessment of MWL are advised to adopt discrete stimuli that occupy the same attentional resource pools as the primary task. For example, if the primary task is a driving one, then the secondary task should be a visuospatial one involving manual response. This ensures that the technique really is measuring spare capacity and not an alternative resource pool.

Physiological measures of workload involve the measurement of physiological responses that may be affected by increased or decreased levels of MWL. Examples of physiological measures include heart rate, heart rate variability, eye movement, and brain activity. The main advantage associated with the use of physiological measures of MWL is that they do not intrude upon primary task performance and they can be applied in the field, as opposed to simulated settings. Disadvantages associated with the use of physiological techniques include the high cost, physical obtrusiveness, and reliability of the technology used, as well as doubts regarding the construct validity and sensitivity of the techniques.

Subjective-rating techniques involve the operators in question providing subjective ratings of their workload during or after task performance. Subjective-rating techniques can be unidimensional (assessing overall level of workload only) or multidimensional (assessing various different dimensions associated with workload) and typically offer a quantitative rating of the level of MWL imposed on an individual during task performance. Young and Stanton (2004) suggested that the data obtained when using unidimensional techniques are far simpler to analyze than data obtained when using multidimensional techniques; however, multidimensional techniques offer a greater level of diagnosticity than unidimensional techniques. The NASA Task Load Index (TLX) (Hart and Staveland, 1988) is the most commonly used subjective rating technique; it utilizes six subscales (mental demand, physical demand, temporal demand, effort, performance, and frustration level) to derive an overall numerical rating of workload.

Subjective-rating assessment techniques are attractive due to their ease and speed of application and the low cost involved. Subjective-rating techniques are also not intrusive to primary task performance and can be used in the field in real-world settings rather than in simulated environments. That said, subjective workload assessment techniques are mainly only used when an operational system is available; therefore, it is difficult to employ them during the design process, as the system under analysis may not actually exist, and simulation can be extremely costly. A host of problems is associated with collecting subjective data post-trial. Often, MWL ratings correlate with performance on the task under analysis. Participants are also prone to forgetting certain parts of the task where variations in their workload may have occurred.

6.6.2.1.1 *Workload Prediction*

More recently, a number of analytical workload prediction models and techniques have been developed (e.g., Gregoriades and Sutcliffe, 2007; Neerincx, 2003; Vidulich, Ward, and Schueren, 1991). According to Gregoriades and Sutcliffe (2006), analytical workload assessment techniques use mathematical models and simulations to simulate human performance and estimate workload. Cognitive task load analysis (CTLA), for example, is used to assess or predict the cognitive load imposed by a particular task (Neerincx, 2003). CTLA uses a model of cognitive task load that describes the effects of task characteristics on MWL. According to the model, cognitive (or mental) task load is comprised of percentage time occupied, level of information processing, and the number of task set switches exhibited during the task in question.

6.6.3 Examples

6.6.3.1 NASA-TLX Procedure and Advice

The NASA-TLX procedure involves participants providing subjective ratings (typically on a scale of 1 (low) to 20 (high) for the following subscales:

- *Mental demand*—How much mental demand and perceptual activity was required (e.g., thinking, deciding, calculating, remembering, looking, searching). Was the task easy or demanding, simple or complex, exacting or forgiving?
- *Physical demand*—How much physical activity was required (e.g., pushing, pulling, turning, controlling, activating). Was the task easy or demanding, slow or brisk, slack or strenuous, restful or laborious?
- *Temporal demand*—How much time pressure did you feel due to the rate or pace at which the tasks or task elements occurred? Was the pace slow and leisurely or rapid and frantic?
- *Effort*—How hard did you have to work (mentally and physically) to accomplish your level of performance?
- *Performance*—How successful do you think you were in accomplishing the goals of the task set by the analyst (or yourself)? How satisfied were you with your performance in accomplishing these goals?
- *Frustration level*—How insecure, discouraged, irritated, stressed, and annoyed vs. secure, gratified, content, relaxed, and complacent did you feel during the task?

The NASA-TLX assessment is presented in Figure 6.8. Guidance on the procedure to adopt when undertaking NASA-TLX workload assessments is presented next:

1. *Define tasks.* The first step involves defining the tasks that are to be subjected to analysis. The type of tasks analyzed depends on the focus of the analysis; for example, when assessing the effects on operator workload caused by a novel design or a new process, it is useful to analyze a set of tasks that are as representative of the devices operations as possible. To analyze a full set of tasks will often be too time consuming and labor

NASA Task Load Index

Mental Demand
How much mental and perceptual activity was required (e.g., thinking, deciding, calculating, remembering, looking, searching)? Was the task easy or demanding, simple or complex, exacting or forgiving?

Low —————————————— **High**

Physical Demand
How much physical activity was required (e.g. pushing, pulling, turning, controlling, activating)? Was the task easy or demanding, slow or brisk, slack or strenuous, restful or laborious?

Low —————————————— **High**

Temporal Demand
How much time pressure did you feel due to the rate or pace at which the tasks or task elements occurred? Was the pace slow and leisurely or rapid and frantic?

Low —————————————— **High**

Effort
How hard did you have to work (mentally and physically) to accomplish your level of performance?

Low —————————————— **High**

Performance
How successful do you think you were in accomplishing the goals of the task set by the analyst (or yourself)? How satisfied were you with your performance in accomplishing these goals?

Low —————————————— **High**

Frustration Level
How insecure, discouraged, irritated, stressed, and annoyed vs. secure, gratified, content, relaxed, and complacent did you feel during the task?

Low —————————————— **High**

FIGURE 6.8 NASA-TLX assessment. (Adapted from Hart, S.G. and Staveland, L.E., in *Human Mental Workload*, P.A. Hancock and N. Meshkati, Eds., Elsevier, Amsterdam, 1988, pp. 139–183.)

intensive, so it is pertinent to use a set of tasks that involve all aspects of the system under analysis.

2. *Conduct a HTA for the tasks under analysis*. Once the tasks under analysis are defined clearly, a hierarchical task analysis should be conducted for each task. This allows the analysts and participants to understand the tasks fully.

3. *Select participants.* Once the tasks under analysis are clearly defined and described, it may be useful to select the participants that are to be involved in the analysis. This may not always be necessary, and it may suffice to simply select participants randomly on the day; however, if workload is being compared across rank or experience levels, then clearly effort is required to select the appropriate participants.
4. *Brief participants.* Before the tasks under analysis are performed, all of the participants involved should be briefed regarding the purpose of the study and the NASA-TLX technique. It is recommended that participants be given a workshop on MWL and MWL assessment. It may also be useful at this stage to take the participants through an example NASA-TLX application, so they understand how the technique works and what is required of them as participants. It may even be pertinent to get the participants to perform a small task and then get them to complete a workload profile questionnaire. This would serve as a pilot run of the procedure and would highlight any potential problems.
5. *Perform task under analysis.* Next, the participants should perform the task under analysis. The NASA-TLX can be administered either during or after the trial, although it is recommended that the TLX be administered after, as online administration is intrusive to primary task performance. If online administration is required, then the TLX should be administered and responded to verbally.
6. *Conduct weighting procedure.* When the task under analysis is complete, the weighting procedure can begin. The weighting software presents to the participants 15 pair-wise comparisons of the 6 subscales (mental demand, physical demand, temporal demand, effort, performance, and frustration level). The participants should be instructed to select from each of the 15 pairs the subscale from each pair that contributed the most to the workload of the task. The software then calculates the total number of times each subscale was selected by the participant. Each scale is rated based on the number of times it is selected by the participant; this is done using a scale of 0 (not relevant) to 5 (more important than any other factor).
7. *Conduct NASA-TLX rating procedure.* Participants should be presented with the interval scale for each of the TLX subscales (this is done via the rating software). Participants are asked to give a rating for each subscale, between 1 (low) and 20 (high), in response to the associated subscale questions. The ratings provided are based entirely on the participants' subjective judgment.
8. *Calculate TLX score.* The TLX software is then used to compute an overall workload score. This is calculated by multiplying each rating by the weight given to that subscale by the participant. The sum of the weighted ratings for each task is then divided by 15 (sum of weights). A workload score of between 0 and 100 is then derived for the task under analysis.

A flowchart depicting the NASA-TLX procedure is presented in Figure 6.9.

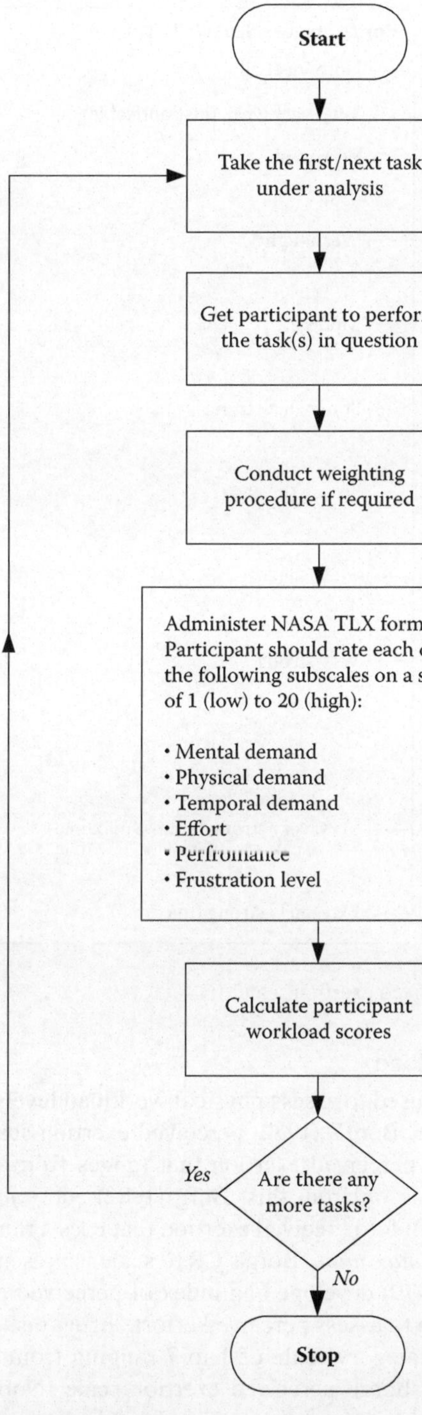

FIGURE 6.9 NASA-TLX flowchart.

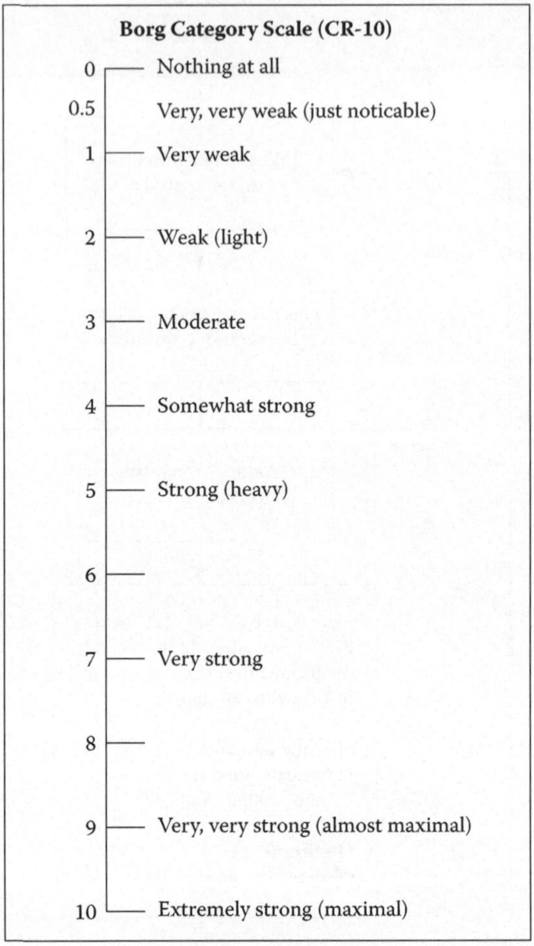

FIGURE 6.10 Borg's perceived exertion scale.

6.6.3.2 Physical Workload

The majority of methods used to assess physical workload levels use scales of perceived effort; for example, Borg's (1982) perceived exertion scale uses a 15-point category scale for rating perceived exertion that ranges from *very, very light* to *very, very hard*. Following on from this, Borg (1982) subsequently developed a category ratio scale (CR10) for perceived exertion that uses a range of 0 to 10, from *nothing at all* to *absolute maximum*. Borg's CR10 scale is presented in Figure 6.10. Hogan and Fleishman (1979) developed an index of perceived physical effort that also used a category scale to assess perceived effort during task performance. The Fleishman index uses a category scale of 1 to 7 ranging from *very, very light* to *very, very* hard. The Pittsburgh perceived exertion scale (Noble and Robertson, 1996) uses a nine-point category scale ranging from *not at all stressful* to *very, very stressful*.

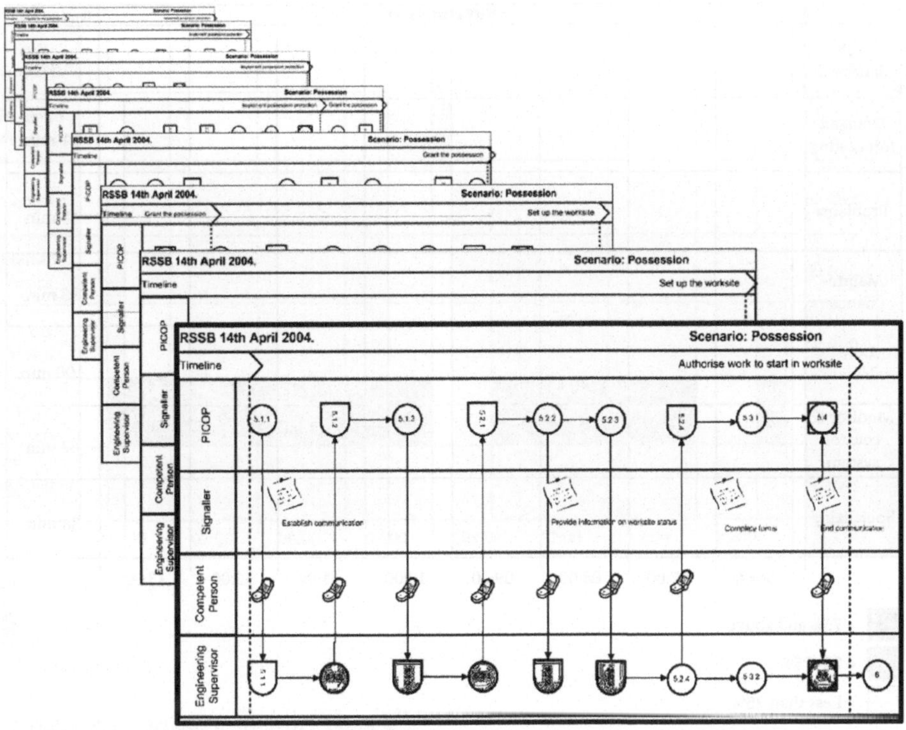

FIGURE 6.11 Operation sequence diagrams.

The level of physical workload imposed on individuals during task performance can also be analyzed and represented using task analysis and charting techniques. Task analysis techniques such as hierarchical task analysis are used to develop a description of task activity, whereas charting techniques are used to depict, pictorially, the task steps undertaken by individuals during task performance. From these analyses, frequency counts for different physical operations can be calculated. Operation sequence diagrams (OSDs) are used to calculate operational loading figures that indicate the frequency with which each individual undertook a particular type of operation during task performance (Stanton, Salmon, Walker, Baber, and Jenkins, 2005). Operational loading figures are typically calculated for a range of operators, such as operation (e.g., perform a physical or cognitive task step), receive (e.g., receive information from another agent or artifact), delay (e.g., delay task performance for some reason, such as waiting for comms/instructions from another agent), decision (e.g., make a decision in relation to the task), transport (e.g., transportation of agents or artifacts), and combined operations (e.g., receive information and perform a task step at the same time). From operational loading figures, an indirect indication of workload is typically derived based on the amount of operations that an individual undertakes during task performance; that is, the more operations individuals must perform during a particular task, the greater their level of workload is taken to be. Examples of OSD outputs are presented in Figure 6.11.

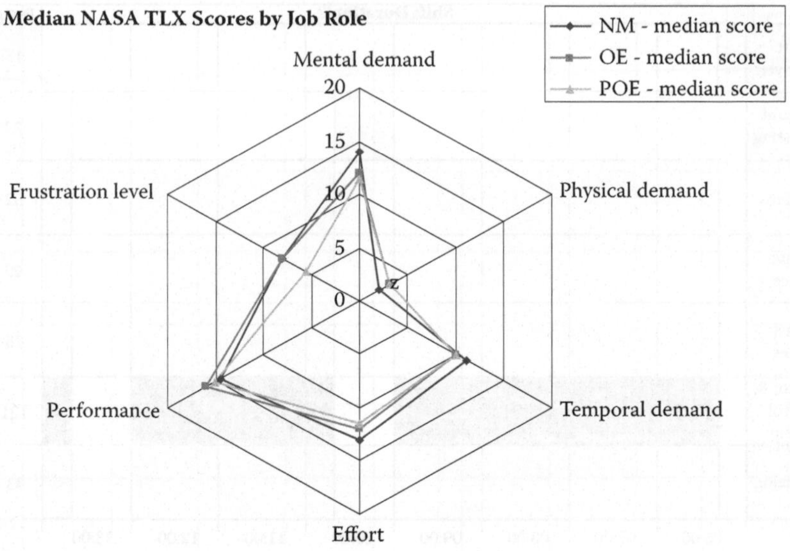

FIGURE 6.14 Median NASA-TLX ratings by role type.

Engineer will take on the same work as the Operations Engineer, and the Shift Control Manager will take on the same work as the Principal Operations Engineer and some of the Network Manager duties.

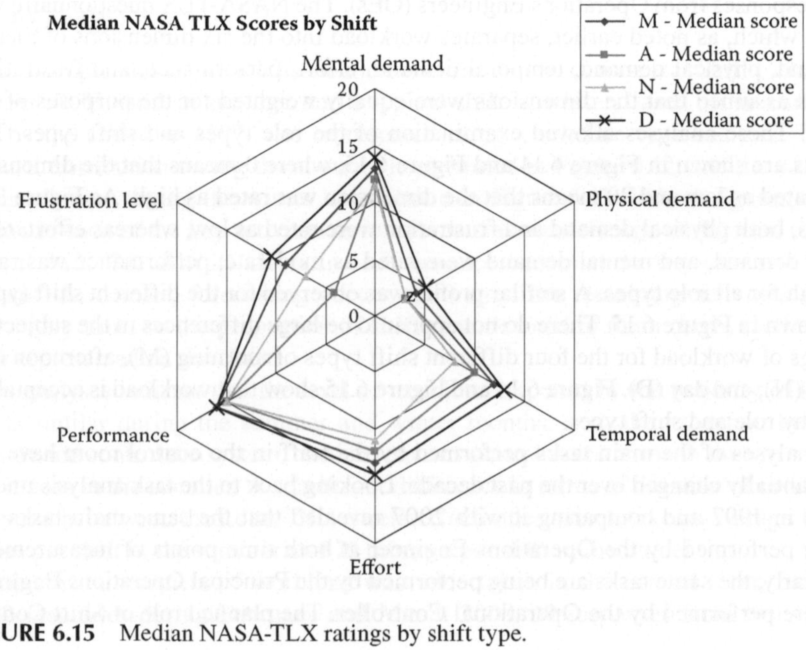

FIGURE 6.15 Median NASA-TLX ratings by shift type.

TABLE 6.1
Comparison of Tasks in the Control Room: 1997, 2007, and Proposed

Role	Main Tasks	1997 (OE/OC)	2007 (OE/POE)	Proposed (SCE/SCM)
Operations Engineer (OE) or Shift Control Engineer (SCE)	Shift handover	✓	✓	✓
	Monitor gas system	✓	✓	✓
	Demand forecasting	✓	✓	✓
	Control of gas system	✓	✓	✓
	Contact external parties	✓	✓	✓
	Conduct maintenance	✓	✓	✓
	Conduct emergency procedures	✓	✓	✓
	Conduct load shedding			✓
Operations Controller (OC), Principal Operations Engineer (POE), or Shift Control Manager (SCM)	Maintain an overview of area	✓	✓	✓
	Overview gas supply and stock	✓	✓	✓
	Oversee systems management	✓	✓	✓
	Conduct Maintenance	✓	✓	✓
	Conduct load shedding	✓	✓	✓
	Conduct emergency procedures	✓	✓	✓
	Development of staff	✓	✓	✓
	Carry out project work	✓	✓	✓

The main changes between 1997 and 2007 had to do with the naming of systems, but they perform the same functions in much the same way. The planned changes will not be very different from the previous two iterations of the roles and tasks. It is anticipated that similar demands will be made on the staff in the control room. The new control system should have specific pages dedicated to demand, intake, and stock profiles, which will make shortfalls and deficits more visible. The demand forecasting will also be largely automated, the products of which will comprise a gas management strategy for the Shift Control Engineer. The point here is that, although the tasks have not changed much for over a decade, some new technologies are being employed to support those tasks. As Table 6.1 shows, the only substantive change being proposed is the additional task of load shedding to be conducted by the Shift Control Engineer. Given that the current workload is within acceptable limits, it is anticipated that this task will not overburden the Shift Control Engineer. This assumption needs to be checked when the new format has been operationalized.

BIBLIOGRAPHY

Bainbridge, L. (1987). Ironies of automation. *Automatica*, 19(6), 775–779.

Borg, G.A. (1982). Psychophysical bases of perceived exertion. *Medicine and Science in Sports and Exercise*, 14, 377–381.

Brabazon, P. and Conlin, H. (2001). *Assessing the Safety of Staffing Arrangements for Process Operations in the Chemical and Allied Industries*, Contract Research Report 348/2001. London: HSE Books.

Brazier, A., Waite, P., and Gait, A. (2004). *Safe Staffing Arrangements—User Guide for CRR348/2001 Methodology: Practical Application of Entec/HSE Process Operations Staffing Assessment Methodology and Its Extension to Automated Plant and/or Equipment*. London: Energy Institute.

Delleman, N.J. and Dul, J. (2006). Evaluation of working postures and movements in relation to machinery. In W. Karwowski (Ed.), *Handbook on Standards and Guidelines in Ergonomics and Human Factors* (pp. 201–206). Mahwah, NJ: Lawrence Erlbaum.

Donald, C.H.M. (1999). Assessing the Human Vigilance Capacity of Control Room Operators. Paper presented at the Institution of Electrical Engineers' People in Control—An International Conference on Human Interfaces in Control Rooms, Cockpits, and Command Centers, University of Bath, Bath, U.K., June 1999.

Endsley, M.R. (1993). A survey of situation awareness requirements in air-to-air combat fighters. *International Journal of Aviation Psychology*, 3, 157–168.

Endsley, M.R. (1995). Towards a theory of situation awareness in dynamic systems. *Human Factors*, 37, 32–64.

Endsley, M.R. and Kaber, D.B. (1999). Level of automation effects on performance, situation awareness, and workload in a dynamic control task. *Ergonomics*, 42(3), 462–492.

Endsley, M.R. and Kiris, E.O. (1995). The out-of-the-loop performance problem and level of control in automation. *Human Factors*, 37(2), 381–394.

Endsley, M.R. and Rodgers, M.D. (1998). Distribution of attention, situation awareness, and workload in a passive air traffic control task: implications for operational errors and automation. *Air Traffic Control Quarterly*, 6(1), 21–44.

Endsley, M.R. and Smolensky, M. (1998). Situation awareness in air traffic control: the picture. In M. Smolensky and E. Stein (Eds.), *Human Factors in Air Traffic Control* (pp. 115–154). New York: Academic Press.

Endsley, M.R., Bolte, B., and Jones, D.G. (2003). *Designing for Situation Awareness*. London: Taylor & Francis.

Gregoriades, A. and Sutcliffe, A. (2006). Automated assistance for human factors analysis in complex systems. *Ergonomics*, 49, 1265–1287.

Gregoriades, A. and Sutcliffe, A. (2007). Workload prediction for improved design and reliability of complex systems. *Reliability Engineering and System Safety*, 93(4), 530–549.

Grier, R.A., Warm, J.S., Dember, W.N., Matthews, G., Galinsky, T.L., Szalma, J.L., and Parasuraman, R. (2003). The vigilance decrement reflects limitations in effortful attention not mindlessness. *Human Factors*, 45, 349–359.

Hancock, P.A. and Verwey, W.B. (1997). Fatigue, workload and adaptive driver systems. *Accident Analysis and Prevention*, 29(4), 495–506.

Hart, S.G. and Staveland, L.E. (1988). Development of a multi-dimensional workload rating scale: results of empirical and theoretical research. In P.A. Hancock and N. Meshkati (Eds.), *Human Mental Workload* (pp. 139–183). Amsterdam: Elsevier.

Hart, S.G. and Wickens, C.D. (1990). Workload assessment and prediction. In H.R. Booher (Ed.), *MANPRINT: An Approach to Systems Integration* (pp. 257–296). New York: Van Nostrand.

Health and Safety Executive (HSE). (2008). *Control Room Design*, Technical Measures Document. London: Health and Safety Executive (http://www.hse.gov.uk/comah/sragtech/techmeascontrol.htm).

Helmreich, R.L. (2000). On error management: lessons from aviation. *British Medical Journal*, 320, 781–785.

Hilburn, B. (1997). Dynamic decision aiding: the impact of adaptive automation on mental workload. In D. Harris (Ed.), *Engineering Psychology and Cognitive Ergonomics* (pp. 193–200). Aldershot: Ashgate.

Hogan, J.C. and Fleishman, E.A. (1979). An index of the physical effort required in human task performance. *The Journal of Applied Psychology*, 64(2), 197–204.

Jacobsson Kecklund, L. and Svenson, O. (1997). Human errors and work performance in a nuclear power plant control room: associations with work-related factors and behavioral coping. *Reliability Engineering and Systems Safety*, 56, 5–15

Kahneman, D. (1973). *Attention and Effort*. Englewood Cliffs, NJ: Prentice Hall.

Lee, J.D. (2006). Human factors and ergonomics in automation design. In G. Salvendy (Ed.), *Handbook of Human Factors and Ergonomics* (pp. 1570–1597). Upper Saddle River, NJ: John Wiley & Sons.

Lehto, M.R. and Buck, J.R. (2008). *Introduction to Human Factors and Ergonomics for Engineers*. Boca Raton, FL: CRC Press.

Marsden, P. and Kirby, M. (2004). Allocation of functions. In N.A. Stanton, A. Hedge, E. Salas, H. Hendrick, and K. Brookhuis (Eds.), *Handbook of Human Factors and Ergonomics Methods* (pp. 338–346). Boca Raton, FL: CRC Press.

Megaw, T. (2006). The definition and measurement of mental workload. In J.R. Wilson and N. Corlett (Eds.), *Evaluation of Human Work* (pp. 525–553). Boca Raton, FL: CRC Press.

Nachreiner, F. (1995). Standards for ergonomics principles relating to the design of work systems and to mental workload. *Applied Ergonomics*, 26(4), 259–263.

Neerincx, M.A. (2003). Cognitive task load analysis: allocating tasks and designing support. In E. Hollnagel (Ed.), *Handbook of Cognitive Task Design* (pp. 281–305). Mahwah, NJ: Lawrence Erlbaum.

Noble, B.J. and Robertson, R.J. (1996). *Perceived Exertion*. Champaign, IL: Human Kinetics.

Reid, G.B. and Nygren, T.E. (1987). The subjective workload assessment technique: a scaling procedure for measuring mental workload. In P.S. Hancock and N. Meshkati (Eds.), *Human Mental Workload* (pp. 185–218). Amsterdam: Elsevier.

Rodgers, M.D. and Nye, L.G. (1993). Factors associated with the severity of operational errors at Air Route Traffic Control Centers. In M.D. Rodgers (Ed.), *An Analysis of the Operational Error Database for Air Route Traffic Control Centers*, DOT/FAA/AM-93/22. Oklahoma City, OK: Human Factors Research Laboratory, Civil Aeromedical Institute, Federal Aviation Administration.

Rouse, W.B. and Morris, N.M. (1986). On looking into the black box: prospects and limits in the search for mental models. *Psychological Bulletin*, 100(3), 349–363.

Sanders, M.S. and McCormick, E.J. (1993). *Human Factors in Engineering and Design*. New York: McGraw-Hill.

Schroeder, D.J. (1982). The loss of prescribed separation between aircraft: how does it occur? In *Proceedings of the Behavioral Objectives in Aviation Automated Systems Symposium* (pp. 257–269). Warrendale, PA: Society of Automotive Engineers.

Schroeder, D.J. and Nye, L.G. (1993). An examination of the workload conditions associated with operational errors/deviations at air route traffic control centers. In M.D. Rodgers (Ed.), *An Examination of the Operational Error Database for Air Route Traffic Control Centers*, Report No. DOT/FAA/AM-93/22. Washington, D.C.: Federal Aviation Administration.

Sebok, A. (2000). Team performance in process control: influence of interface design and staffing levels. *Ergonomics*, 43(8), 1210–1236.

Sharit, J. (2006). Human error. In G. Salvendy (Ed.), *Handbook of Human Factors and Ergonomics* (pp. 708–760). Upper Saddle River, NJ: John Wiley & Sons.

Spath, D., Braun, M., and Hagenmeyer, L. (2006). Human factors and ergonomics in manufacturing and process control. In G. Salvendy (Ed.), *Handbook of Human Factors and Ergonomics* (pp. 1597–1626). Upper Saddle River, NJ: John Wiley & Sons.

Stager, P. and Hameluck, D. (1990). Ergonomics in air traffic control. *Ergonomics*, 33, 493–499.

Stanton, N.A., Salmon, P.M., Walker, G.H., Baber, C., and Jenkins, D.P. (2005). *Human Factors Methods: A Practical Guide for Engineering and Design*. Aldershot: Ashgate.

Stanton, N.A., Young, M.S., and McCaulder, B. (1997). Drive-by-wire: the case of mental workload and the ability of the driver to reclaim control. *Safety Science*, 27(2–3), 149–159.

Tsang, P.S. and Vidulich, M.A. (2006). Mental workload and situation awareness. In G. Salvendy (Ed.), *Handbook of Human Factors and Ergonomics* (pp. 243–268). Upper Saddle River, NJ: John Wiley & Sons.

Vidulich, M.A., Ward, G.F., and Schueren, J. (1991). Using the subjective workload dominance (SWORD) technique for projective workload assessment. *Human Factors*, 33(6), 677–692.

Welford, A.T. (1978). Mental work-load as a function of demand, capacity, strategy and skill. *Ergonomics*, 21, 151–167.

Wickens, C.D. (1984). Processing resources in attention. In R. Parasuraman and R. Davies (Eds.), *Varieties of Attention* (pp. 63–101). New York: Academic Press.

Wickens, C.D. (1992). *Engineering Psychology and Human Performance*, 2nd ed. New York: Harper Collins.

Wickens, C.D. (2002). Situation awareness and workload in aviation. *Current Directions in Psychological Science*, 11, 128–133.

Wierwille, W.W. and Eggemeier, F.T. (1993). Recommendation for mental workload measurement in a test and evaluation environment. *Human Factors*, 35, 263–281.

Wilson, J.R. and Rajan, J.A. (1995). Human–machine interfaces for systems control. In J.R. Wilson and E.N. Corlett (Eds.), *Evaluation of Human Work: A Practical Ergonomics Methodology* (pp. 357–405). London: Taylor & Francis.

Young, M.S. and Stanton, N.A. (2001). Mental workload: theory, measurement, and application. In W. Karwowski (Ed.), *International Encyclopaedia of Ergonomics and Human Factors*, Vol. 1 (pp. 818–821). Boca Raton, FL: CRC Press.

Young, M.S. and Stanton, N.A. (2002). Attention and automation: new perspectives on mental underload and performance. *Theoretical Issues in Ergonomics Science*, 3(2), 178–194.

Young, M.S. and Stanton, N.A. (2004). Taking the load off investigations of how adaptive cruise control affects mental workload. *Ergonomics*, 47(9), 1014–1035.

Young, M.S. and Stanton, N.A. (2005). Mental workload. In N.A. Stanton, A. Hedge, E. Salas, H. Hendrick, and K. Brookhuis (Eds.), *Handbook of Human Factors and Ergonomics Methods* (pp. 390–400). Boca Raton, FL: CRC Press.

7 Automation

7.1 OUTLINE

An automatic device is able to operate independently of human control, and *automation* refers to the replacement of human manual control, planning, and problem solving with automatic devices and computers (Bainbridge, 1982). BS/EN/ISO 11064-1:2001 recognizes that "driven by demands for safer, more reliable and efficient operations, innovations in information technology have led to the increased use of automation and centralized supervisory control" in control room environments. Automation is a powerful trend that is typically driven by the availability and performance of technical innovations. Despite this trend, it can be noted that "even highly automated systems … need human beings for supervision, adjustment, maintenance, expansion, and improvement" (Bibby, Margulies, Rijnsdorp, and Withers, 1975). In fact, even today's most advanced Supervisory Control and Data Acquisition (SCADA) systems still rely on humans; moreover, the best examples harness the unique capabilities of both. The process of automation and the tenor of the relevant standards are very much oriented around this idea of joint human and technical optimization.

The purpose of this chapter is to provide insights and methods to enable relevant standards to be adhered to, and exceeded, and for a framework to be put in place that allows human factors to be embedded in the design, procurement, and operation of systems that embody automation. Automation is a very large topic; indeed, in some respects it encompasses several other chapters in this book. The current chapter, therefore, focuses on relevant standards and the available human factors considerations and techniques that can directly support them.

7.2 KEY ISSUES

The topic of automation involves the following:

- A dichotomy between tasks performed by humans and those performed by machines/automation
- An appreciation for the technological drivers behind automation and the sometimes competing needs of joint optimization
- The notion of allocation of function and the human factors criteria that can be used to decide what tasks should be performed by personnel and what tasks should be automated
- Recognition that contemporary systems involving high levels of automation and computerization often have more degrees of freedom in their design than previous hardware based systems

7.3 STANDARDS AND GUIDELINES

The following standards relate to automation:

- BS/EN/ISO 11064-1:2001 (Ergonomic Design of Control Centers. Part 1. Principles for the Design of Control Centers) deals directly with the design, expansion, refurbishment, and technological upgrading of control centers. It provides a set of principles to be applied in order to emphasize and maximize the strengths, features, and capabilities of both humans and machines in a complementary fashion. To that end it deals explicitly with the issues of automation and allocation of function.
- BS/EN/ISO 13407:1999 (Human-Centered Design Processes for Interactive Systems) also provides guidance on allocation of function. It cites the benefits of a high-level human factors approach as being not just about technical optimization but also increased productivity, enhanced quality of work, reductions in support and training costs, and improved user satisfaction.
- BS/EN/ISO 6385:2004 (Ergonomic Principles in the Design of Work Systems) relates, in part, to specific task analysis methods relevant to determining what tasks are amenable to automation and what tasks are best matched to human personnel.
- BS/EN/ISO 9241-5:1999 (Ergonomic Requirements for Office Work with Visual Display Terminals. Part 2. Guidance on Task Requirements) makes reference to specific human capabilities, again useful in the context of allocation of function.
- IEC 60964:2007 (Design for Control Rooms of Nuclear Power Plants) refers, in part, to functional analysis methods and the determination of technical capabilities of a system.
- BS IEC 61839:2000 (Nuclear Power Plants—Design of Control Rooms—Functional Analysis and Assignment) provides details on how to carry out a specific hierarchical functional analysis method.
- EEMUA guidelines provide wide-ranging and explicit design guidance for advanced automated control room technology.

7.4 HUMAN FACTORS LITERATURE

7.4.1 History of Automation

The process of automation has largely been driven by the capabilities of available technology. Four key stages have been identified by human factors research: local manual control, local automatic control, centralized control, and SCADA. If these developments were plotted against three key variables—area of responsibility, level of autonomy, and distance from process under control—automation can be seen to trace a distinct historical trajectory, each stage bringing with it corresponding changes in the nature of work that humans undertake within such systems. The four key stages of automation are briefly outlined below:

1. *Local manual control*—In situations with local manual control, the operator stood with the piece of equipment that required controlling and physically manipulated valves and levers in response to readings taken from instruments. Under these conditions (widespread in the early days of industrialization), operators received no assistance from the machine; they were colocated with the process under control, they worked alone, and they were responsible for only a small part of the total process (Kragt, 1992).
2. *Local automatic control*—Pneumatic technology enabled a shift from local manual control to local automatic control. The operator was able to stand back and monitor the displays as the machine undertook the manual tasks of physically manipulating the process; however, the operator was still colocated with the process under control, still worked alone, and was still responsible for a relatively small part of the total process (Kragt, 1992).
3. *Centralized control*—Local automatic control created the conditions necessary for remote and centralized forms of control in which all the displays and equipment could be moved to a central location. Thus, previously disparate individuals now came together in one location to oversee machines that carried out manual tasks. These individuals no longer worked alone but as a team. A corollary of this is that coordination and communication needs rose, and the area of responsibility became wider.
4. *SCADA*—More advanced forms of automation, particularly those based on emerging computer technology, enabled increased autonomy and control abilities to be bestowed upon machines. As a result, not only did communication and coordination needs increase, as did the distance between humans and the process under control, but also the area of their responsibility, which became significantly larger.

The central theme is not just one of changing technological capability but also a corresponding change in the nature of the role of humans.

7.4.2 Ironies of Automation

Automated systems are still human–machine systems. Generally speaking, "the more advanced a control system is, so the more crucial may be the contribution of the human operator" (Bainbridge, 1982). The ironies of automation argument is presented in a stereotypical form here:

- The first irony is that when attempts are made to eliminate the human factor in the operation of the system, the human factor does not disappear; rather, it shifts from operation to system design. System design, therefore, becomes the primary source of human intervention in the system and thus a primary source of error.
- The second irony concerns the tasks that are performed in these systems. Automation relieves the operator of some tasks but then creates new ones, the ones that cannot be automated. The result is that "the operator can be left with an arbitrary collection of tasks and little thought may have been given to providing support for them" (Bainbridge, 1982).

- The third irony concerns manual takeover. Humans are not required in the routine running of the operation except to monitor the process. Automation is thus very adept at relieving the human operator of tasks during routine situations, but is often of least help in difficult situations, which typically require the operator to take over control (Norman, 1990). Vigilance tasks during routine system operation are not well matched to human capabilities. Moreover, not taking an active role in system operation can prevent the user from acquiring or maintaining precisely the skills required in order to assume control in abnormal circumstances. So, by "automating the process the human operator is given a task which is only possible for someone who is in online control" (Bainbridge, 1982).

7.4.3 Out-of-the-Loop Performance Problems

Automation removes the human from the control loop. The effect of this is to reduce the level of human–system interaction, which in turn can give rise to reduced situational awareness. This can pose a significant problem, as operators have been shown to be slower to respond to events that require their intervention (e.g., Kaber and Endsley, 1997). A particular class of incident is reported widely in the human factors literature. These are cases of the automation gradually reaching the limits of its compensatory powers, up until the point when it can no longer cope and the problem situation is suddenly handed over to the human operator, who has reduced awareness of the conditions that have led to the failure because the automation was running the system up until that point. Examples are given of aircraft autopilots gradually compensating for a failing engine, only to hand over to the pilots when the aircraft is in a critical condition (e.g., Wickens, 1992).

Another example is the case of adaptive cruise control, a car technology that in normal circumstances controls the speed of the vehicle depending on prevailing traffic situations. Studies have shown that drivers are much slower to respond to situations that require intervention compared to drivers who remained in manual control. Not just that, but the response of those in manual control was more efficacious in terms of avoiding a crash (Stanton and Young, 1998).

Out-of-the-loop performance problems highlight the potentially difficult and paradoxical performance that arises from systems that are not jointly optimized. First, during the period leading to events that require human intervention, the system state can become uncoupled from the operator's understanding of that state. The result can often be inappropriate corrective actions. Second, the difficulty that human operators have in performing vigilance tasks such as monitoring leads to, among other things, a situation of mental underload. Mental underload can lead to delays in corrective actions.

7.4.4 Sociotechnical Considerations

The standards relating to control center automation highlight the human experience within such systems as being an equally critical determinant of performance as that of the purely technical aspects. Per BS/EN/ISO 13407:1999, this concern goes beyond basic health and safety to encompass "human capabilities, skills, limitations, and needs." It has been seen above that automation can potentially change the nature of tasks performed by human operators. Many of the ironies and out-of-the-loop performance problems are accompanied by a corresponding degree of boredom, meaninglessness, stress, and dissatisfaction for role incumbents. These subjective phenomena are injurious to system performance.

Hackman and Oldman (1980) proposed five core job characteristics that speak toward human needs at work. For a job to be inherently satisfying, meaningful, and engaging, it should exhibit:

- Skill variety (requiring the deployment of a multiplicity of skills)
- Task identity (in which the entire cycle of an operation is experienced from beginning to end, a so-called *whole task*)
- Task significance (the task should have meaningful impacts on the system)
- Autonomy (so leadership resides at the level of the group)
- Feedback (or knowledge of results; see Annett and Kay, 1957)

Skill variety, task identity, task significance, autonomy, and feedback—all of these factors can in some respects readily apply to the hands-on experience of manually controlling a system; however, they can apply equally to the case of highly automated systems. The operator of an automated system, for example, typically has a wider area of responsibility that potentially requires more varied skills. Likewise, automation has the potential for task identity and cases where entire operating cycles (or subcycles) can be experienced. Also, the wider area of responsibility and degree of centralized control can increase task significance, and the contemporary emphasis on teamwork can bestow a new form of autonomy. There is often no technical reason why automation *per se* is not inherently incapable of meeting all five of these core job characteristics; however, to do so, appropriate human-centered design needs to take place.

A well-known example of so-called sociotechnical design in which these objectives have been met derives from the industrial setting of vehicle manufacture. In this case, a radically different automation strategy was adopted that serves to highlight what joint optimization might look like in practice.

The defining feature of the highly automated vehicle assembly plant was its shift toward an organization based on smaller groups and an implementation of automation that worked in harmony with this social organization. The most radical change was that the production-line method of assembly, although technically effective, was abandoned. It was replaced by autonomous groupwork undertaken by well-qualified personnel. The parts for the vehicles were organized as if they were kits, with each member of the team completing a portion of the kit and the team as a whole effectively building an entire vehicle (independently of other teams). The natural task groupings bestowed a form of autonomous responsibility on the group; there was a whole task to be undertaken, as well as the requisite skills within the group to undertake it from beginning to end.

Advanced automation remained but was pressed into the service of handling production material (Sandberg, 1995, p. 147). The effects of joint optimization were observed in that "model changes ... needed less time and less costs in tools and training" compared to a similar plant that was organized around the traditional production line principle (Sandberg, 1995, p. 149). In fact, "the learning in this work organization is impressive. Being engaged in all aspects of work makes the production comprehensible and the employees become, as part of their job, involved in the customer's demands and in striving after constant improvement. Work intensity is high" (Sandberg, 1995, p. 148); indeed, somewhat higher than a more traditionally organized facility.

In terms of system state vs. operator awareness of that state, an interesting effect is that "whereas the former organization had been maintained in a steady state only by the constant and arduous efforts of management, the new one proved to be inherently stable and self correcting" (Trist, 1978, p. 53). The organization therefore started to behave like an open system, one that could achieve a steady state based on a constant throughput of inputs and outputs and could maintain this steady state despite considerable changes in the environment. People within the system were in the control loop and highly aware of local system states.

This radical example of joint optimization is presented to illustrate that, even in industries associated with high levels of automation, philosophies concerning how to deploy it can vary considerably. Human factors best practice is concerned with remedying the ironies and paradoxes of automation and in seeking joint optimization. The fundamental question is not one of what *can* be automated, rather what *should* be automated and specifically *how*.

7.4.5 Dynamic Allocation of Function

So far, the allocation of function between humans and machines could be assumed as being a one-shot effort; allocation of function, once defined, remains static. Dynamic allocation of function is an alternative that "makes it possible to adjust the division of labor between the human and the automation over time" (Lee, 2006, p. 1581). This approach relies on two distinct automation philosophies:

- *Adaptable automation*—Operators have the freedom to choose the level of automation they desire or require. Another name for this is *soft automation* (where the user retains authority and control), compared to *hard automation* (where the user is forced to rely on automation with no override capability).
- *Adaptive automation*—The automation itself embodies some form of meta-awareness enabling it to sense internal and external states and perform adjustments accordingly. One example is the case of high task demands, when the system will adjust to a situation of greater automated assistance, and *vice versa* for low demand situations.

Where facilities exist in the automation to be able to undertake either type of automation, the guidance provided in BS/EN/ISO 11064-1:2001, combined with the human factors best practice to follow, will be helpful.

It is important to note that attractive, dynamic allocation of function is not a panacea. Improper design can quite often magnify the detrimental effects, of which there are three to consider. First, adaptive and adaptable automation still requires a degree of in-the-loop monitoring to occur which may place extra demands on the operator. Second, adaptive automation requires the ability to sense various parameters that govern the level of automation; however, some critical parameters may not be amenable to sensing. Third, the evolvable nature of work systems and the adaptable nature of humans within them can render dynamic allocations inappropriate or subject to unexpected adaptations. Without great care in the design and implementation of these automation philosophies, the system can take on the appearance of behaving erratically, increasing the probability of mode errors and offering to the user a confusing and inconsistent interface (Lee, 2006). The guidance, recommendations, and human factors best practice discussion that follow all contribute to resolving issues like these.

7.5 GUIDANCE AND RECOMMENDATIONS

BS/EN/ISO 11064-1:2001 is comprised of nine general considerations and principles of ergonomic design and a more explicit framework for an ergonomic design process. An overview of the salient features of both is provided below:

- *General considerations and principles of ergonomic design*—Table 7.1 presents the overarching principles involved in the ergonomic design of control centers. Note that human factors principles are well embedded within them.
- *Framework for an ergonomic design process*—Figure 7.1 illustrates the framework in the form of a flowchart comprised of 5 phases and 11 individual steps. This represents a methodological sequence of procedures for an ergonomic design process. Emphasis is given to Phase B (Analysis and Definition), as it is within this stage that issues central to automation, such as allocation of function, are dealt with.

Phase A (clarification) is about achieving a good working understanding of the purpose, context, resources, and constraints that surround the implementation, modification, and assessment of control center automation. Per BS/EN/ISO 11064-1:2001, this includes taking into account existing situations that could be used as a reference. Note that there is an insistent focus on a systems perspective in which the role of the control center and its relationships with other relevant subsystems must be identified and documented. Phase B (analysis and definition) is the most cogent in terms of automation issues. It is comprised of five steps: (1) defining the behavior and functions of the system, (2) using that information to allocate functions to humans or machines, (3) defining task requirements, (4) designing jobs, and (5) verifying the results.

Step 1 of Phase B is concerned with defining the behavior and functions of the system or automation. This should be undertaken for all conditions (steady-state operation, normal transient operation, abnormal or emergency operation, and maintenance). BS/EN/ISO 11064-1:2001 presents four generic methods by which the behavior and functions of the automated system can be explored and documented:

TABLE 7.1
General Considerations and Principles of Ergonomic Design

Principle 1 Apply a human-centered design approach.	The combination of humans and machines is considered as an overall system to be jointly optimized. The human characteristics to be considered not only include basic physical capabilities and limitations but also emphasize the unique cognitive strengths of humans (such as perceptual ability, problem solving, and decision making). In addition to the immediate and obvious ergonomic requirements, more subtle psychological demands may require special attention (e.g., sociotechnical considerations).
Principle 2 Integrate ergonomics in engineering practice.	A project should be organized in such a way that an integration of technical and ergonomic expertise is encouraged. This process is often referred to as human factors integration (HFI).
Principle 3 Improve design through iteration.	Establishing the validity of an individual element of the design in isolation does not guarantee that the assembled system will be validated or effective. It should be noted that users, either consciously or unconsciously, may adapt their behavior to modifications and that such behavioral changes may not be consistent with good ergonomics practice.
Principle 4 Conduct situational analysis.	A situational analysis of existing or similar situations is recommended. In this way, the functions of the future system can be thoroughly understood.
Principle 5 Conduct task analysis.	The tasks delegated to individual control room operators and to other significant users of the control center should be formally understood.
Principle 6 Design error-tolerant systems.	Human error cannot be totally eliminated; an important tool is the use of risk assessment.
Principle 7 Ensure user participation.	Experienced users can offer valuable empirical contributions to the control center design.
Principle 8 Form an interdisciplinary design team.	An interdisciplinary design team should be formed to oversee and influence all phases of the design project. This team may include system and process engineers, ergonomists, architects, and industrial designers. This team should be available during all appropriate phases of the project.
Principle 9 Document ergonomic design basis.	Internal documents should be developed that reflect the ergonomic design basis for the project—for example, fundamental reasoning or significant task analysis findings.

Source: Permission to reproduce table from BS/EN/ISO 11064 is granted by the British Standards Institution.

1. Postulated walk-/talk-throughs of operational modes
2. Definition of operational safety and reliability requirements
3. Top-down functional process diagrams
4. Plant topologies

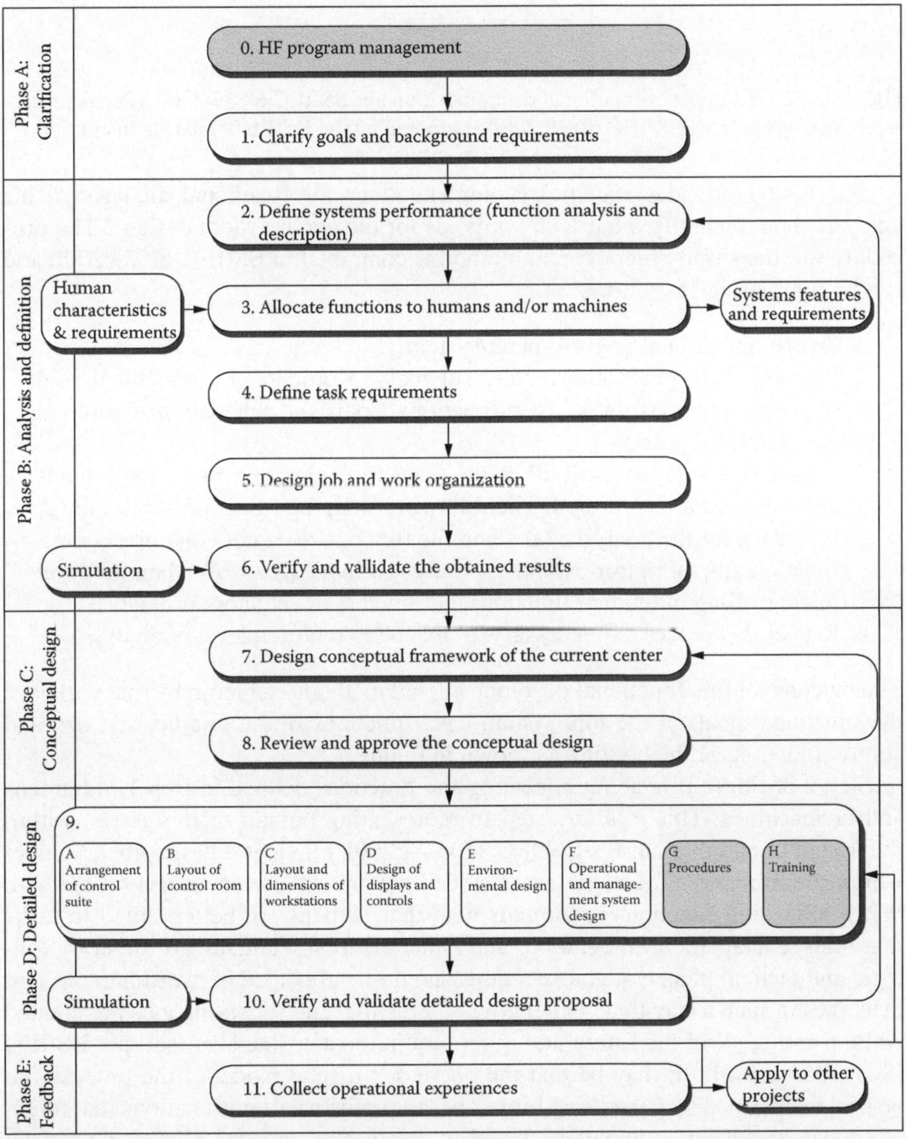

FIGURE 7.1 Framework for the ergonomic design of control centers according to BS/EN/ISO 11064-1:2001. Phase B deals with allocation of function. (Permission to reproduce figure from BS/EN/ISO 11064-1:2001 is granted from the British Standards Institution.)

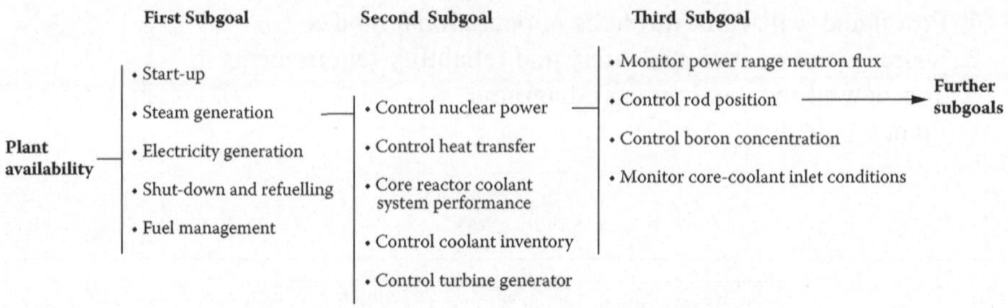

FIGURE 7.2 Example of functional decomposition per BS/IEC 61839:2000. (Permission to reproduce figure from BS/IEC 61839:2000 is granted by the British Standards Institution.)

IEC 60964 provides specific recommendations for functional analysis: "This analysis should identify a hierarchy of goals for the control room design." The procedure for this strictly hierarchical method is contained in BS/IEC 61839:2000 and can be summarized as follows:

1. Define operational goals of plant/system.
2. Develop these goals into further subgoals to produce a hierarchical goal structure (relationship between functional goals and subgoals structured in a hierarchical order).
3. Pursue this analysis until all of the plant functions to achieve these goals and subgoals are identified. The following stopping rule represents a lower boundary for the analysis: (a) when the function does not contain a control function any more (for example, purely mechanical), or (b) when the level of controlling individual functions, parameters or actuators is reached.
4. Repeat this process, if necessary, to reach the required level of analysis.

The outcome of this functional decompositional method is a hierarchy that will have the functional goals at the top, system-level functions in the middle, and detailed control functions at the bottom, as shown in Figure 7.2.

Step 2 of Phase B is about allocating the functions defined in step 1 to humans and or machines. This is a key step in representing human factors issues within the design of automation. BS/EN/ISO 11064-1:2001 provides a basic process under which preliminary allocations are founded on the following overview method: When allocating functions to humans note that humans are better suited to tasks that require adaptation, integration, and generalization. Humans are superior strategic and tactical planners, and functions should be allocated between humans and machines in such a way that challenging, interesting, and satisfying jobs are created while meeting all of the safety and other requirements. Note also that, per BS/EN/ISO 11064-1:2001, "It may be that the operator's mental model of the process can be best maintained by permitting him or her to perform certain functions that might otherwise have been automated." In other words, the standard echoes the human factors maxim that "just because something can be automated does not necessarily mean that it should be." Distilling this overview approach into a set of heuristics reveals the following four points to consider when allocating tasks to humans:

- Avoid repetitive and monotonous tasks.
- Allow overall authority to be maintained (i.e., soft automation).
- Enable an adequate understanding of the state of the automation to be achieved (and ensure adequate levels of situational awareness).
- Retain a feeling of usefulness.

In the case of dynamic allocation, there is an additional requirement to consider how these heuristics apply to the same functions but in different situations; for example, a task could conceivably be repetitive and monotonous in one situation but encourage a feeling of usefulness in another.

Figure 7.3 presents a flowchart depicting the basic allocation process contained within BS/EN/ISO 11064-1:2001. The process recognizes that some tasks have to be allocated either to a machine or to a human, so to that extent they are mandatory allocations. This leaves a set of preliminary allocations in which the criteria above are more fully engaged in order to decide who (or what) is optimal for undertaking a particular task. This then leaves cases in which both humans and machines could perform the task, meaning that the decision process shifts to one based on who (or what) would be preferable. The basic allocation process, therefore, progresses from who has to do a task to who is best suited and, finally, to who would be preferable. Solutions are based on iteration. The outputs of step 2 are as follows:

- A set of functions to be performed by humans
- A set of functions to be performed by machines
- A set of functions to be performed by humans and machines in different situations or under varying criteria (in the case of dynamic allocation of function)
- Sets of interactions between humans and machines
- An auditable record to show how allocation decisions have been made
- A record of tasks that cannot be performed adequately by humans and machines and the interventions that are required to remedy this situation

Step 3 of Phase B is concerned with the definition of task requirements. When the functions of the system or automation have been defined and these functions have been mapped to humans or machines, BS/EN/ISO 11064-1:2001 recommends that the fundamental properties of the tasks assigned to humans should be fully explored. According to the standard, "The fundamental task elements to be considered include manual and cognitive activities, task frequency, duration, complexity, communication requirements, environmental conditions, and any other unique factors required for one or more humans to perform a given task." In this regard, many of the other chapters in this book apply. To be compliant with the standard, a formal record of all relevant task elements, based on the systematic breakdown described above, must be created. Studies, walk- and talk-throughs, and other data-gathering methods are all entirely appropriate in diagnosing what different task elements mean for human–system performance. Likewise, the task analysis can be extended further to be used as a tool for proposing initial engineering solutions. In summary, the outputs of this step are as follows:

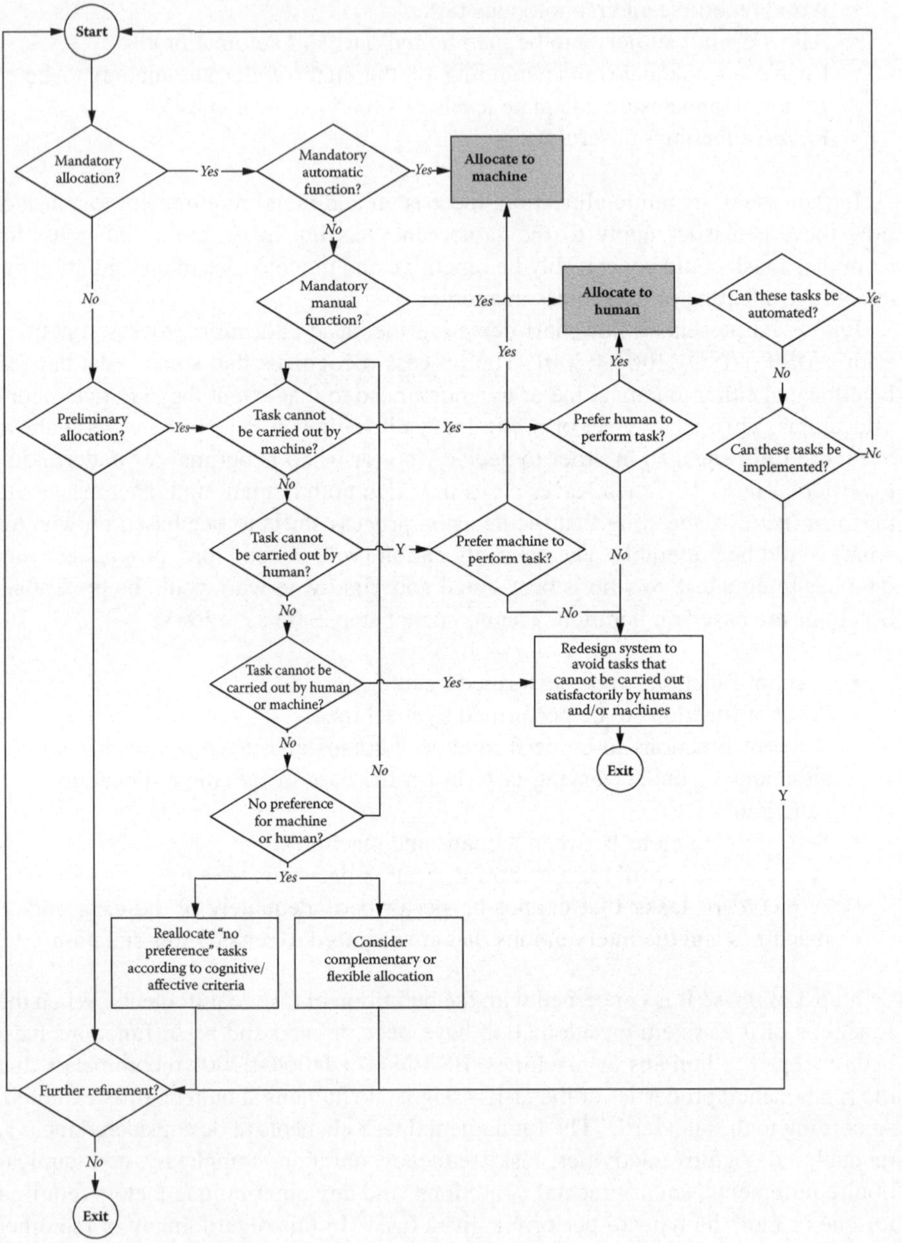

FIGURE 7.3 Flowchart for allocating functions to humans and machines.

- Task analysis
- Systematic and exhaustive record of relevant task elements
- Results of additional data-gathering activities (where undertaken)
- Initial engineering solutions (where appropriate)

Step 4 of Phase B turns the definition of the tasks allocated to humans in the system into specific jobs and work organizations. The allocation criteria above must be deployed again so the collection of interacting tasks that comprise a job (and the interacting jobs that comprise an organization) still match operators' physical characteristics, cognitive and analytical abilities, organizational and leadership skills, and social system factors. A job assignment criteria checklist has to be developed to facilitate this process, and BS/EN/ISO 11064-1:2001 suggests a number of such individual criteria:

- Workload
- Job sharing
- Information and data requirements
- Predictability of the controlled system
- Required tools, physical space, and facilities
- Conditions where tasks are to be performed

The systems perspective is evident in the requirement to consider not only the specific jobs assigned but also how those jobs interrelate. To that end, work assignments are also iterative and likely to require a period of ongoing and repeated modification. In summary, the outputs of this step are as follows:

- Jobs assigned to each operator
- Work organization (structure and number of operators)
- Communications requirements
- Requirements for operating procedures
- Requirements for training
- Requirements for information and control.

Step 5 of Phase B is to verify and validate the obtained results. It can be noted that the iterative, cyclical design process operates at several intermediate levels. It is at this level that all previous iterations are validated and verified globally so all conflicts are identified and resolved before proceeding to Phase C (conceptual design).

7.6 ASSESSMENT AND EVALUATION

BS/EN/ISO 11064-1:2001 has a considerable amount of human factors best practice already embedded within it. It is, of course, possible to benefit still further, particularly in terms of the methods (and sequence of methods) that can be used to undertake functional analyses, task analyses, actual allocation of function, and the diagnosis of cognitive and affective criteria. The standard touches on the concept of *dynamic allocation of function* and this is explored further in this section.

7.6.1 Functional Analysis

BS/EN/IEC 61839:2000 states that other function representations may be more suitable than the hierarchical method proposed. It would, therefore, be consistent with relevant standards to deploy a method such as Cognitive Work Analysis

(CWA) (Jenkins et al., 2008; Vicente, 1999) not necessarily as a replacement but as an adjunct. It should be noted that CWA is a comprehensive, in-depth method in its own right, so only a brief description of its possibilities in this context can be provided here.

Cognitive Work Analysis can be applied, like hierarchical functional analysis, to operations that are ostensibly predictable in their behavior and in which the options for completing a task are well understood and comparatively limited. The value of an approach like CWA is that *unlike* hierarchical functional analysis it can also be deployed in situations where task performance is subject to influences and disturbances that cannot always be foreseen. So, whereas the former is good at describing what *should* happen, the latter can describe what *could* happen. In other words, it can help to describe the space of possibilities afforded by the physical objects in the environment, thus lending itself to describing systems that are not currently in existence (i.e., systems at a conceptual or hypothetical level of development) and avoiding the pitfall of becoming bogged down in existing practices.

Cognitive Work Analysis is comprised of five individual methods and the first, work domain analysis (WDA), is the means by which this functional analysis can be performed. The work domain analysis phase involves describing or modeling the system in which the activity under analysis takes place. A work domain analysis is used to identify the functional purpose and structure of the work domain in terms of the overall system goals, the processes adopted, and the artifacts used within the system. In modeling a system in this way, the system constraints that modify activity within the work domain are specified. These constraints can of course greatly influence the selection of automation strategies and allocation of function.

The use of an abstraction hierarchy (AH) provides a graphical representation of the work domain where the AH represents the system at a number of levels. At the highest level, the AH captures the system's reason for being, its functional purpose, while at the lowest level it captures the physical objects in the system's environment. In between are levels of abstraction that relate more specifically to functions and which can form the basis for an alternative approach to function analysis. The five different layers in an AH are defined briefly in Table 7.2.

In summary, conventional hierarchical functional analysis tends to focus on the *what* of a system: What is it and what does it do? The abstraction hierarchy, on the other hand, takes a much wider perspective by looking at the *why* and the *how.* Whereas traditional functional analysis methods excel at what *should* happen, this complementary approach is useful at looking at what *could* happen: Could automation serve this purpose better than a human or *vice versa*?

7.6.2 Mandatory Assignments

In the basic procedure for allocation of function contained in BS/EN/ISO 11064-1:2001 is a class of allocations referred to as "mandatory." These are allocations that for whatever reason have to be carried out by either humans or machines. Further components of CWA can be used in support of this analysis, specifically the *contextual activity template* and *social organizational and cooperation analysis.*

TABLE 7.2
Detailed Description of the Different Levels in an Abstraction Hierarchy

Level of Abstraction	Prompts
Functional purposes	For what reasons does the work system exist?
	What are the highest level objectives or ultimate purposes of the work system?
	What services does the work system provide to the environment?
Values and priority measures	What criteria can be used to judge whether the work system is achieving its purposes?
	What criteria can be used to compare the results or effects of the purpose-related functions on the functional purposes?
	What are the performance requirements of various functions in the work system?
	How is the performance of various functions in the work system measured or evaluated and compared?
Purpose-related functions	What functions are required to achieve the purposes of the work system?
	What functions are performed in the work system?
	What are the functions of individuals, teams, and departments in the work system?
	What functions are performed with the physical resources in the work system?
Object-related processes	What can the physical objects in the work system do or afford?
	What processes are the physical objects in the work system used for?
	What are the functional capabilities and limitations of physical objects in the work system?
Physical objects	What are the physical objects or physical resources in the work system—both manmade and natural?
	What physical objects or physical resources are necessary to enable the processes and functions of the work system?

The contextual activity template (CAT) is a way of representing activity in work systems that are characterized by both work situations and work functions. Work situations are situations that can be decomposed based on recurring schedules or specific locations. Work situations can be represented along a horizontal axis with work functions shown along a vertical axis. This arrangement creates a contextual activity template as shown in Figure 7.4. The circles indicate the work functions, and the bars show the extent of the table in which the activity typically occurs. The dotted boxes around each circle indicate all of the work situations in which a work function *can* occur (as opposed to *must* occur). By these means, the constraints of the system can be captured. Figure 7.4 shows three activities plotted on the CAT. These activities can take place in three different situations. Activity A can only take place in situation 1, activity B can take place in any of the situations (although typically it only takes place in situation 1), and activity C, meanwhile, can take place in any situation and typically does.

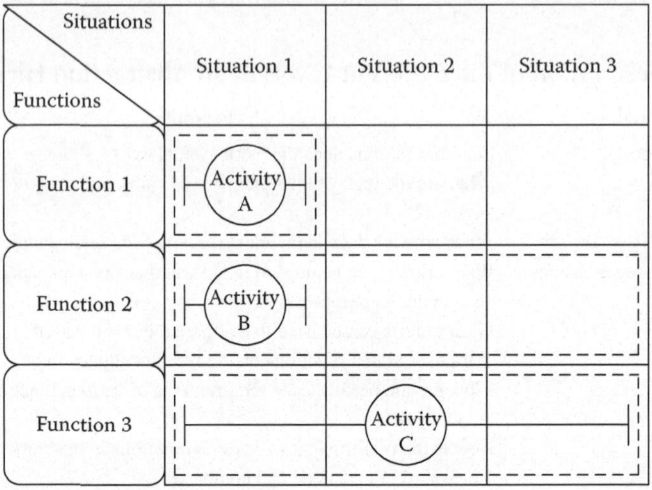

FIGURE 7.4 Contextual activity template.

This type of analysis provides further insight for allocation of function decisions. Activities A and C might indicate situations where automation has to occur (and where it typically does occur). This is a useful check in terms of mandatory assignments. Activity B, on the other hand, shows a similar pattern for situation 1 (perhaps automation has to, and typically does, occur here) but not for situations 2 and 3. Activity *could* occur in these situations (but typically does not), yet the allocation of function decisions might be quite different in these cases. This, then, is one route to systematically considering dynamic allocation of function—that is, a form of automation that is situation dependent.

7.6.3 Preliminary Assignments

The step "define task requirements" comes after "allocate function" and "define system performance" (as shown in Figure 7.1). From a human factors point of view, a task analysis typically represents the first step to be undertaken, not one of the last. Just as functions can be subject to analysis, so can the tasks that move a system from function to function. A method that integrates functions and tasks in this way is called Task Analysis for Error Identification (TAFEI) (Baber and Stanton, 1994). Again, this is a reasonably comprehensive method for which only a brief summary can be provided.

Task Analysis for Error Identification was developed to allow errors to be predicted, but in this context it can serve as a useful way of modeling the interaction between user and the system together, rather than separately. TAFEI assumes that people use systems in a purposeful manner, such that the interaction may be described as a cooperative endeavor. The technique makes the further assumption that actions are constrained by the state of the system at any particular point in the interaction and that the system offers information to the user about its functionality.

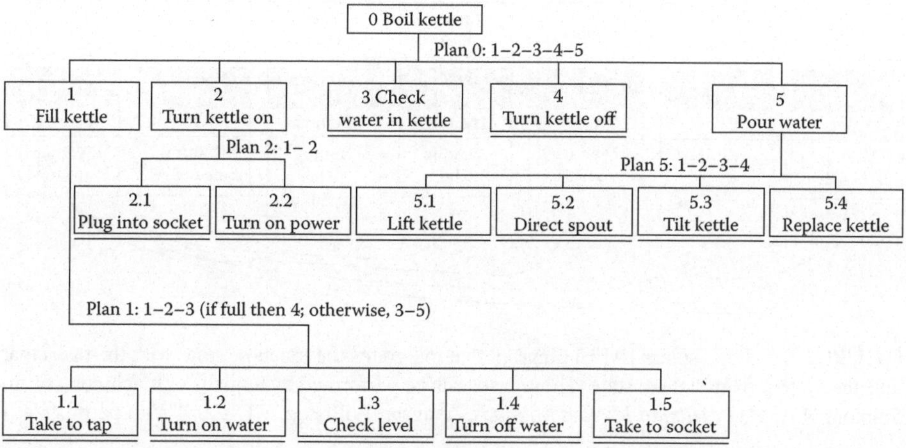

FIGURE 7.5 Example of a hierarchical task analysis (HTA). (From Stanton, N.A. et al., *Human Factors Methods*, Ashgate, Williston, VT, 2005. With permission.)

Thus, the interaction between users and devices progresses through a sequence of states. At each state, users select the action most relevant to their goal. TAFEI proceeds as follows.

The first step is to construct a hierarchical task analysis (HTA) of the system to model the human side of the interaction (an example is shown in Figure 7.5). As TAFEI is best applied to analyses based on scenarios, it is often wise to consider just one specific goal at a time, as described by the HTA (e.g., a specific, closed-loop task of interest). This helps to keep the analysis to a manageable size and progressing through a series of steps. When this goal has been selected, the analysis proceeds to constructing state–space diagrams (SSDs) that model the system side of the interaction. SSDs are constructed to represent the behavior of the system. A SSD essentially consists of a series of states that the system passes through from a starting state to the goal state. For each series of states, there will be a current state and a set of possible exits to other states. At a basic level, the current state might be off, with the exit condition "switch on" taking the system to the on state. Thus, when the system is off, it is waiting for an action (or set of actions) to take it to the on state. Hierarchical functional analysis (e.g., as specified in BS/IEC 61839:2000) can be used to develop an exhaustive set of such states for the system under analysis.

The next step is to map the HTA onto the SSDs, specifically the plans contained in the HTA that state the sequence of actions to be performed to take the system from one state to another. If a state transition is activated by the system itself (via automation, for example), this can also be indicated on the SSD, using the letter "M" on the TAFEI diagram to indicate machine. The integration of HTA and SSDs results in a TAFEI diagram, which is shown in Figure 7.6.

The final step is to interrogate the TAFEI diagram by creating a transition matrix. All possible states are entered as headers, as shown in Table 7.3. The cells represent state transitions (e.g., the cell at row 1, column 2 represents the transition between state 1 and state 2). With support from the allocation heuristics and CWA-based

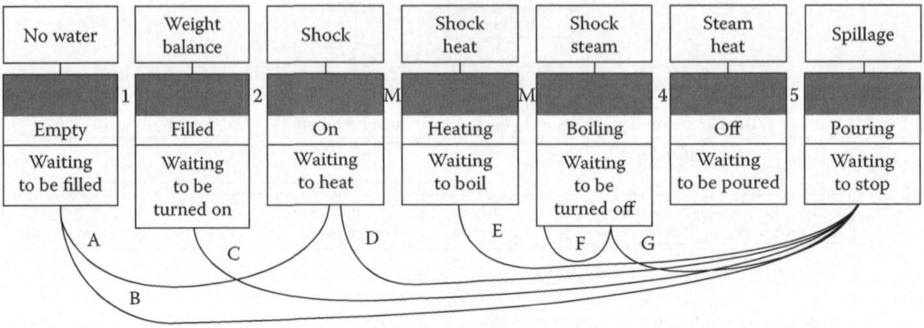

FIGURE 7.6 State–space TAFEI diagram that integrates the machine state with the tasks that take the system from state to state. Those tasks can be performed by humans or machines. (From Stanton, N.A. et al., *Human Factors Methods*, Ashgate, Williston, VT, 2005. With permission.)

methods described above, the matrix can be used as a tool to systematically critique every single state transition in terms of mandatory, preliminary, and affective-based assignments. The matrix, therefore, provides a complete state space and test bed upon which any number of alternative automation strategies and job designs (i.e., the tasks that enable states to be changed) can be systematically and exhaustively explored.

7.6.4 Allocation of Function

An explicit allocation of function methodology designed by Marsden and Kirby (2005) provides a way to meet the twin goals of human factors best practice and BS/EN/ISO 11064-01:2001. Marsden and Kirby's method is comprised of three main stages:

1. *Task analysis*—As above, the first step is to undertake a hierarchical task analysis of the system in question. A well-executed task analysis will have high-level goals and subgoals that are solution or activity independent; that

TABLE 7.3
Transition Matrix Using a Tea Kettle as an Example

From State \ To State	Empty	Filled	On	Heating	Boiling	Off	Pouring
Empty	✓			✓	✓	✓	
Filled		✓		✓	✓	✓	
On			✓	✓	✓	✓	
Heating						✓	
Boiling							
Off							
Pouring							

Source: Stanton, N.A. et al., *Human Factors Methods*, Ashgate, Williston, VT, 2005. With permission.

is, they specify *what* is to be achieved not *how* it is to be achieved. This is important, because otherwise it "assumes a particular way of carrying out the task and preempts the allocation of function, prematurely ruling out alternative allocations of function and creative solutions" (Marsden and Kirby, 2005, p. 34-3).

2. *Stakeholder analysis*—The objective in this stage is to maintain the analysis at the level of *what* tasks are required to be carried out (not *how*) and to "identify the sources of job satisfaction and dissatisfaction of any stakeholder whose job might conceivably be affected" (Marsden and Kirby, 2005, p. 34-3). Human factors best practice is achieved by this step, as it is an explicit means by which the core job characteristics alluded to above can be embedded in the analysis. This involves "identifying both the current knowledge and skills of the stakeholders and their potential to gain new knowledge and skills" (p. 34-3). It also requires consideration of how important the following factors are to job incumbents:
   - Having a challenging job to do
   - Having the opportunity to exercise specialist skills and benefit from the status this can bring
   - Being able to develop new skills
   - Keeping the job as simple as possible
   - Removing tedious aspects of the job
   - Having a variety of work to do
   - Enjoying (or avoiding) interaction with other people
   - Working (or not working) as part of a team
   - Knowing what is going on
   - Deriving pride from making a contribution to something successful
   - Enjoying the challenge of dealing with difficult problems

   The nature of this assessment is such that asking participants these questions directly is not necessarily a guarantee of ensuring valid answers. Further human factors techniques may be required to provide the required level of insights (see Section 7.6.6, Core Job Characteristics, for additional guidance).
3. *Undertake and analyze allocations*—With this, and any mandatory assignments and heuristics (such as those already present in BS 11064-1:2001 or those presented in the next section) in mind, the analyst should work through the task analysis and annotate each higher level goal or subgoal as follows:
   - H, only humans should perform this task.
   - H-C, shared between human and computer or automation (human in control)
   - C-H, shared between human and computer or automation (computer or automation in control)
   - C, only automation or computer should perform task

The systemic nature of the problem requires a review of the impact of these allocations by considering the following (Marsden and Kirby, 2005, p. 34-4):

1. The extent to which an allocation is compatible with or conflicts with the job satisfaction criteria
2. The potential for human error, particularly where the task is safety critical or there are security issues
3. The need for the human to be sufficiently involved in the task to stay alert and have sufficient knowledge of the situation to be able to act when needed
4. Any likely change in the cost and use of resources, including time if speed of task completion is important

Amendments based on this consideration should be reflected in the annotated task analysis and the process iterated as many times as is practical to ensure that any unexpected or systemic effects are captured. Following this iterative cycle, and with a task model whose H, H-C, C-H, and C annotations reflect the requirement to achieve joint sociotechnical optimization, further detail can be added to the model. This involves a shift in emphasis from *what* tasks have to be performed to considering more explicitly *how*. Practically, this involves working at a lower level of resolution in the analysis, ascribing to each low-level goal or task in the analysis its own H, H-C, C-H, and C annotation. The objective is to arrive at a set of low-level tasks that contain only H or C annotations, stating explicitly and in detail who is charged with doing what. Figure 7.7 provides a graphical summary of this detailed and systematic method.

7.6.5 Fitts' List

In BS/EN/ISO 11064-1:2001 preliminary function allocations are performed using an overview approach comprised of four basic heuristics. Human factors best practice offers an extension to this approach in the form of Fitts' List (Fitts, 1951). The expanded collection of heuristics is anchored around information-processing stages and the relative performance of humans or automation in performing them. Fitts' List is shown in Table 7.4.

In the context of preliminary assignments, heuristic approaches like this provide a good starting point, but it is important to note that some authors are highly critical. Woods and Hollnagel (2006), for example, noted that, "The idea that new technology can be introduced as a simple substitution of machines for people—preserving the system through improving the results—is a persistent oversimplification fallacy" (p. 118). According to Lee (2006, p. 1580/1), the disadvantages of this approach can be summarized as:

- Limited ability to cope with interconnections between functions
- The fact that sociotechnical systems evolve and coevolve
- Diminishing number of situations in which human abilities exceed those of the automation

Nevertheless, heuristics remain an established part of the relevant standards and an accepted, expedient approach. The limitations inherent in them are overcome to some extent by embedding them within an iterative design cycle, such that

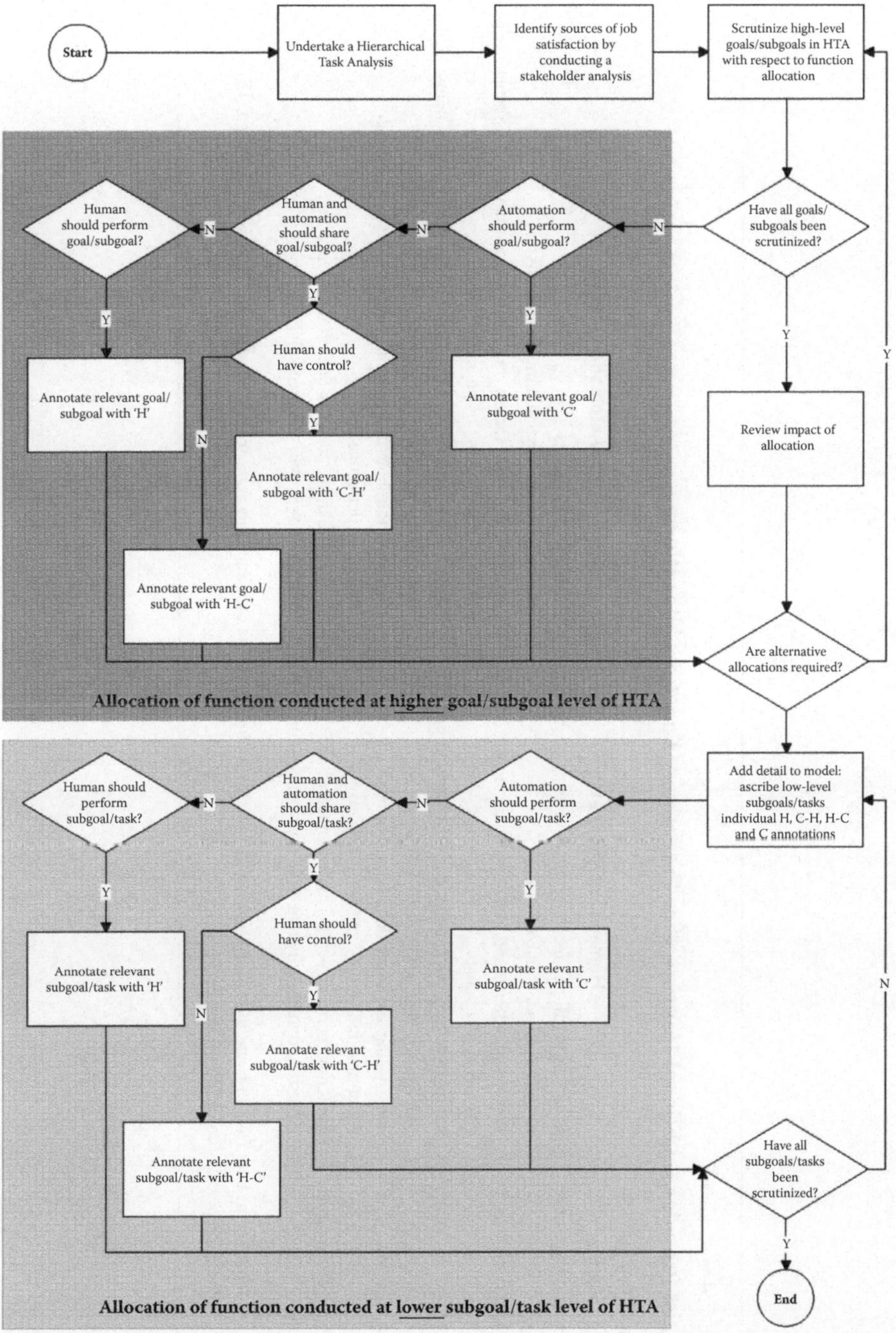

FIGURE 7.7 Flowchart illustrating the process of applying Marsden and Kirby's (2005) allocation of function analysis.

TABLE 7.4
Fitts' List Provides Additional Heuristics for Preliminary Function Allocations

Information-Processing Stage	Humans Are Better at	Automation Is Better at
Information acquisition	Detecting small and meaningful amounts of change	Monitoring processes
	Detecting a wide range of stimuli	Detecting signals beyond human capability
Information analysis	Perceiving patterns and making generalizations	Ignoring extraneous factors and making quantitative assessments
	Exercising judgment	Consistent application of precise criteria
	Recall of related information and development of innovative associations between items	Storing information for long periods and recalling specific items with perfect accuracy
Action selection	Improvising and using flexible procedures	Repeating the same procedure in precisely the same manner
	Reasoning inductively and correcting errors	Reasoning deductively
Action implementation	Switching between actions as demanded by the situation	Performing many complex operations at once
	Adjusting dynamically to a wide range of conditions	Responding quickly and precisely

Source: Adapted from Lee, J.D., in *Handbook of Human Factors and Ergonomics*, 3rd ed., G. Salvendy, Ed., John Wiley & Sons, New York, 2006, pp. 1570–1596. With permission of John Wiley & Sons.

interconnections and evolution can be observed and modified (again, it is important to note that in some sections of the human factors community even this is not deemed adequate). For current purposes, however, it should be noted that heuristics alone are not sufficient for performing allocation of function; they must be part of an iterative, cyclical design cycle.

7.6.6 Core Job Characteristics

In the context of an iterative design cycle (as advocated in BS/EN/ISO 11064-1:2001), additional strategies can be deployed to provide feedback of a sociotechnical or affective nature into the design process. Human factors best practice in this regard centers around various self-report questionnaire methods, administered to individuals who have had an opportunity to use the system in the context of trials, simulations, or other phased delivery methods. Numerous job satisfaction questionnaires exist and have been widely applied, from the Minnesota Satisfaction Questionnaire (with up to 100 questions; see Weiss et al., 1967) to the Job Descriptive Index (requiring simple yes/no answers; see Smith, Kendall, and Hulin, 1969). A concept called *cohesion* (Siebold and Kelly, 1988) also maps well onto the core job characteristics mentioned earlier.

7.7 EXAMPLE

The worked example that expresses some of the human factors best practice discussed above is based on an energy distribution system—specifically, the part of the total system that deals with the storage of gas. In this scenario, the task is concerned with alarms that warn control room operators of abnormal states.

7.7.1 Marsden and Kirby's Allocation of Function Analysis

The first example reflects human factors best practice by illustrating the use of a freely available HTA tool. This tool enables many of the basic requirements of BS/EN/ISO 11064-1:2001 to be met, including the definition of task requirements, using the outputs of the HTA, combined with heuristics, in order to perform a preliminary allocation of function, then going further to help meet the requirements for job design.

Specifically, the tool greatly facilitates the "define task requirements" stage of the ergonomic design of control center framework in BS/EN/ISO 11064-1:2001. As Figure 7.8 shows, various extensions are employed to systematically and exhaustively undertake preliminary task allocations using the heuristics described earlier. Figure 7.8 shows the analysis at an overall level (due to space) with a job design questionnaire that has allowed the overall task (as well as each component task) to be systematically assessed in relation to information flow, when tasks are performed, their temporal relation, constraints, delays, errors, workload, knowledge, and skills. These map onto the fundamental task elements discussed in BS/EN/ISO 11064-1:2001. The tool is flexible enough for additional user-defined analysis categories to be added, such as the BS/EN/ISO 11064-1:2000 job assignment criteria.

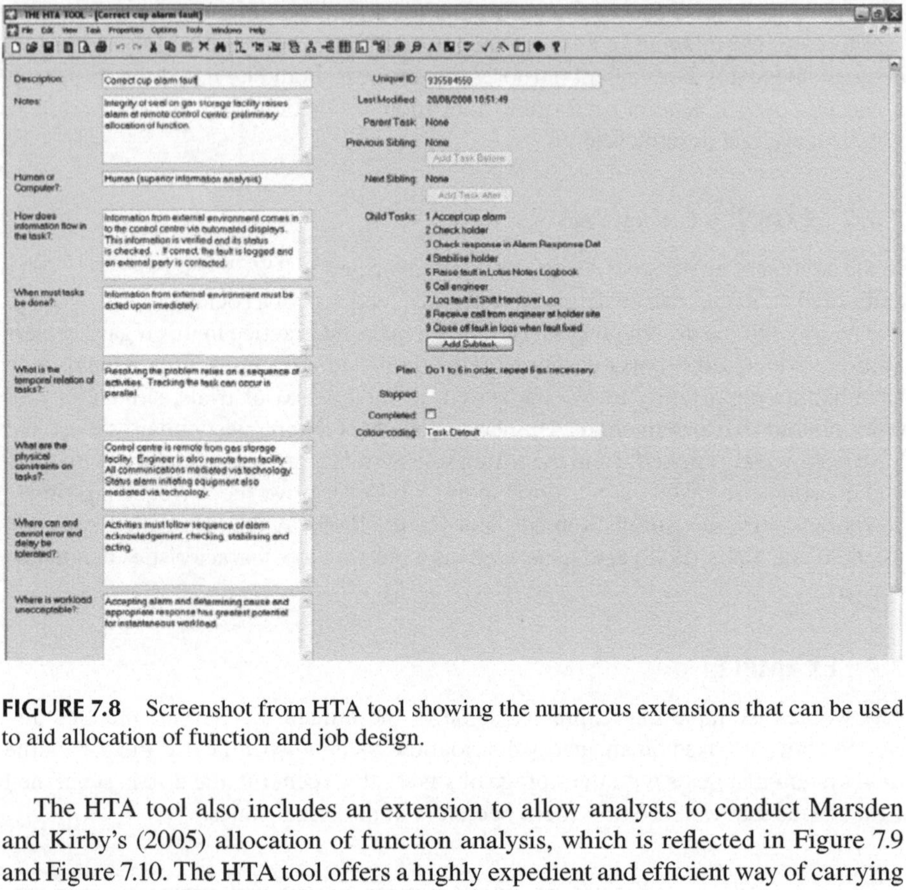

FIGURE 7.8 Screenshot from HTA tool showing the numerous extensions that can be used to aid allocation of function and job design.

The HTA tool also includes an extension to allow analysts to conduct Marsden and Kirby's (2005) allocation of function analysis, which is reflected in Figure 7.9 and Figure 7.10. The HTA tool offers a highly expedient and efficient way of carrying out Marsden and Kirby's (2005) allocation of function analysis and, more generally, a number of aspects relating to BS/EN/ISO 11064-1:2001's framework for ergonomic control centers. Of equal importance is the subsidiary need to document the outcomes of any ergonomic analysis, which the HTA tool also readily facilitates.

7.7.2 Constraints-Based Approach to Functional Analysis

The second example derived from the same scenario shows how work domain analysis (under the auspices of CWA) can complement the hierarchical functional analysis method advocated in IEC 60964:2007. Figure 7.11 shows that the fundamental functional purpose of the system (its reason for being) is to "manage the status of a gas storage facility." To fulfill this purpose, the system has to diagnose deviations from acceptable states and determine appropriate action to take. Note that at this stage, and all subsequent stages, no mention is made of humans or machines; the description is purely systemic. Stepping down to the next two layers of the decomposition space reveals the system's functions, expressed in an abstract, technological, and actor-independent way. Generalized functions such as "making system operators aware of potential faults" and "facilitating correct alarm responses" are those that are required

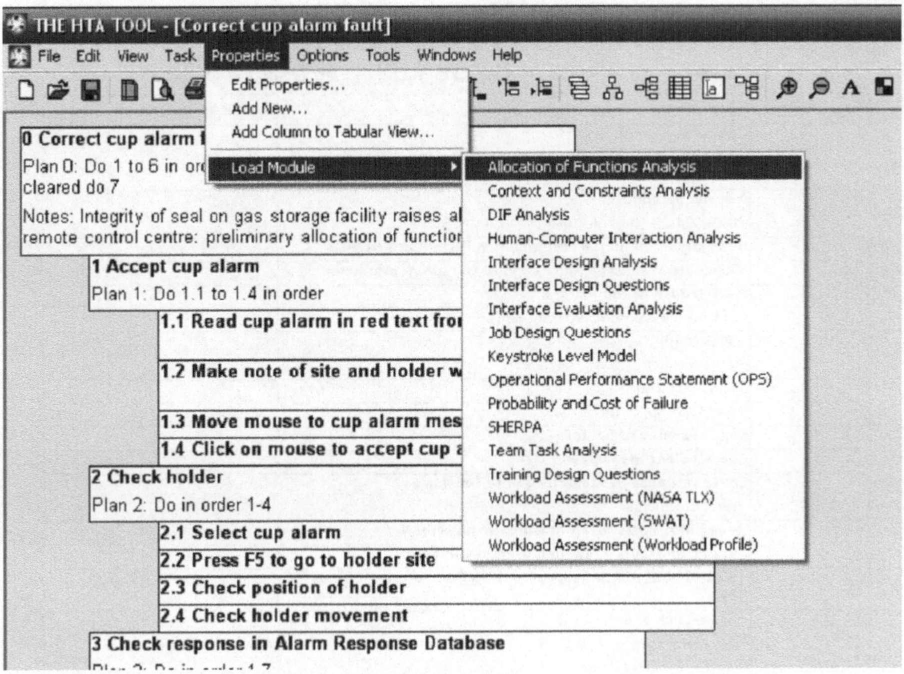

FIGURE 7.9 Screenshot from HTA tool showing the allocation of function module being loaded.

to achieve the purposes of the work system and to satisfy external constraints. Physical functions are what the objects in the work domain (e.g., equipment, artifacts, tools) actually do, or can afford. In this case, they "detect abnormal conditions in the storage vessel," "communicate the presence of an alarm," and "provide an audit trail for alarm handling activities." The abstract nature of the decomposition space enables the consideration of what *could* happen; for example, detection of abnormal conditions in the storage vessel could be a task performed by humans (who according to Fitts' List are good at perceiving patterns and making generalizations). Equally, the task could be performed by automation (which is good at making quantitative assessments based on the consistent application of precise criteria). The situations and constraints, and other factors, can help to shape allocation decisions like these in a multifaceted way.

The central point of this method is to help avoid a situation whereby the analysis becomes unnecessarily tied to how existing systems work and who *should* or *normally does* perform a task. By raising the level of abstraction a level or two, the quality of critique can be enhanced significantly.

7.7.3 Dynamic Allocation of Function

Cognitive Work Analysis also helps to facilitate dynamic allocation of function. This feature can be demonstrated with a contextual activity template derived from the same scenario. Figure 7.12 plots the anticipated modes of system operation specified in

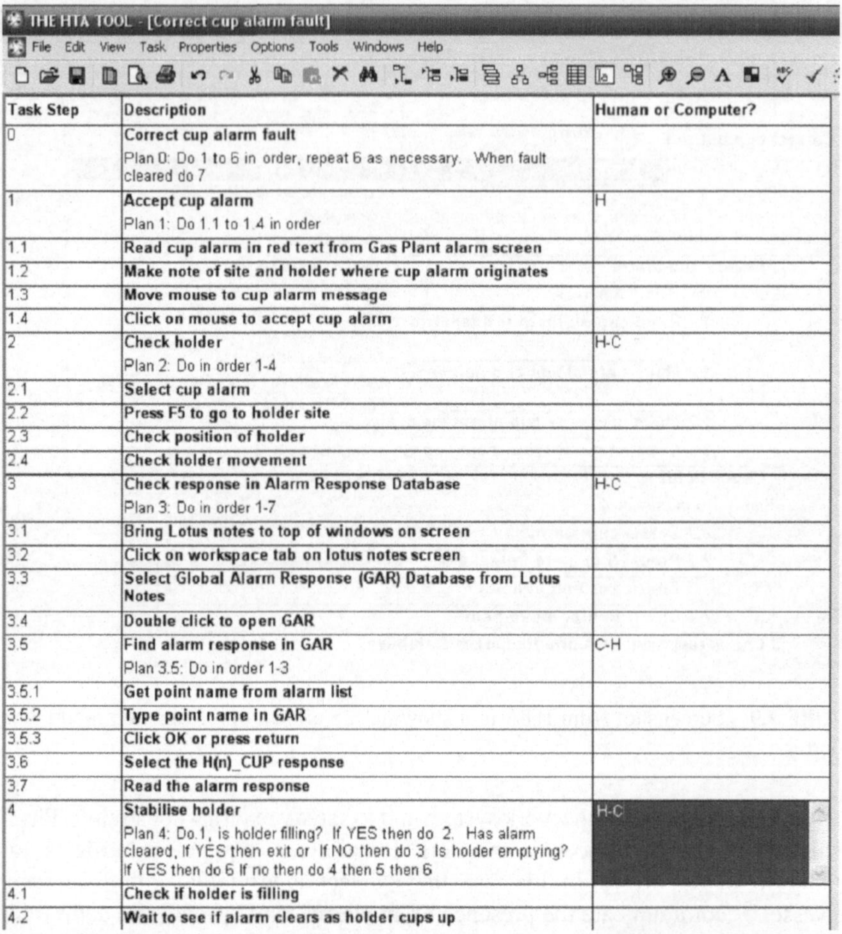

Task Step	Description	Human or Computer?
0	**Correct cup alarm fault** Plan 0: Do 1 to 6 in order, repeat 6 as necessary. When fault cleared do 7	
1	**Accept cup alarm** Plan 1: Do 1.1 to 1.4 in order	H
1.1	**Read cup alarm in red text from Gas Plant alarm screen**	
1.2	**Make note of site and holder where cup alarm originates**	
1.3	**Move mouse to cup alarm message**	
1.4	**Click on mouse to accept cup alarm**	
2	**Check holder** Plan 2: Do in order 1-4	H-C
2.1	**Select cup alarm**	
2.2	**Press F5 to go to holder site**	
2.3	**Check position of holder**	
2.4	**Check holder movement**	
3	**Check response in Alarm Response Database** Plan 3: Do in order 1-7	H-C
3.1	**Bring Lotus notes to top of windows on screen**	
3.2	**Click on workspace tab on lotus notes screen**	
3.3	**Select Global Alarm Response (GAR) Database from Lotus Notes**	
3.4	**Double click to open GAR**	
3.5	**Find alarm response in GAR** Plan 3.5: Do in order 1-3	C-H
3.5.1	**Get point name from alarm list**	
3.5.2	**Type point name in GAR**	
3.5.3	**Click OK or press return**	
3.6	**Select the H(n)_CUP response**	
3.7	**Read the alarm response**	
4	**Stabilise holder** Plan 4: Do.1, is holder filling? If YES then do 2. Has alarm cleared, If YES then exit or If NO then do 3 Is holder emptying? If YES then do 6 If no then do 4 then 5 then 6	H-C
4.1	**Check if holder is filling**	
4.2	**Wait to see if alarm clears as holder cups up**	

FIGURE 7.10 Screenshot from HTA tool showing Marsden and Kirby's (2005) high-level function allocations (H, human to carry out subgoal; H-C, shared allocation with human in control; C-H, shared allocation with computer in control).

BS/EN/ISO 11064-1:2001 on the horizontal axis and generalized functions from the work domain analysis along the vertical axis. The dotted box shows what situations the various functions can occur in, and the circle/whisker shows where it typically does occur. The gray shading shows human allocation; white is machine allocation.

From the diagram, it is clear that the heuristics driving these allocation decisions are sensitive to the situations in which functions are occurring; for example, in steady-state operation perhaps the automation can be relied upon to make system operators aware of potential faults. In these conditions, it might be sufficient for human operators to be relatively passive receptors and to have information pushed onto them; however, in abnormal situations, it might be necessary to augment this function with human attributes. Rather than being passive receptors of information,

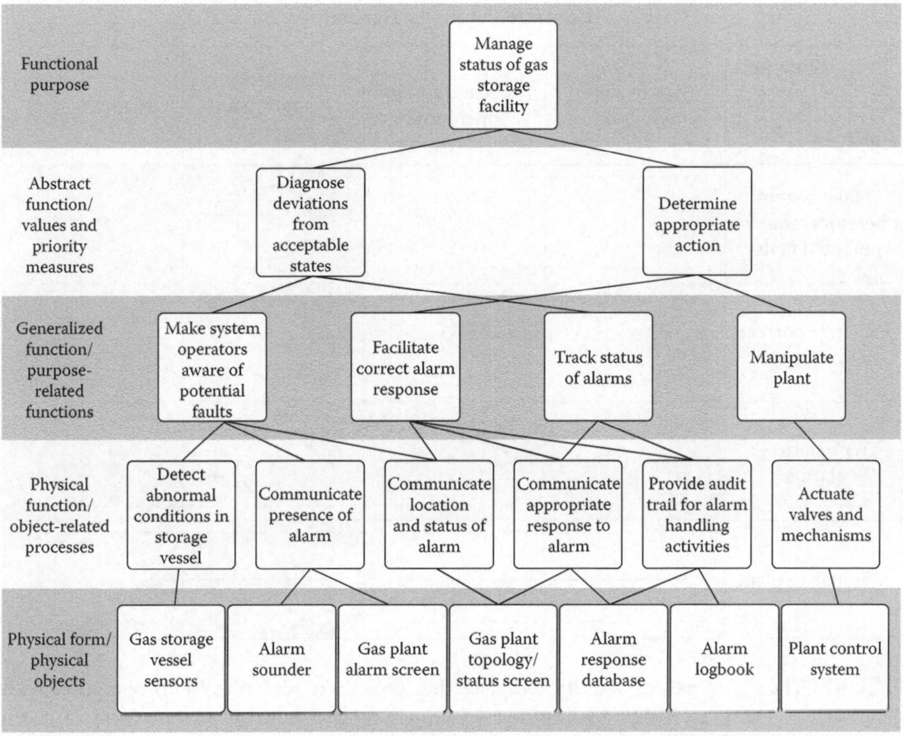

FIGURE 7.11 Abstraction hierarchy showing generalized and physical functions.

potential fault states and their etiology should perhaps be pulled from the situation. The contextual activity template provides a vehicle to undertake such an analysis in a systematic and auditable fashion.

In summary, the HTA tool is presented as a means by which a significant evaluation of basic requirements and preliminary allocations can be undertaken. The two CWA-based approaches are presented as means to extend that analysis beyond basic requirements and increase the quality of technical critique—in particular, positioning the analysis so dynamic allocation of function can be facilitated.

The topic of automation is a mainstay of human factors and a very large area of research in its own right. Because of this, many of the issues and concerns raised by automation overlap with other sections in this book. In so far as the relevant standards are concerned, and despite a lot of human factors best practice already embedded within it, several important considerations and useful approaches have been presented. Principal among them is not the question of what *can* or *could* be automated, but what *should* be automated. This gives rise to sets of heuristics that enable allocations to be made and an iterative design process that tests the efficacy and interconnections between these allocations. Of course, even highly automatic systems are sociotechnical systems (at the very least, humans have to design and maintain them), thus they have to be jointly optimized. The human factors best practice described above is designed to help achieve this aim.

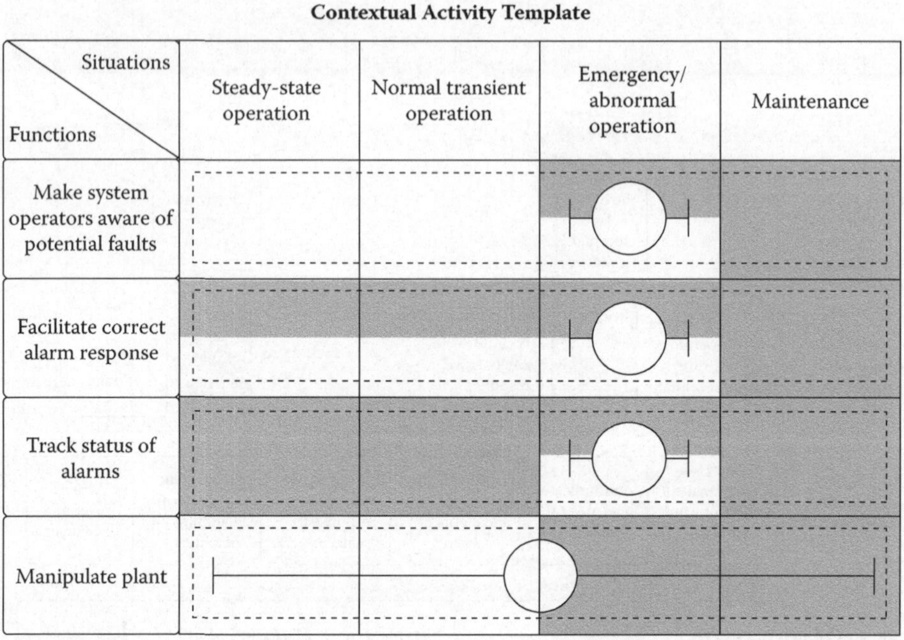

FIGURE 7.12 Contextual activity template that crosses modes of system operation with generalized functions. Gray shading denotes functions that humans can perform in given situations; white shading shows machine functions.

BIBLIOGRAPHY

Allen, R.E. (1984). *The Pocket Oxford Dictionary of Current English*. Oxford: Clarendon Press.

Annett, J. and Kay, H. (1957). Knowledge of results and skilled performance. *Occupational Psychology*, 31(2), 69–79.

Baber, C. and Stanton, N.A. (1994). Task analysis for error identification: a methodology for designing error tolerant consumer products. *Ergonomics*, 37, 1923–1941.

Bainbridge, L. (1982). Ironies of automation. In J. Rasmussen, K. Duncan, and J. Neplat (Eds.), *New Technology and Human Error*. New York: John Wiley & Sons.

Bibby, K.S., Margulies, F., Rijnsdorp, J.E., and Withers, R.M.J. (1975). Man's role in control systems. In *Proceedings of IFAC 6th Triennial World Congress*, Boston, MA.

Fitts, P.M. (1951). *Human Engineering for an Effective Air Navigation and Traffic Control System*. Washington, D.C.: National Research Council.

Hackman, J.R. and Oldman, G. (1980). *Work Redesign*. New York: Addison-Wesley.

Jenkins, D.P., Stanton, N.A., Salmon, P.M., and Walker, G.H. (2008). *Cognitive Work Analysis: Coping with Complexity*. Aldershot: Ashgate.

Kaber, D.B. and Endsley, M.R. (1997). Out-of-the-loop performance problems and the use of intermediate levels of automation for improved control system functioning and safety. *Process Safety Progress*, 16(3), 126–131.

Kragt, H. (1992). Introduction to enhancing industrial performance. In Kragt, H. (Ed.), *Enhancing Industrial Performance* (pp. 1–23). London: Taylor & Francis.

Lee, J.D. (2006). Human factors and ergonomics in automation design. In Salvendy, G. (Ed.), *Handbook of Human Factors and Ergonomics*, 3rd ed. (pp. 1570–1596). New York: John Wiley & Sons.

Marsden, P. and Kirby, M. (2005). Allocation of functions. In N. Stanton, A. Hedge, K. Brookhuis, E. Salas, and H. Hendrick (Eds.), *Handbook of Human Factors and Ergonomics Methods* (pp. 338–346). Boca Raton, FL: CRC Press.

Norman, D.A. (1990). The "problem" with automation: inappropriate feedback and interaction, not "over-automation." *Philosophical Transactions of the Royal Society of London B*, 327, 585–593.

Sandberg, A. (1995). *Enriching Production: Perspectives on Volvo's Uddevalla Plant as an Alternative to Lean Production*. Aldershot: Avebury.

Siebold, G.L. (2000). The evolution of the measurement of cohesion. *Military Psychology*, 11(1), 5–26.

Siebold, G.L. and Kelly, D.R. (1988). *Development of the Combat Platoon Cohesion Questionnaire*, Tech. Rep. No. 817. Alexandria, VA: U.S. Army Research Institute for the Behavioral and Social Sciences.

Smith, P.C., Kendall, L.M., and Hulin, C.L. (1969). *Measurement of Satisfaction in Work and Retirement*. Chicago, IL: Rand McNally.

Stanton, N.A. and Young, M.S. (1998). Vehicle automation and driving performance. *Ergonomics*, 41(7), 1014–1028.

Stanton, N.A., Salmon, P.M., Walker, G.H., Baber, C., and Jenkins, D.P. (2005). *Human Factors Methods*. Williston, VT: Ashgate.

Trist, E.L. (1978). On socio-technical systems. In W.A. Pasmore and J.J. Sherwood (Eds.), *Sociotechnical Systems: A Sourcebook* (pp. 43–57). San Diego, CA: University Associates.

Vicente, K.J. (1999). *Cognitive Work Analysis: Toward Safe, Productive, and Healthy Computer-Based Work*. Boca Raton, FL: CRC Press.

Weiss, D.J., Dawis, R.V., England, G.W., and Lofquist, L.H. (1967). *Manual for the Minnesota Satisfaction Questionnaire*. Minneapolis, MN: University of Minnesota Industrial Relations Center.

Wickens, C.D. (1992). *Engineering Psychology and Human Performance*. New York: Harper Collins.

Woods, D.D. and Hollnagel, E. (2006). *Joint Cognitive Systems: Patterns in Cognitive Systems Engineering*. Boca Raton, FL: Taylor & Francis.

8 Supervision

8.1 OUTLINE

The supervision of process control teams is a key factor in ensuring plant efficiency and safety. Previous research in various domains has identified the key role that effective supervision plays in the efficiency of subordinates' work performance. Further, various studies have identified poor or inadequate supervision as a key contributory factor in accidents and incidents within complex sociotechnical systems (e.g., BOMEL Consortium, 2003; Brazier, Gait, and Waite, 2004). For example, Brazier et al. (2004) highlighted the Texaco oil refinery, Hickson & Welch, Nobel, and Piper Alpha explosions as examples of accidents in which inadequate supervision was evident.

Various definitions of the concept have been presented (e.g., Allen, 1998; Lardner and Miles, 1998; McGoldrick, 1994; Mintzberg, 1979; Ward et al., 2004; Weston, Grimshaw, and Norton, 1989). For example, Mintzberg (1979) suggested that supervision is the "co-ordination by someone taking responsibility for the work of others including planning, scheduling, allocating, instructing, and monitoring actions." Allen (1998) suggested that supervisors "share information, trust others, surrender authority, and understand when to intervene. They participate in setting objectives, defining roles, and managing processes, such as time, disagreements and change." Brazier et al. (2004, p. 2) formally defined supervision as:

> … a management function that can be delivered by one or more individuals within and/or external to a team. It involves controlling, influencing and leading a team, and includes activities such as directing the work of others, allocating workload, planning and scheduling tasks, instructing and monitoring actions. People with supervisory roles are expected to maintain discipline, to take responsibility and be held accountable for the actions of the team.

Of course, the exact nature of the supervisory activities involved in managing a particular system is likely to be domain specific. Based on a synthesis of the literature, however, Ward et al. (2004) suggested that (at a high level, at least) supervision involves the following activities: directing the work of others, allocating workload, planning and scheduling, instructing and monitoring actions, maintaining discipline, taking responsibility, and ultimately being held accountable for the work undertaken.

The nature of supervision within complex sociotechnical systems is changing. Increased automation, enhanced technological capability, increased operator skill, evolving working practices, and changing team structures are all currently impacting the way in which supervision is delivered. As a result, organizations are experiencing a new set of problems associated with their supervisory approach. For example, Ward et al. (2004) observed that the current movement from traditional, hierarchical teams (where roles are explicitly defined and lines of control and communication are clear) toward self-managed, multi-skilled teams has led to various problems, including a lack of leadership and clarity over roles and responsibilities and poor communication external to the team. The Health and Safety Executive (HSE, 2007) stated that, as organizations have evolved, the way in which supervision is delivered and who delivers it have become less and less clear. The consequence of these changes in the way that supervision works is that organizations now must have a clear understanding of how they currently deliver supervision and weaknesses in their approach to supervision (Ward et al., 2004).

8.2 KEY ISSUES

The key issues related to the concept of supervision concern how it is delivered by an organization, what roles and functions the supervisors fulfill, what the weaknesses of the supervisory approach are, how current supervision levels can be assessed, and what measures can be taken to enhance the supervision levels that that are provided to staff. Also, factors such as the competence of the supervisor and the supervisor's relationship with his or her managers and subordinates should be considered. A summary of the key issues is given below.

8.2.1 Supervision Type

The type or form of supervision used within an organization is a key consideration, and a clear understanding of the form of supervision in place allows organizations to understand the associated weaknesses and to take appropriate countermeasures. Ward et al. (2004) identified six forms of supervision related to the type of teams used within the chemical and allied industries. These included traditional hierarchical teams, teams with a team leader, teams with shared leadership, teams with a coach or mentor, self-managed teams, and multi-skilled teams. It is notable that attention is currently being given to the move away from traditional hierarchical teams toward the use of self-managed and multi-skilled teams. Each form of supervision has its own strengths and weaknesses, and it is critical that the organization in question recognizes the weaknesses of its approach and takes suitable countermeasures.

8.2.2 Supervisor's Role, Responsibilities, and Functions

The nature of the supervisor's role, responsibilities, and functions (i.e., what it is that the supervisor is required to do) should be clearly thought out and explicitly defined. For supervision to function effectively, it is critical that the organization, managers, supervisor, and supervisor's subordinates all have a clear understanding of the supervisor's role and what his or her responsibilities are.

8.2.3 Staff Roles and Responsibilities

The staff hierarchy in place should be clearly defined, including a breakdown of the roles and responsibilities of each staff member and a clear definition of the lines of reporting that are currently in place. Subordinates must have a clear understanding of their own roles and responsibilities in addition to a clear understanding of what the supervisor's role is and what his or her responsibilities are.

8.2.4 Supervision Assessment

Organizations should have valid and reliable means of assessing the forms of supervision that they have in place. Ward et al. (2004) argued that, due to the changing nature of supervision, organizations must understand their own approach to supervision and its weaknesses so they can take effective countermeasures. Brazier and Ward (2004) presented a supervision assessment methodology that allows organization to evaluate their supervisory arrangements and the subsequent impact on health and safety (see Section 8.6, Assessment and Evaluation).

8.2.5 Weaknesses Associated with Supervisory Approach

As described above, in order to be able to approach flaws in their supervisory approach, it is critical that organizations fully understand their current supervisory arrangements and the flaws associated with it. Weaknesses associated with different forms of supervision include a lack of team member interaction with the supervisor, lack of team member autonomy, overloading of supervisor, lack of team member development, lack of clarity over roles and responsibilities, and a lack of leadership (Ward et al., 2004). It is also critical for organizations to identify when their supervisory system is not functioning effectively.

8.2.6 Supervision Weakness Countermeasures

When an organization has identified the weaknesses associated with its current approach to supervision, it should take appropriate countermeasures designed to mitigate or remove them. This might involve taking a whole new approach to supervision and implementing a new supervisory system or using measures such as increasing worker involvement, making roles and responsibilities more clear, appropriate task delegation by supervisors, supervisors taking responsibility for developing team member skills, and encouraging team member interactions (Brazier et al., 2004).

8.2.7 Supervisor Competence

The competence of the supervisors working within an organization is of critical importance to efficient and effective supervision. Organizations need to clearly define what competencies are required for each supervisory role, develop training and assessment programs to ensure that supervisors develop and maintain these competencies, and develop approaches for dealing with instances in which supervisor competence is not up to standard (see Chapter 2). A description of the competencies required for each supervisory role can also be used to drive the selection or recruitment of supervisors and to inform the development of appropriate supervisor training interventions.

8.2.8 Supervisor Workload

The level of workload placed on the supervisor is of importance, and instances of supervisors becoming overloaded should be avoided (see Chapter 6).

8.2.9 Relationship between Supervisor and Subordinates

The relationship between supervisors and their subordinates plays a key role in the performance of teams. Factors such as communication with the supervisor, trust in the supervisor, and the behaviors that the supervisor exhibits have all been found to influence subordinates' performance in some way or another.

8.2.10 Relationship between Management and Supervisor

The relationship between supervisors and their managers plays a key role in the supervisor's performance and in subordinates' perceptions of the supervisor and his or her efficiency.

8.2.11 Communication

Communication plays a critical role in the success or failure of an organizations approach to supervision. This includes communication among team members, communication between team members and the supervisor, and communication between the supervisor and higher levels of management. Communication among all elements should be encouraged where appropriate (see Chapter 5) by providing the necessary communication channels.

8.2.12 Interaction between Team Members and Management

The level of interaction between team members and management should be considered carefully. If the nature of the team's supervision is such that there is a lack of interaction between team members and management then the organization should take measures to increase team member–management interaction levels.

8.3 STANDARDS AND GUIDELINES

8.3.1 *Successful Health and Safety Management*

Successful Health and Safety Management (HSE, 1997) states that "adequate supervision complements the provision of information, instruction, and training to ensure that the health and safety policy of an organization is effectively implemented and developed."

8.3.2 Different Types of Supervision and the Impact on Safety in the Chemical and Allied Industries: Assessment Methodology and User Guide

Brazier and Ward (2004) have developed a methodology that allows organizations in the chemical and allied industries to evaluate their supervisory arrangements and the subsequent impacts on health and safety. The approach is interview based and uses a series of questions designed to assess the adequacy of an organization's supervisory arrangements (see Section 8.6, Assessment and Evaluation, for a full description of the methodology).

8.4 HUMAN FACTORS LITERATURE

8.4.1 Supervision

As stated above, various definitions of supervision have been presented in the literature. For the purposes of this document, Mintzberg's (1979) assertion that supervision is the "co-ordination by someone taking responsibility for the work of others including planning, scheduling, allocating, instructing, and monitoring actions" is appropriate. Komaki (1986) suggested that, despite the importance of defining what it is that constitutes effective supervision, defining the concept has remained something of an enigma, and evidence that particular behaviors are critical to effective supervision is meager at best. Based on a review of the literature, Ward et al. (2004) concluded that the main elements of supervision are directing the work of others, allocating workload, planning and scheduling, instructing and monitoring actions, maintaining discipline, taking responsibility, and ultimately being held accountable for the work done. Brazier et al. (2004) identified the following key supervisory activities:

- Planning routine work
- Allocating routine work to team members
- Making decisions under normal conditions
- Making decisions under abnormal conditions
- Monitoring a team's performance
- Planning activities specifically related to health and safety
- Allocating activities specifically related to health and safety
- Monitoring health and safety performance
- Ensuring compliance with health and safety rules and procedures
- Leading the team during normal situations
- Leading the team during abnormal situations
- Facilitating communications within the team and between the team and management
- Ensuring teamwork and developing the team
- Facilitating workforce involvement
- Applying disciplinary procedures

In their review of the literature on supervision, Ward et al. (2004) concluded that supervision is actually a management team function that can be delivered by more than one individual either within or external to a team, rather than an exclusively individual function, and that most organizations fail to fully understand the relationship between supervision and health and safety. Brazier and Ward (2004) suggested that supervision involves the following activities:

- Defining overall team workload
- Allocating day-to-day work priorities
- Allocating manpower for daily tasks
- Controlling and monitoring work patterns and shift rotas
- Communicating operational information to and from the team
- Problem solving and decision making
- Identifying competence requirements for work tasks
- Assessing training requirements of team members
- Measuring team performance
- Carrying out appraisal of team members
- Implementing first-level discipline
- Initiating and leading incident investigations
- Maintaining/updating procedures
- Providing leadership in emergency situations

Komaki, Zlotnick, and Jensen (1986) developed the Operant Supervisory Taxonomy Index (OSTI), which is a tool for classifying supervisory behaviors. The OSTI uses an operant-based taxonomy of supervisory behavior that contains behaviors related to the following categories of supervisory behavior: (1) performance consequences, (2) performance monitors, (3) performance antecedents, (4) own performance, (5) work-related, (6) non-work-related, and (7) solitary. The operant taxonomy of supervisory behavior is presented in Figure 8.1. This model suggests that supervisory behaviors include work-related and non-work-related interactions. The non-work-related interactions include interactions related to the supervisor's own performance and to the supervisor's subordinates' performance and interactions that are not about performance or work related. Work-related interactions focusing on subordinates' performance include performance antecedents, performance monitors, and performance consequences. Performance antecedents include providing instructions about performance, performance monitors include collecting information about subordinates' performance, and performance consequences indicate knowledge of performance and its outcomes.

A study of shift supervisor decision making in the nuclear power domain by Carvalho, Santos, and Vidal (2005) found that during micro-incidents (incidents that interfere with the normal operating mode) shift supervisor decision making was primarily (80%) based on pattern recognition, the basis for which appeared to be derived from experience and training rather than standard operating procedures. Supervisors select courses of action that are generated via the recognition of situational characteristics that are similar to previous situations in which the supervisor has been involved.

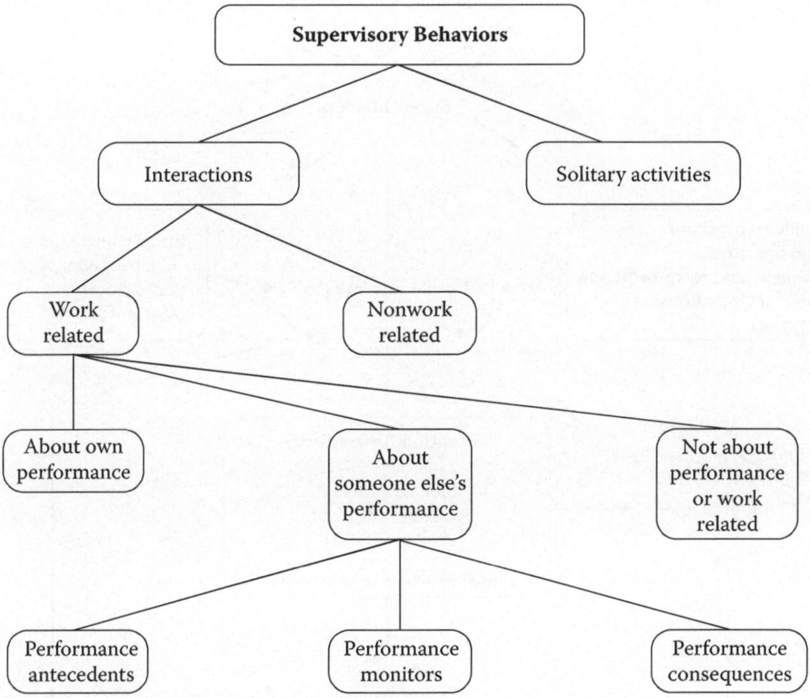

FIGURE 8.1 Operant taxonomy of supervisory behavior. (Adapted from Komaki, J.L. et al., *Journal of Applied Psychology*, 71(2), 260–269, 1986.)

This is known as recognition-primed decision (RPD) (Klein, 2000). The number of options available to supervisors during nuclear power plant micro-incidents is limited by organizational constraints, including hardware (e.g., interfaces, automation), software (incongruent procedures), and liveware (e.g., underlying assumptions). Based on their research, Carvalho et al. (2005) developed a simplistic model of supervisor decision making during micro-incidents that is presented in Figure 8.2. They concluded that supervisors (at least within the nuclear power domain) have ultimate responsibility for control room operation and safety and serve as the main control room communication channel to the other parts of the plant and outside of the plant. They also observed that the supervisor has many roles, both technical and administrative.

8.4.2 Different Forms of Supervision

The literature indicates that different forms of supervision are available to organizations, each with its own strengths and weaknesses. According to Brazier et al. (2004), different forms of supervision are appropriate for different circumstances depending on the nature of operations undertaken, structure and culture of the organization, and nature of the teams involved. Ward et al. (2004) developed a supervision continuum (see Figure 8.3) to demonstrate how differences in the structure of teams impact the

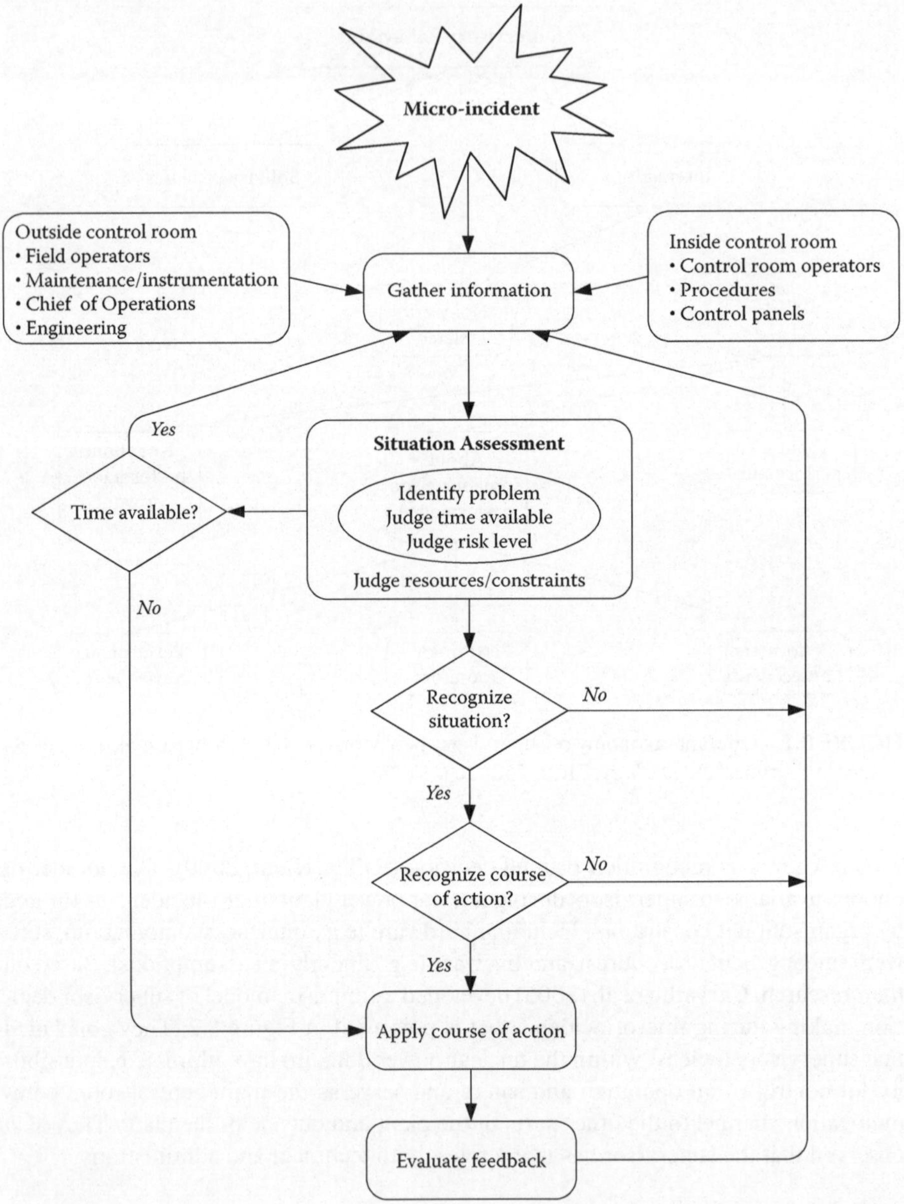

FIGURE 8.2 Supervisor decision making during micro-incidents. (Adapted from Carvalho, P.V.R. et al., *International Journal of Industrial Ergonomics*, 35, 619–644, 2005. With permission.)

type of supervision approach adopted. They identified six different forms of supervision and their inherent strengths and weaknesses (see Table 8.1). They also observed that few teams in the chemical and allied industries represent a truly traditional or self-managed structure but rather are characteristic of a combination of both.

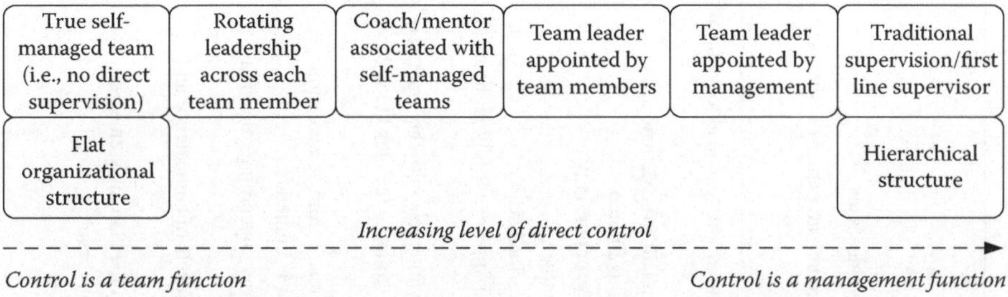

FIGURE 8.3 Supervision continuum. (Adapted from Ward, R. et al., *Different Types of Supervision and the Impact on Safety in the Chemical and Allied Industries: Literature Review*, Health and Safety Executive, London, 2004. Reproduced under the terms of the Click-Use Licence.)

8.4.3 Note on Leadership Styles

There is a huge body of literature available surrounding the concept of leadership (e.g., Bass, 1990, 1998, 1999; Burns, 1978). Although the scope of this document makes it impossible to summarize the literature on leadership here, it is worth discussing briefly the concepts of transformational and transactional leadership, as they are directly related to supervision and subordinate performance. *Transformational leadership* has been defined as "the leader moving the follower beyond immediate self-interests through idealized influence (charisma), inspiration, intellectual stimulation, or individualized consideration" (Bass, 1999, p. 11). Transformational leaders raise the morale and motivation of their subordinates, whereas transactional leaders cater to their followers' immediate self-interests (Bass, 1999). *Transactional leadership* refers to "the exchange relationship between leader and follower to meet their own self-interests" (Bass, 1999, p. 10). The transformational leader emphasizes what you can do for your country, whereas the transactional leader emphasizes what your country can do for you. Organizational changes have led to the need for leaders to become more transformational and less transactional remain effective. Bass (1999) suggested that transformational leadership is clearly the optimum and that members of transformational teams are high performing, care about each other, intellectually stimulate each other, inspire each other, and identify with the team's goals.

8.4.4 *Status Quo*

Despite its obvious importance to complex system safety and efficiency, the concept of supervision has received surprisingly little attention within the human factors literature. The concept does however emerge within the human error and teamwork literature. Within the teamwork literature it is often referred to as a key teamwork behavior and within the human error literature as a causal factor in organizational accidents and incidents; it is also apparent that only limited guidance relates to how to provide, assess, and improve supervision within complex sociotechnical systems.

At the current time, it is acknowledged that the nature of supervision is changing; for example, Ward et al. (2004) noted a shift from traditional hierarchical team structures toward the concept of self-managed teams. This shift toward the use of

TABLE 8.1
Different Supervision Types and Their Strengths and Weaknesses

Team Type	Supervisory Characteristics	Main Strengths	Main Weaknesses
Traditional, hierarchical	Supervisor is responsible for the work of others and for planning, allocation of work, instructing, and monitoring actions. Supervisor maintains discipline and is accountable for the work undertaken by his subordinates.	Lines of reporting are clear. Role definition and understanding of responsibilities are clear. Communications channel between team and management is provided by the supervisor. Supervisor is responsible for team performance.	Decision making and leadership are entirely in the hands of the supervisor. Management typically only interacts with the supervisor. Team members may lack autonomy (can lead to reduced job satisfaction). Inefficient operations may result from demarcation. Team member development is limited. Team members may lack the opportunity to communicate with one another.
Team with team leader	Supervisor acts as a single leader but also typically works as a team member and has team duties in additional to a leadership role. Supervisor may be nominated by other team members or by management (depending on the organizations culture).	Lines of reporting are clear. Role definition and understanding of responsibilities are clear. Communications channel between team and management is provided by the supervisor. Supervisor is responsible for team performance.	Supervisor may become overloaded. Decision making and leadership are entirely in the hands of the supervisor. Management typically only interacts with the supervisor. Team members may lack autonomy (can lead to reduced job satisfaction). Inefficient operations may result from demarcation. Team member development is limited.
Team with shared leadership	Supervisory role is shared by a number of team members, possibly by way of rotation of leadership role across different team members.	Supervisor is still responsible for team performance. Increased levels of responsibility are experienced by different workers. Approach is good for worker development.	There may be a lack of clarity concerning roles and responsibilities. Team members need high levels of competence and supervisory skills. Informal hierarchies within the team may emerge. Increased training needs may result.

Team with coach or mentor	Mentor facilitates teamworking and may be responsible for factors such as developing the team and providing them with support to progress. Mentor is not typically part of the team itself.	Workforce involvement in factors such as decision making increases. Levels of responsibility are increased. Team interaction is enhanced.	Roles and responsibilities may not be clear. Management may only interact with mentor. Team member development may be limited.
Self-managed	A self-managed team is one that operates without any direct supervision. Team members typically have increased responsibility, autonomy, involvement in decision-making processes, and mutual accountability and are responsible for monitoring their own task performance. Self-managed teams usually adopt a flat organizational structure.	Employee involvement and autonomy are increased. Fewer layers exist between team and management. Workforce involvement in factors such as decision making is increased. Levels of responsibility are increased. Team interaction is enhanced.	Roles and responsibilities may not be clear. Team members need high levels of competence and supervisory skills. Management may not fully commit to self-managed teams. There may be a lack of leadership. There may be reduced levels of interaction between the team and management. Informal hierarchies within the team may emerge. Increased training needs may result.
Multi-skilled	A multi-skilled team typically has increased levels of competency which enables them to carry out tasks previously carried out by another function.	Employee involvement increases. Levels of employee skill and competence increase.	Roles and responsibilities may not be clear. There may be a lack of support from management. Long-term skill maintenance is necessary. Lack of leadership within the team may result. There may be reduced levels of interaction among team members. There is a greater potential for team members to become overloaded through higher expectations of management. Loss of competence is possible. Supervisors may not be entirely familiar with the multiple skills of the team members.

Source: Adapted from Ward, R. et al., *Different Types of Supervision and the Impact on Safety in the Chemical and Allied Industries: Literature Review*, Health and Safety Executive, London, 2004.

self-managed, multi-skilled teams is impacting the nature of supervision and creating new problems, such as a lack of leadership, reduced interaction between team members, and poor communication external to the team. Most organizations have moved from hierarchical management structures to flatter structures which has led to a lack of clarity in the way in which supervision is delivered. In summarizing the research undertaken to date on self-managed teams, Brazier et al. (2004) reported that it seems that the implementation of self-managed teams can have either a positive or neutral effect on health and safety. Although communications within the team and the involvement of team members in the decision-making process generally improve, it has also been found that communications between teams and with senior management can deteriorate. Brazier et al. (2004) also reported that the literature suggests that self-managed teams can experience problems achieving appropriate levels of leadership, training, and clarity of roles and responsibilities. They concluded, however, that traditional forms of supervision do not compare favorably with self-managed teams with respect to their impact on health and safety performance.

8.4.5 Teams and Teamwork

Supervision is intrinsically related to the well-discussed topic of teams and teamwork. Salas (2005) defined a *team* as consisting of two or more people dealing with multiple information resources who work to accomplish some shared goal. Teamwork is a complex process and as such is comprised of various processes. *Teamwork* is defined as "the ability of team members to work together, communicate effectively, anticipate and meet each other's demands, and inspire confidence, resulting in a coordinated collective action (Salas et al., 2000). At a simplistic level, a distinction between teamwork and taskwork has been made. Teamwork refers to those instances where individuals interact or coordinate behavior to achieve tasks that are important to the team's goals (i.e., behavioral, attitudinal, and cognitive responses coordinated with fellow team members), whereas taskwork (i.e., task-oriented skills) describes those instances where team members are performing individual tasks (in light of their individual roles within the team) separate from their team counterparts. According to Glickman et al. (1987), team tasks require a combination of taskwork and teamwork skills to be effective.

Many researchers have attempted to define the different processes or behaviors underlying effective teamwork and supervision (or leadership). Cannon-Bowers et al. (1995), for example, identified eight core dimensions that are generalizable across most (if not all) teams: adaptability, shared situational awareness, performance monitoring and feedback, leadership and team management, interpersonal relations, coordination, communication, and decision making. Salas, Burke, and Cannon-Bowers (2000) presented these dimensions along with their associated definitions and subskills. Leadership and team management refer to the "ability to direct and coordinate the activities of other team members, assess team performance, assign tasks, motivate team members, plan and organize and establish a positive atmosphere" (Salas et al., 2000, p. 342).

8.4.6 Study of Supervision

The supervision of employees undertaking work activities is not a new concept, and research has been conducted focusing on different forms of supervision and supervisory behaviors and their effects on employee performance. Many classical studies in the field of occupational psychology, for example, have demonstrated that the quality of supervision is directly related to worker productivity (e.g., Mowry, 1957). A series of seminal studies focusing on leadership undertaken at the Personnel Research Board of Ohio State University investigated the description and assessment of leadership and its effects on employee performance, one of the outputs of which was the Supervisory Behavior Description Questionnaire (SBDQ) (Fleishman, 1953).

Rinefort and Van Fleet (1993) investigated the effects of reduced supervision within the chemical, paper, and wood product manufacturing domains. Based on an analysis of the average number of employees per supervisor and the cost of supervision per employee in 140 organizations, they found that those organizations with greater numbers of first-level supervisors had fewer work injuries. In postulating reasons for this, they suggested that such organizations had sufficient supervisors to effectively coordinate work activities and to provide communication and other forms of assistance.

Mearns, Flin, Fleming, and Gordon (1997), among other things, investigated the role of supervisors in facilitating safe working practices in teams or crews within the offshore oil domain. They concluded that more effective supervisors (in terms of safety performance) value their subordinates and the work activities they are required to perform to a greater extent than did less effective supervisors and that more effective supervisors appear to have a supportive style of management and are concerned for their subordinates' welfare. More effective supervisors were found to visit the worksite more often; they engaged with subordinates and encouraged participation in decision making. Less effective supervisors were found to focus more on productivity and deadlines and felt under pressure to get the job done. In light of their findings, Mearns et al. (1997) subsequently recommended that supervisors should be trained in key management-related skills, including skills linked to leadership, communications, understanding, and respect for others' opinions. They also suggested that Crew Resource Management (see Chapter 2) is a suitable form of training for offshore oil supervisors.

In light of the small sample size used by Mearns et al., Fleming (1999) attempted to corroborate their findings via a follow-up study designed to investigate if specific safety leadership behaviors carried out by offshore supervisors directly influenced subordinates' safety-related behavior. The results of this study suggested that workgroup safety can be optimized by supervisors clearly valuing their subordinates, involving them in decision-making processes, visiting the worksite frequently, encouraging group participation, and communicating their personal positive attitude about safety (all of which supported the initial findings of Mearns et al.). In conclusion, Fleming argued that supervisors sit at the culture–behavior interface and that any safety climate changes will require supervisor support to cross this interface. Further, the findings suggested that interpersonal aspects of supervisor safety management should be addressed during supervisor training programs and that the two-way relationship between supervisors and their subordinates is likely to influence the supervisor's ability to change his or her own safety leadership behavior.

Brazier et al. (2004) undertook a study of operating, maintenance, and contractor teams from nine organizations within the chemical and allied industries to understand how they delivered supervision. When they examined how responsibility for the key supervisor activities described above was distributed throughout the team, they found that only one of 29 teams examined conformed to a traditional supervisory structure (i.e., all supervisor tasks allocated to one individual). Operating teams were more traditionally structured, whereas the maintenance teams were more representative of self-managed teams. It was also notable that none of the teams was completely self-managed. Contractor teams were generally given less responsibility, with more external input being given, than the operating and maintenance teams analyzed.

Based on this study and a literature review, Brazier et al. (2004) identified the following five primary safety assessment principles relating to supervision:

- Supervision is a critical management function, and this must be reflected in an organization's safety management system.
- Supervision has a key influence on the way teams perform; therefore, supervision must be delivered appropriately for the way in which the team in question functions.
- Individuals with a supervisory role must have the necessary competence and opportunity to fulfill their responsibilities.
- Supervision of contractors is also critical and must be properly managed.
- Organizational arrangements must ensure good leadership in emergencies.

When Bono, Jackson-Foldes, Vinson, and Muros (2007) examined the effects of supervisors and managers on employee emotions, they found that supervisors and managers have a powerful impact on employee emotions. Bono et al. (2007) posited that employees may feel anxious when interacting with supervisors, as supervisors are individuals who directly validate performance (which leads to employees becoming anxious about their own performance during supervisor interactions) and because individuals have a need for autonomy that tends to be limited by supervisors. Glaso and Einarsen (2006) found four affective factors relevant to the supervisor–subordinate relationship, of which three were negative (frustration, violation, and uncertainty).

8.4.7 Consequences of Inadequate Supervision

Inadequate supervision has a number of consequences for the organization involved, ranging from process performance decrements to fatal accidents and incidents. In a study investigating the causal factors in fatal incidents, the BOMEL Consortium (2003) found supervision to be one of the most significant organizational factors involved. Ward, Brazier, and Lancaster (2004) reported that the BOMEL study found that supervisors were generally providing inadequate levels of supervision and guidance to workers, particularly in the hazardous industries. Brazier et al. (2004) also highlighted a series of high-profile fatal incidents in the chemical and allied industries where inadequate supervision was identified as a causal factor.

One of the consequences of inadequate supervision is human error. A recent shift in the theoretical perspective on human error (from a person approach to a newer systems perspective approach) and industrial accidents and incidents have led practitioners to treat human error as a systems failure and consider the presence of latent (or error-causing) conditions, such as inadequate supervision and its role in errors made at the sharp end by operators. These approaches suggest that errors made by operators at the sharp end are a consequence of these latent conditions residing with the system, as opposed to the earlier person-based approaches that viewed errors as being the result of psychological factors within an individual, such as forgetfulness, inattention, poor motivation, carelessness, negligence, and recklessness (Reason, 2000). Reason's (1990) systems perspective (or Swiss cheese model, as it is more commonly referred to) is the most commonly embraced systems perspective approach to human error and organizational accidents. Within Reason's model, latent conditions and unsafe acts combine to produce organizational accidents and incidents. Latent conditions are inadequate conditions or failures residing throughout a system and include poor designs, inadequate supervision, manufacturing defects, maintenance failures, inadequate training, clumsy automation, inappropriate or ill-defined procedures, inadequate equipment, and procedural short-cuts, to name only a few. Unsafe acts, on the other hand, represent those errors that are made by human operators at the sharp end of system operation that have an immediate impact on system safety.

Reason's model focuses on the interaction between latent conditions and errors and their contribution to accidents within organizations. According to the model, complex systems consist of various levels, each of which contributes to production (e.g., decision makers, line management, productive activities, and defenses). Each layer has defenses that are designed to prevent the occurrence of occupational accidents. Examples of defenses include protective equipment, rules and regulations, training, checklists, and engineered safety features. It is notable that effective supervision is a critical factor in the success or failure of these defenses. Holes or weaknesses in these defenses, created by the latent conditions and unsafe acts, create windows of opportunity for accident trajectories to breach the defenses and cause an accident. Organizational accidents occur when the holes in the system defenses line up in such a way that the accident trajectory can penetrate each of the different layers. Reason's systems perspective model is presented in Figure 8.4.

Board decisions and policy-level decisions made by system designers and higher level management can create latent conditions within the system. Examples of fallible board decisions and policy include the selection of inadequately designed equipment, the vetoing of system improvement measures, and the use of policies that place time pressures on agents within the system. The next level of Reason's model is the line management level, where line management problems arise from incompetent management and the fallible board decisions and policy from the preceding level in the model. Line management problems represent those instances where management is either inadequate or inappropriate. Examples of line management problems include inadequate management or supervision and the use of inadequate or inappropriate training and procedures. Psychological precursors of unsafe acts are latent states that create the potential for unsafe acts. According to Reason (1990), the precise nature of unsafe acts is defined through the complex combination of a number

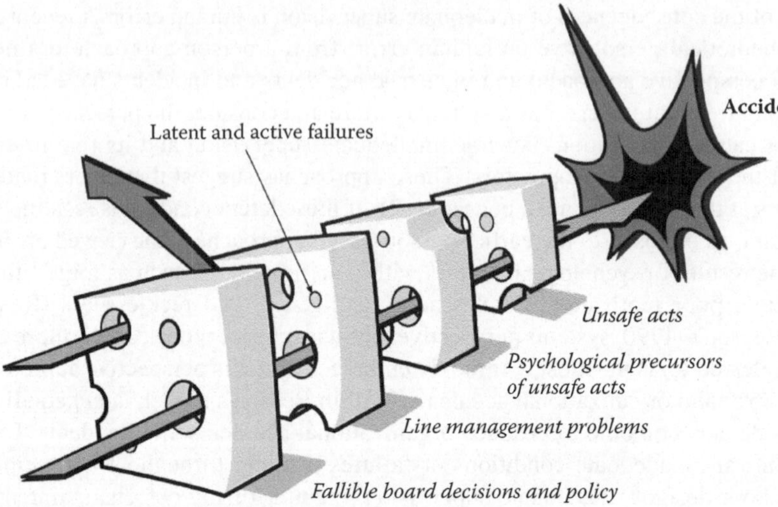

FIGURE 8.4 Systems perspective model of accident causation in complex systems. (Adapted from Reason, J., *British Medical Journal*, 320, 768–770, 2000.)

of factors, including the task being performed, the environmental conditions, and the presence of hazards. Each precursor can contribute to a great number of unsafe acts, depending on the associated conditions. Examples of these precursors include poor motivation, negative attitudes, and a failure to perceive hazards.

The methods used to analyze organizational accidents and incidents also indicate the role that supervision plays in organizational health and safety. When accidents and incidents do happen, accident analysis and investigation are used to reconstruct accidents and identify the human and system contributions, allowing practitioners to specify exactly what happened and why and then use the findings to ensure that similar accidents do not occur again. According to Roed-Larsen, Valvisto, Harms-Ringdahl, and Kirchsteiger (2004), accident investigation is the most widely used method for clarifying the basic, contributing, and immediate causes of accidents and identifying appropriate measures to prevent the occurrence of similar events in the future. It is notable that various accident analysis and investigation techniques make explicit reference to inadequate supervision as a causal factor in industrial accidents and incidents; for example, the Human Factors Analysis and Classification System (HFACS) (Wiegmann and Shappell, 2003) approach to analyzing aviation accidents utilizes an unsafe supervision taxonomy containing supervisory error modes. Wiegmann and Shappell (2003) suggested that the role of supervisors is to provide their personnel with the opportunity to succeed, and this is achieved by providing guidance, training, leadership, oversight, and incentives. Supervisors influence the operational conditions under which operators work and the way in which activity is performed. The unsafe supervision level within the HFACS taxonomy contains four categories: inadequate supervision, planned inappropriate operations, failure to correct a known problem, and supervisory violations. *Inadequate supervision* refers to those instances when efficient supervision was not provided. Examples of inadequate supervision

TABLE 8.2
Examples of Unsafe Supervision

Inadequate Supervision
Failure to provide proper training
Failure to provide professional guidance or oversight
Failure to provide current publications, adequate technical data, or procedures
Failure to provide adequate rest period
Lack of accountability
Perceived lack of authority
Failure to track qualifications
Failure to track performance
Failure to provide operational doctrine
Overtasked or untrained supervisor
Loss of supervisory situational awareness

Planned Inappropriate Operations
Poor crew pairing
Failed to provide adequate brief time or supervision
Benefit outweighed by risk
Failure to provide adequate opportunity for crew rest
Excessive tasking or workload

Failure to Correct a Known Problem
Failure to correct inappropriate behavior or identify risky behavior
Failure to correct a safety hazard
Failure to initiate corrective action
Failure to report unsafe tendencies

Supervisory Violations
Unqualified crew authorized for flight
Failure to enforce rules and regulations
Violation of procedures
Unnecessary hazard authorized
Willful disregard for authority by supervisors
Inadequate documentation
Fraudulent documentation

Source: Adapted from Wiegmann, D.A. and Shappell, S.A., *A Human Error Approach to Aviation Accident Analysis: The Human Factors Analysis and Classification System*, Ashgate, Burlington, VT, 2003.

include failure to provide proper training, failure to provide professional guidance or oversight, and failure to provide adequate rest periods. *Planned inappropriate operations* refer to those instances where the operational tempo or scheduling of aircrew puts individuals at unacceptable risk, jeopardizes crew rest, and affects performance. Examples of planned inappropriate operations include poor crew pairing and failure to provide adequate opportunity for crew rest. *Failure to correct a known problem* refers to those instances when a supervisor is aware of inadequacies within the system, such as inadequate equipment, training, or individuals, but does not attempt to rectify the problem. Examples of failure to correct a known problem include failure to correct inappropriate behavior or identify risky behavior and failure to report unsafe tendencies. *Supervisory violations* refer to those instances when rules and regulations are willfully disregarded by supervisors. Examples of supervisory violations include authorizing an unqualified crew flight and violating procedures (see Table 8.2).

Other accident analysis and investigation approaches that make explicit reference to supervision as a potential causal factor include AcciMaps (Svedung and Rasmussen, 2002), which diagrams company management and technical, operational, and management failure levels, and the Incident Cause Analysis Method (ICAM) (BHP Billiton, 2001), which addresses inadequate supervision and poor supervisor or worker ratio error modes.

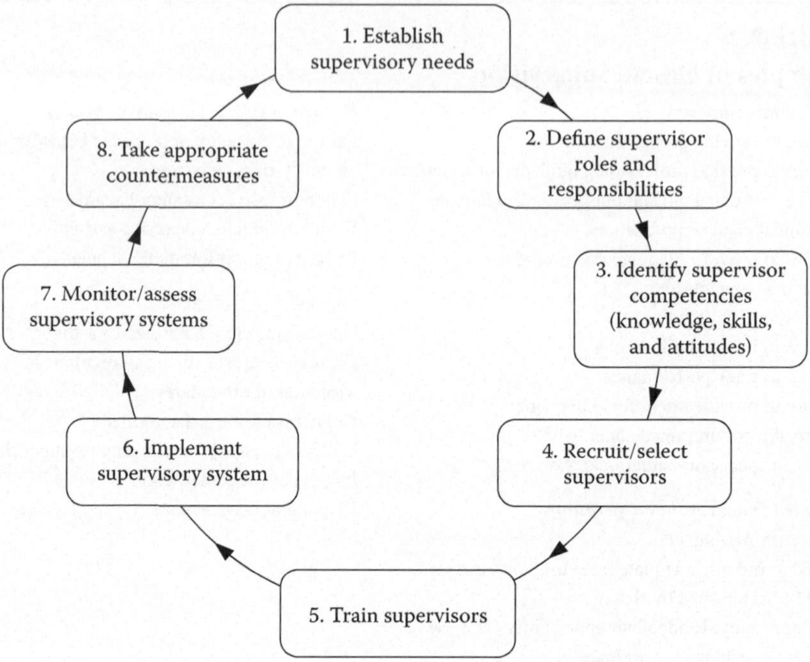

FIGURE 8.5 Supervisory management framework.

8.5 GUIDANCE AND RECOMMENDATIONS

The importance of the use of appropriate forms of supervision and its key role in the safety and efficiency of process control are outlined above. Further, the importance of understanding and acting on the weaknesses associated with an organization's approach to supervision is also emphasized. It is recommended that, in managing their approach to supervision, process control organizations need to clearly understand their supervisory needs, take measures to satisfy these supervisory requirements, assess the performance of their chosen supervisory approach, and take any countermeasures required to increase the efficiency of their approach. This process should occur whenever new technology or processes are introduced into the process control system. A supervision management framework is presented in Figure 8.5.

At step 1, the organization should establish its supervisory needs. This might involve the use of task analysis approaches to identify where in a process supervision is required and discussions between managers within the organization. When the parts of the process that require supervision have been identified, the roles and responsibilities of the supervisors should be clearly defined. Again, task analysis procedures will be useful for this purpose. Following on from this, the competencies, expressed as knowledge, skills, and attitudes (KSAs) required for the supervisory role, should be clearly defined (see Chapter 2 for guidance on this process). This can then be used to drive the selection (from existing staff pool) or recruitment of appropriate staff to fill the supervisory roles in question and inform the development of appropriate training programs for the supervisors. When they are trained, the supervisory system should be implemented, and the supervisory system should be continually monitored and assessed to determine how well it is working and if any particular weaknesses are associated with it. Following this, countermeasures designed to address any of the problems found should be implemented. Based on their literature review, site visits, and supervision assessment study, Brazier et al. (2004) developed a series of countermeasures designed to tackle flaws in the various supervisory approaches that they identified. These countermeasures are presented in Table 8.3, Table 8.4, and Table 8.5. Based on the literature reviewed for their report, they formulated a series of general guidelines for supervision within complex sociotechnical systems:

1. Identify supervisor competencies. Managers should have a very clear understanding of what competencies (KSAs) are required for the supervisory role in question.
2. Be sure the appropriate individual is selected for the supervisory role. This is critical. The choice should be based on the individual's experience, competencies, personality, and overall suitability for the role. It is important to note that existing job performance may not be sufficient evidence for an individual's suitability for a supervisory role.
3. Provide appropriate training for the supervisors. The training should be based on the competencies required for the role and should be ongoing.
4. Clearly define and communicate the supervisor's role and responsibilities. It is critical that everybody within an organization, including the managers, the supervisor, and the supervisor's subordinates, have a clear understanding of the supervisor's roles and responsibilities.
5. Continually monitor and assess the supervisory system to ensure that it is working efficiently and to highlight any problems associated with it. Countermeasures should be taken immediately to address any problems identified (see Table 8.3, Table 8.4, and Table 8.5 for examples).
6. Ensure and encourage communication among supervisors, subordinates, and management. Communication plays a key role in the success or failure of a particular supervisory system. Organizations should encourage adequate communication among all levels through the use of procedures and clear communication channels.
7. Ensure that lines of reporting among management, supervisors, and subordinates are clearly defined and expressed for all involved.
8. Encourage high levels of interaction among supervisors, subordinates, and management. It is important that all levels of the system continue to interact with one another, and interaction between management and a supervisor's subordinates is particularly critical.
9. Ensure team member development. Regardless of the form of supervision in place, team member development should continue to occur. Team members should be involved in decision-making and leadership tasks and should be given sufficient levels of autonomy.

TABLE 8.3
Team Countermeasures

Problem	Countermeasure
Supervisor is team's main point of contact, so direct management communication with the team is limited.	Make arrangements for managers to meet team members. Develop an open communicating culture. Be sure that the supervisor is an effective channel of communication.
Remote managers find it difficult to understand how the team actually functions.	Encourage high levels of involvement from team members in analyzing situations, making decisions, and implementing change.
Flat organizational structures (where individuals have more autonomy and freedom) may result in personal agendas being followed rather than company or team objectives.	Make objectives explicitly clear, and communicate them throughout the organization.
Too many procedures make locating information needed difficult.	Identify where procedures are necessary, and make sure they are easy to find and navigate and are presented in an appropriate format.
Teams working shifts may find it difficult to communicate with day workers and management.	Be sure that day workers understand that they have a responsibility to engage with shift workers.
Teams lacking supervisor may perceive supervisory function to be devalued.	Be sure that the organization has a clear understanding of the importance of effective supervision.
Dictatorial supervision leads to poor communication and lack of ownership.	The personality of supervisors is a critical factor. Supervisors should be able to engage with team members effectively and know when and how to provide leadership.

Bureaucracy may lead to supervisors becoming overloaded with paperwork and unable to undertake supervisory activities.	Criteria detailing the proportion of time to be spent on different supervisory activities should be developed and clearly specified.
Hierarchical organizations encourage individuals to restrict their activities to a relatively small portion of the total work undertaken.	Encourage teamwork so individuals understand the common goals of the team.
Selection of existing team members for supervisory roles often results in individuals being selected based on their ability to do their current job rather than on them satisfying supervisory job requirements.	The competencies (KSAs) required for the supervisory role should be identified and clearly articulated, and selection should take into account these factors.
Teams containing self-managed individuals may lead to supervisors perceiving that they no longer have responsibility for the individuals actions.	Clearly define supervisor responsibilities, and make sure they are fully understood by both the supervisor and the team.
Small teams are less able to maintain technical expertise in all aspects of the work undertaken.	Provide technical support by individuals who can communicate in a form that the team members require.
Multi-skilled teams may be unable to maintain a depth of technical knowledge for all aspects of the work undertaken.	Provide technical support by individuals who can communicate in a form that the team members require.
Externally (from the team) developed plans may be impractical and fail to account for the resources required and available.	Be sure resource requirements for activities are assessed and that planners understand what resources are available.

Source: Adapted from Brazier, A. et al., *Different Types of Supervision and the Impact on Safety in the Chemical and Allied Industries*, Health and Safety Executive, London, 2004.

TABLE 8.4
Individual Countermeasures

Problem	Countermeasure
Individuals with multiple roles may not understand how much time they should devote to supervisory activities.	Provide clear guidance on how much time an individual should devote to each of their roles.
Individuals with a combined supervisory and operational role may be restricted in the amount of interaction with their team members.	Identify the operational tasks carried out by people with supervisory responsibility, and ensure that they do not have a detrimental impact on their supervisory role.
Individuals working in flat organizational structures may perceive that they are in a weak position, unable to respond to pressure from management.	Be sure that the personality and behavior of managers are open and receptive and that undue pressure is not placed on individuals.
Individuals with multiple roles may not understand how much time they should devote to supervisory activities.	Provide clear guidance on how much time individuals should devote to each of their roles.
Individual working in multi-skilled teams may not have competency in all aspects of the jobs that they supervise.	Be sure that team competence is sufficient in all areas and that technical support is available.
Individuals working in multi-skilled teams may feel that their supervisors are unable to answer certain technical questions and so may make unauthorized decisions.	Clearly and explicitly define tiers of responsibility.
Unrealistic expectations are placed on supervisors of hierarchical teams to achieve behavioral change and continuous improvement.	Be sure everybody understands what is required to achieve behavioral change.
In flat organizational structures, there may be unrealistic expectations that evolution will improve team performance in all areas.	Be sure that the direction is set for teams, and monitor all aspects of performance to ensure that improvements are being made.
Individuals promoted into supervisory roles are unlikely to have sufficient experience for the role.	In addition to selecting people based on their suitability, provide adequate training prior to commencing the role and that extra support is provided initially.

Source: Adapted from Brazier, A. et al., *Different Types of Supervision and the Impact on Safety in the Chemical and Allied Industries*, Health and Safety Executive, London, 2004.

TABLE 8.5
Safety Management System Countermeasures

Problem	Countermeasure
Supervision may not provide sufficient priority during major hazards.	Safety management system should provide direction so supervisory activity covers all risks appropriately.
Flat organizational structures have more general role descriptions which often means that supervision does not feature prominently.	Safety management system should define how supervisory functions are to be identified and allocated to individuals or job roles. Supervisory activities should be included within task procedures.
Demarcation of responsibilities occurs within hierarchical organizations.	Safety management system should state that all individuals are responsible for their own behavior and for monitoring others to prevent unsafe acts being undertaken. This should also be reinforced through training, appraisal, and review.
Multi-skilled teams may deal with problems themselves rather than confer with higher authority.	Safety management system should clearly define the chain of command and identify tiers of responsibility.
Different types of organizations require different forms of supervision.	Safety management system should reflect that supervisors, team members, and management must all have a personality consistent with the style of the organization.
Evolving organizations make it difficult to ensure that all elements of supervisory function are delivered adequately at all times.	Safety management system should include arrangements for dealing with organizational changes.
Organizational culture impacts supervision and how it is delivered.	Safety management system should reflect the importance of generating an open culture.
Devolved supervision makes performance assessment difficult as it covers a large number of people.	Safety management system should clearly define organizational and team performance indicators that address the supervisory function.
On multiple-team sites, conflict between teams can occur and be difficult to resolve.	Safety management system should ensure that priorities and responsibilities are clearly defined and understood. A single person onsite at all times (e.g., shift manager) with overall responsibility may be useful.

Source: Adapted from Brazier, A. et al., *Different Types of Supervision and the Impact on Safety in the Chemical and Allied Industries*, Health and Safety Executive, London, 2004.

10. Optimize supervisor workload. It is imperative that the supervisor does not become overloaded in a way that impacts his or her supervisory role. Conversely, a supervisor with too little to do may cause subordinates to question the supervisor's role.
11. Avoid demarcation at all costs.

8.6 ASSESSMENT AND EVALUATION

The importance of an organization's ability to assess its own approach to the supervision of teams is clear. Despite this, there is a lack of structured approaches available for organizations to do so. Most research in this area has focused either on the impact of supervisory behaviors on subordinate behavior, performance, motivation, and attitudes to work or on the type of supervisory approach used. This has involved the use of various approaches such as the Leadership Behavior Description Questionnaire (Hemphill, 1950), the Supervisory Behavior Description Questionnaire (Fleishman, 1953), and the Operant Supervisory Taxonomy Index (Komaki et al., 1986). Other more standardized approaches such as observation and task analytic approaches such as cognitive task analysis (Carvalho et al., 2005) have also been employed. Inherent within these approaches is an inability to make a judgment on the strengths and weaknesses of the form of supervision assessed. In response to this, Brazier and Ward (2004) developed an assessment approach that allows organizations in the chemical and allied industries to evaluate their supervisory arrangements and the subsequent impacts on health and safety. The approach is interview based and uses a series of questions designed to assess the adequacy of an organization's supervisory arrangements. The questions are decomposed into the following three sections:

1. *Safety management system questions* are designed to assess the organization's safety management system with regard to supervision.
2. *Team assessment questions* focus on the supervisory arrangements from the point of view of the different teams working within the organization.
3. *Individual assessment questions* are designed to assess those individuals working within the organization who have supervisory responsibilities.

The process involves analysts holding interviews or discussions with safety management system personnel, teams, and then individuals who have supervisory responsibilities. A flowchart depicting the methods procedure is presented in Figure 8.6. The questions underlying the method can be found in Brazier and Ward (2004), and they also have provided an assessor's guide, which contains prompts and guidance on how to interpret the answers provided by respondents.

Brazier et al. (2004) included a summary of analyses using the methodology that were undertaken in the chemical and allied industries. Eight companies (six involved in chemical manufacture, two involved in oil or gas processing) were assessed by two analysts using the methodology described above. All of the participants considered supervision to be critical to health and safety; however, only one of the organizations assessed felt that supervision was specifically defined within their safety

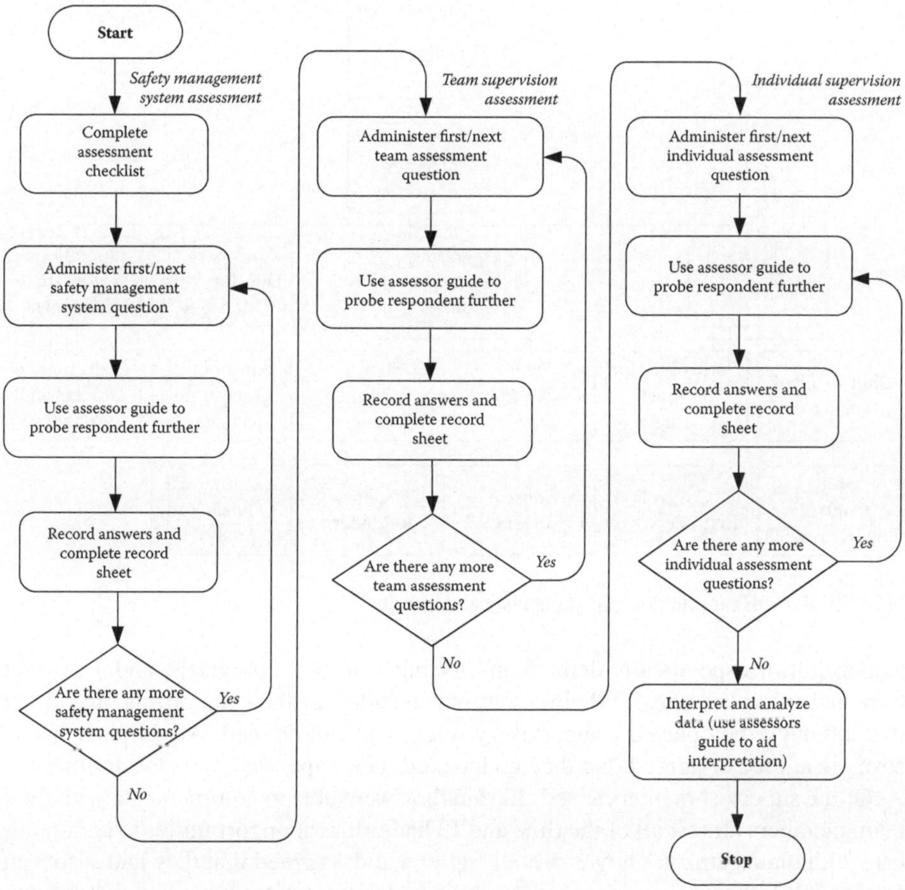

FIGURE 8.6 Supervision assessment methodology flowchart.

management system. Further, only three of the organizations regularly and formally considered supervisory performance, and only six of the organizations (three strongly agreed and three agreed) did enough to improve supervisory performance when problems were identified.

From an analysis of teams within these organizations, 10 out of the 11 teams analyzed had a team leader appointed by management (the remaining team had a traditional supervisor). All of the teams assessed felt that the majority of supervisory activities were undertaken within the team. Only 5 teams felt that they had a comprehensive set of procedures and instructions covering supervisory activities, 6 teams felt that procedures and instructions were covered in some form, and 4 felt that they had no procedures or instructions covering supervisory activities. Finally, the majority of the teams either strongly agreed or agreed that everyone in the team understood how supervision was delivered.

From the individual assessment, 15 participants out of 26 interviewed were identified as having a supervisory role. Nine out of the 15 felt that they had their supervisory

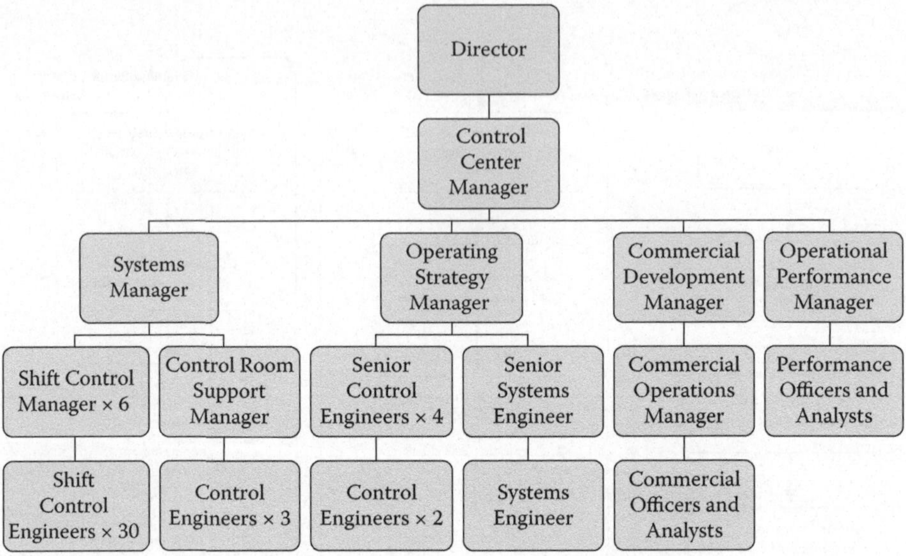

FIGURE 8.7 Example domain supervisory hierarchy.

responsibilities specifically defined and included in their appraisals, and 4 felt that they had them specifically defined but not included in their appraisals, and 2 felt that although they played a supervisory role it was not defined. All 15 participants strongly agreed or agreed that they understood their supervisory responsibilities.

Of the supervisors interviewed, 13 said they were able to communicate with their team members almost all of the time and 13 had frequent opportunities to communicate with management. Only 3 strongly agreed and 4 agreed that they had sufficient time to fulfill their supervisory responsibilities. In terms of training, 8 of the supervisors felt that they had received extensive training and experience, and 7 felt that they were sufficiently experienced but had not received any training for their supervisory role. Overall, 12 strongly agreed and 3 agreed that they were sufficiently competent for their supervisory role. Finally, 6 strongly agreed and 7 agreed that they received sufficient support from others to fulfill their supervisory responsibilities.

8.7 EXAMPLE

The supervisory arrangements to be used in the proposals were analyzed. This involved a human factors specialist observing the day-to-day work activities and conducting interviews with control room personnel. On the basis of this, relevant elements of the supervision assessment approach described above were completed for the example system. The team hierarchy presented in Figure 8.7 occurs within the context of a wider system organization, which is represented in Figure 8.8. Elements of Brazier and Ward's (2004) supervision assessment methodology were used to analyze the supervisory arrangements in place within the example system. The supervisory arrangements within the system in this case can be formally defined as traditional

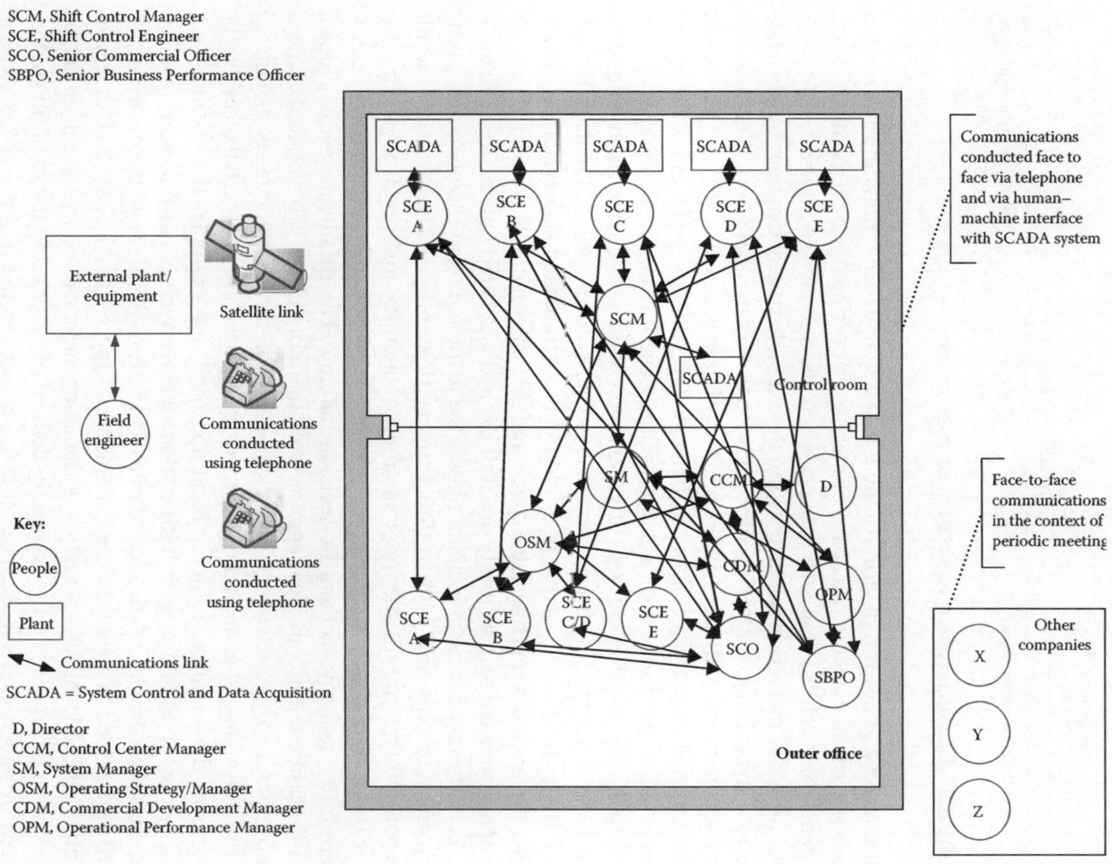

FIGURE 8.8 Network communications structure (links outside of the control room and outer office have been removed).

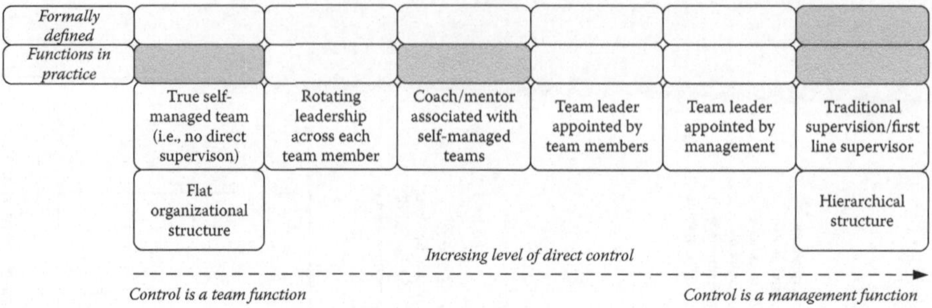

FIGURE 8.9 Example system supervisory arrangement classification.

and hierarchical; however, in terms of the actual way in which it functions in practice, elements of self-management, a coach or mentor supporting the team members, and a traditional, hierarchical approach to supervision are evident (see Figure 8.9).

The form of supervision in place within the example system can therefore be classified as a hybrid one, consisting of a traditional, hierarchical team structure with supervisors in place at all levels of the hierarchy but in practice involving elements of self-management and team coaches associated with self-managed teams. The hybrid nature of this supervisory approach allows it to benefit from the advantages associated with the three different forms of supervision while also counterbalancing some of the flaws associated with each approach to supervision. As outlined by Brazier et al. (2004), the main strengths of a traditional, hierarchical supervisory structure are that the lines of reporting are explicit and that each individual's role within the team and associated responsibility are well defined and clearly understood. In addition, the self-managed team element allows the team and team members to operate with more autonomy and without direct supervision and provides additional levels of responsibility, team interaction, mutual accountability, and involvement in decision making. The coach/mentor approach also increases team member autonomy, responsibility, and involvement in decision making. The disadvantages associated with the traditional, hierarchical approach are therefore effectively counterbalanced by the other two approaches. Conversely, some of the disadvantages associated with the self-managed team and coach/mentor approach are counterbalanced by the strengths of the traditional, hierarchical approach; for example, the possible lack of clarity over roles and responsibilities that emerges within self-managed teams is counteracted by the clear definition of roles and responsibilities and the clear lines of reporting provided by the traditional, hierarchical approach. Also, the lack of leadership and reduced levels of interaction with management caused by the self-managed team approach is remedied by the clearly defined hierarchy and levels of interaction between managers and subordinates that are evident in the traditional hierarchical approach.

The different supervisory roles within the example system and the numbers of personnel under direct supervision of each role are presented in Table 8.6. A breakdown of the roles and responsibilities of the various supervisors within the example system is presented in Table 8.7. Analysis of the proposed supervisory arrangements

TABLE 8.6
Supervisor Roles and Subordinates

Team Members/Roles	Number in Team (Per Shift)	Does This Position Have Supervisory Responsibility?	How Many People Does This Position Supervise?
Director	67	Yes	1
Control Center Manager	66	Yes	4
Systems Manager	40	Yes	10
Shift Control Manager	5	Yes	5
Operating Strategy Manager	8	Yes	5
Control Room Support Manager	4	Yes	3
Senior Control Engineer	1	Yes	1
Commercial Development Manager	6	Yes	1
Commercial Operations Manager	5	Yes	2
Senior Commercial Officer	2	Yes	2
Operational Performance Manager	7	Yes	4
Senior Business Performance Officer	1	Yes	1

shows that all of the supervisory roles have been identified. Further, the planned supervisory system for the 24/7 control room has both a formal hierarchical structure for day working and a flatter structure for other hours. This arrangement is in keeping with best practice for supervision of control room staff.

BIBLIOGRAPHY

Allen, G. (1998). *Teambuilding*. http://ollie.dcccd.edu/mgmt1374/book_contents/4directing/teambldg/teambldg.htm.

Bass, B.M. (1990). *Bass and Stogdill's Handbook of Leadership*. New York: Free Press.

Bass, B.M. (1998). *Transformational Leadership*. Mahwah, NJ: Lawrence Erlbaum.

Bass, B.M. (1999). Two decades of research and development in transformational leadership. *European Journal of Work and Organizational Psychology*, 8(1), 9–32.

BHP Billiton. (2001). *Minicam Investigation Guide*, Issue 2, October 2001.

BOMEL Consortium. (2003). *The Factors and Causes Contributing to Fatal Accidents 1996/97 to 2000/01*, Summary Report, HSE Task ID BOM\0040, C998\01\117R, Rev. B. London: Health and Safety Executive.

Bono, J.E., Jackson-Foldes, H., Vinson, G., and Muros, J.P. (2007). Workplace emotions: the role of supervision and leadership. *Journal of Applied Psychology*, 92(5), 1357–1367.

Brazier, A. and Ward, R. (2004). *Different Types of Supervision and the Impact on Safety in the Chemical and Allied Industries: Assessment Methodology and User Guide*. London: Health and Safety Executive.

Brazier, A., Gait, A., and Waite, P. (2004). *Different Types of Supervision and the Impact on Safety in the Chemical and Allied Industries*. London: Health and Safety Executive.

Burke, C.S. (2004). Team task analysis. In N.A. Stanton, A. Hedge, K. Brookhuis, E. Salas, and H. Hendrick (Eds.), *Handbook of Human Factors and Ergonomics Methods* (pp. 526–535). Boca Raton, FL: CRC Press.

Burns, J.M. (1978). *Leadership*. New York: Harper & Row.

TABLE 8.7
Supervisor Roles and Responsibilities

Tasks	Responsible Supervisors
Defining overall team workload	Control Center Manager
Allocating day-to-day work priorities	Systems Manager, Operating Strategy Manager, Commercial Development Manager, Operational Performance Manager
Allocating manpower for daily tasks	Systems Manager, Operating Strategy Manager, Commercial Development Manager, Operational Performance Manager
Controlling and monitoring work patterns and shift rotas	Systems Manager, Operating Strategy Manager, Commercial Development Manager, Operational Performance Manager, Shift Control Managers
Communicating operational information to and from the team	Systems Manager, Shift Control Managers
Problem solving and decision making	Shift Control Managers
Identifying competencies required for work tasks	Systems Manager, Operating Strategy Manager, Commercial Development Manager, Operational Performance Manager, Shift Control Managers
Assessing training requirements of team members	Systems Manager, Operating Strategy Manager, Commercial Development Manager, Operational Performance Manager, Shift Control Managers
Measuring team performance	Systems Manager, Operating Strategy Manager, Commercial Development Manager, Operational Performance Manager, Shift Control Managers
Carrying out appraisal of team members	Systems Manager, Operating Strategy Manager, Commercial Development Manager, Operational Performance Manager, Shift Control Managers
Implementing first-level discipline	Systems Manager, Operating Strategy Manager, Commercial Development Manager, Operational Performance Manager, Shift Control Managers
Initiating and leading incident investigations	Systems Manager
Maintaining and updating procedures	Systems Manager, Operating Strategy Manager, Commercial Development Manager, Operational Performance Manager, Shift Control Managers
Providing leadership in emergency situations	Shift Control Managers to Systems Manager to Control Center Manager

Cannon-Bowers, J.A., Tannenbaum, S.I., Salas, E., and Volpe, C.E. (1995). Defining competencies and establishing team training requirements. In R.A. Guzzo and E. Salas (Eds.), *Team Effectiveness and Decision Making in Organizations* (pp. 333–381). San Francisco, CA: Jossey-Bass.

Carvalho, P.V.R., Santos, I.L., and Vidal, M.C.R. (2005). Nuclear power plant shift supervisor's decision making during micro incidents. *International Journal of Industrial Ergonomics*, 35, 619–644.

Fleishman, E.A. (1953). The description of supervisory behavior. *Journal of Applied Psychology*, 37(1), 1–6.

Fleming, M. (1999). *Effective Supervisory Safety Leadership Behaviors in the Offshore Oil and Gas Industry*. London: HSE Books.

Glaso, L. and Einarsen, S. (2006). Experienced affects in leader–subordinate relationships. *Scandinavian Journal of Management*, 22, 49–73.

Glickman, A.S., Zimmer, S., Montero, R.C., Guerette, P.J., Campbell, W.J., Morgan, B.B., and Salas, E. (1987). *The Evolution of Team Skills: An Empirical Assessment with Implications for Training*, Technical Report NTSC-87-016. Arlington, VA: Office of Naval Research.

Health and Safety Executive (HSE). (1997). *Successful Health and Safety Management*, HSG65. London: HSE Books.

Health and Safety Executive (HSE). (2007). *Human Factors: Supervision*. London: Health and Safety Executive (http://www.hse.gov.uk/humanfactors/comah/supervision.htm).

Hemphill, J.K. (1950). *Leader Behavior Description*. Columbus, OH: Ohio State University.

Klein, G. (2000). *Sources of Power: How People Make Decisions*. MIT Press: Cambridge, MA.

Komaki, J.L. (1986). Toward effective supervision: an operant analysis and comparison of managers at work. *Journal of Applied Psychology*, 71, 270–279.

Komaki, J.L., Zlotnick, S., and Jensen, M. (1986). Development of an operant-based taxonomy and observational index of supervisory behavior. *Journal of Applied Psychology*, 71(2), 260–269.

Lardner, R. and Miles, B. (1998). Supervision, self-management and safety. *The Chemical Engineer*, 650, 28–31.

McGoldrick, D.A. (1994). First-Line Supervisors in the Offshore Oil Industry. Paper presented at the Society of Petroleum Engineers Health, Safety, and Environment in Oil and Gas Exploration and Production Conference, Jakarta, Indonesia, January 1994.

Mearns, K., Flin, R., Fleming, M., and Gordon, R. (1997). *Human and Organizational Factors in Offshore Safety*. London: HSE Books.

Mintzberg, H. (1979). *The Structuring of Organizations*. Upper Saddle River, NJ: Prentice Hall.

Mowry, H.W. (1957). A measure of supervisory quality. *Journal of Applied Psychology*, 41(6), 405–408.

Reason, J. (1990). *Human Error*. Cambridge, U.K.: Cambridge University Press.

Reason, J. (1997). *Managing the Risks of Organizational Accidents*. Burlington, VT: Ashgate.

Reason, J. (2000). Human error: models and management. *British Medical Journal*, 320, 768–770.

Rinefort, F.C. and Van Fleet, D.D. (1993). Safety issues beyond the workplace: estimated relationships between work injuries and available supervision. *Employee Responsibilities and Rights Journal*, 6(1), 1–8.

Roed-Larsen, S., Valvisto, T., Harms-Ringdahl, L., and Kirchsteiger, C. (2004). Accident investigation practices in Europe: main responses from a recent study of accidents in industry and transport. *Journal of Hazardous Materials*, 111, 7–12.

Salas, E. (2005). Team methods. In N.A. Stanton, A. Hedge, K. Brookhuis, E. Salas, and H. Hendrick (Eds.), *Handbook of Human Factors and Ergonomics Methods* (pp. 43.1–43.4). Boca Raton, FL: CRC Press.

Salas, E., Burke, C.S., and Cannon-Bowers, J.A. (2000). Teamwork: emerging principles. *International Journal of Management Reviews*, 2(4), 339–356.

Svedung, I. and Rasmussen, J. (2002). Graphic representation of accident scenarios: mapping system structure and the causation of accidents. *Safety Science*, 40, 397–417.

Ward, R., Brazier, A., and Lancaster, R. (2004). *Different Types of Supervision and the Impact on Safety in the Chemical and Allied Industries: Literature Review*. London: Health and Safety Executive.

Weston, R., Grimshaw, M., and Norton, D.C. (1989). *Supervision of Technical Staff. An Introduction for Line Supervisors*. Suffolk, U.K.: St. Edmundsbury Press.

Wiegmann, D.A. and Shappell, S.A. (2003). *A Human Error Approach to Aviation Accident Analysis: The Human Factors Analysis and Classification System*. Burlington, VT: Ashgate.

9 Shift Patterns

9.1 OUTLINE

Control room activities have a long history of 24-hour operation which has necessarily required operators to work in shifts. While shiftwork quite often brings with it individual compensation in terms of better pay, time off, flexible working, and so on, numerous issues and risks for the individual as well as the system as a whole require careful consideration. The Health and Safety Executive (HSE) defines shiftwork as usually meaning "a work activity scheduled outside standard daytime hours, where there may be a handover of duty from one individual or work group to another" (HSE, 2006, p. 3). The accepted definition of shiftwork has two dimensions:

- The first dimension covers the time period in which the work takes place.
- The second dimension refers to the handover of work from one individual or group to another.

Standard daytime hours are regarded as the period between 7:00 a.m. and 7:00 p.m., which is normally split in two by a lunch break. Shiftwork, therefore, is work activity that routinely occurs outside of this 12-hour daytime period.

It should be noted that, while shiftwork is commonly associated with individuals who have to work during the night, working at other times can be equally if not more disruptive to the individuals concerned and ultimately to the performance of the system they are working within. Many authors have argued for an inclusive definition of shiftwork, such as: "We shall define any work taken on a regular basis outside [the defined daytime period] as shiftwork and any individuals involved in that work as shiftworkers. Thus, we shall include part-timers, full-timers, evening shiftworkers, early morning shiftworkers, as well as the rotating and fixed shiftworkers" (Monk and Folkard, 1992, p. 2). The HSE view of shiftworking is similarly inclusive and covers, in addition, several other categories of shiftwork (HSE, 2006, p. 3):

- Work during the afternoon, night, or weekend, typically with periods of the work scheduled outside standard daytime hours
- Extended work periods of 12 hours or more, often associated with compressing the working week
- Rotating hours of work
- Split shifts, where work periods are divided into two distinct parts with several hours' break in between
- Overtime
- Standby or on-call duties

A definition of shiftwork according to time period thus:

- Is inclusive
- Refers to a daytime working norm of 7:00 a.m. to 7:00 p.m.
- Is orientated around the impact that such work has on workers

The typical conception of work undertaken during normal daytime hours (7:00 a.m. to 7:00 p.m.) is that it begins at the start of the period, stops at the end of the working day, and is resumed in the next period. Shiftwork is different. Rather than being paced by a normal working day, it is instead continuous in nature, paced by 24-hour operation.

The continuous nature of shiftwork brings with it another aspect unique to this method of working: the requirement for a handover of duty from one individual or work group to another. Shiftwork, in addition to its out-of-hours nature, is "a pattern of work where one employee replaces another on the same job within a 24-hour period" (HSE, 2006, p. 3).

The purpose of the handover is to let an incoming group of workers take over the task. In control room settings, a compatible state of situational awareness needs to be developed by the incoming team, normally with the help of the outgoing team. This allows the new team to perceive and comprehend the status of the system appropriately. The quality of the handover, therefore, is also an essential part of system performance and safety.

9.2 KEY ISSUES

There is little doubt that shiftwork is demanding and brings with it risks to individual well-being. A number of specific examples include "disruption of the internal body clock; fatigue; sleeping difficulties; disturbed appetite and digestion; reliance on sedatives and/or stimulants; social and domestic problems" (HSE, 2006, p. 5). These risks must be appropriately understood and managed because they all have the potential to adversely affect system performance and safety. Individual risks and issues have been grouped into the following four headings (Monk and Folkard, 1992):

- Sleep and the biological clock
- Social challenges
- Stress and strain
- Health consequences

These factors link and interact with each other to create performance and safety consequences; however, not everyone is the same, and the attenuating effect of individual differences should be represented, as should mediating strategies instigated by both the individual worker and the organization. Given that all of these factors interact, a systems perspective is an appropriate conceptual response.

9.3 STANDARDS AND GUIDELINES

The following standards and guidelines pertain to shiftworking:

- Working Time Regulations (1998, amended 2003)
- Health and Safety at Work etc. Act (1974)
- Management of Health and Safety at Work Regulations (1999)

9.3.1 Working Time Regulations

The Working Time Regulations have a significant influence on shiftworking arrangements. They are legally enshrined, in full, throughout most of the European Union, although the situation in the United Kingdom is somewhat different. The current situation is that U.K. workers can voluntarily opt out of the provisions of the act if they so wish. This means that in some situations they can work for longer periods. Note, however, that employers are not able to coerce or otherwise require workers to opt out; the spirit of this arrangement is that it is at the worker's discretion and any opt-outs are fully cancelable (normally with 7 days' notice). In the absence of any voluntary opt-outs, and while there are some exclusions from the regulations (principally in the transport and aviation sectors), they apply to control room settings in all industry sectors. The Working Time Regulations make the following provisions:

- 48 hours per week (on average)* is the maximum that a worker can be required to work, although workers can choose to work more if they want to provided that an opt-out is signed. This opt-out is cancelable at the worker's discretion (usually with 7 days' notice). Note that working lunches or travel time in the course of working duties are included in the 48 hours.
- 8 hours (on average) is the maximum that a night worker is required to work in a 24-hour period, and no opt-out is available. For demanding work, the 8-hour average becomes an 8-hour maximum.
- Night workers have the right to receive free health assessments before such work is undertaken and then throughout.
- All workers have the right to 11 hours of rest a day.
- All workers have a right to a day off each week.
- All workers have a right to an in-work rest break if the working day is longer than 6 hours.
- All workers have a right to 4 weeks of paid leave per year.

9.3.2 Health and Safety Legislation

While the Working Time Regulations lay out a number of practical parameters and boundaries that govern the working arrangements of shiftworkers, the Health and Safety at Work etc. Act (1974) and the Management of Health and Safety at Work Regulations (1999) describe the wider legislative background as well as the

* The average is normally calculated over a 17-week reference period, although this can be increased through local agreement.

processes for implementing the Working Time guidelines. Briefly: "Under the HSW Act all employers … have a duty, so far as is reasonably practicable, to protect the health, safety, and welfare at work of their employees. They also have a duty, so far as is reasonably practicable, to ensure that others are not exposed to health and safety risks through their undertaking" (HSE, 2006, p. 2).

Shiftworking, as an activity, has the potential to exert a considerable influence on the health, safety, and welfare of shiftworkers, meaning that such risks and issues are required by law to be considered. In addition, there is a mutual responsibility. Employees, too, have "a duty to take reasonable care of their own health and safety and that of other people who may be affected by their activities at work. This duty implies that employees should take positive steps to understand the risk factors in their work, such as the causes of fatigue; comply with safety rules and procedures; and make sure that nothing they do nor fail to do at work puts anyone at risk" (HSE, 2006, p. 2). Managing the risks arising from shiftworking, therefore, relies on strategies for both the worker and the organization. This is reflected in the guidance that follows.

9.4 HUMAN FACTORS LITERATURE

9.4.1 The Body Clock

Humans are rhythmic creatures whose cycle of wakefulness and sleep generally tracks daytime hours. This daily pattern is referred to as circadian rhythm. The physiology of the human body is attuned to this cyclical pattern, adjusting itself to promote maximum wakefulness during daytime hours while configuring itself for sleep during night time. Of the many physiological parameters that can be used to measure circadian rhythms, the most common is body temperature. As shown in Figure 9.1, a circadian rhythm "is characteristic in showing a comparatively steep rise from early to mid-morning, followed by a more gradual rise to the mid-evening peak, and then a fairly precipitate decline just prior to bedtime" (Monk and Folkard, 1992, p. 8). Monk and Folkard (1992) defined three key characteristics of human circadian rhythms:

1. Circadian rhythms are internally generated, thus "the outputs of the circadian system are not reactions to changes in posture or environment but come from within the individual" (p. 11). As well as being endogenous they are also persistent and self-sustaining, meaning that the cycle of wakefulness and sleep continues in the absence of external cues.
2. The most important characteristic of the human circadian system for the shiftworker is that much of it is resistant to abrupt changes in schedule. This means that the circadian rhythm will continue to cycle in a way that is not adapted to current working patterns for some time, possibly weeks or even months. Note that circadian rhythms never fully adapt, even in cases where the operator is working on a permanent night shift. The practical effect of this is that the operator's circadian rhythm can be calling for sleep when

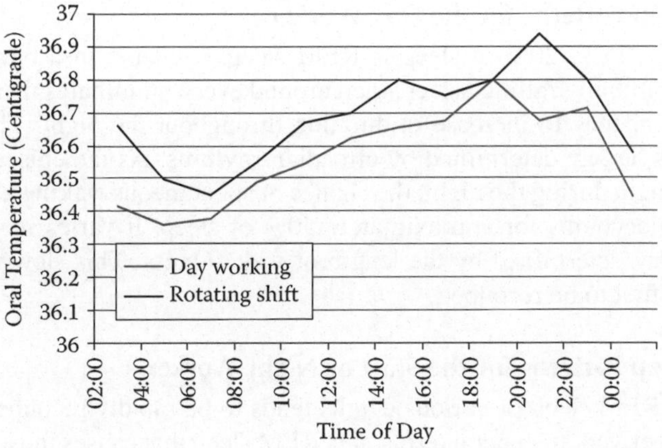

FIGURE 9.1 Circadian rhythm as measured by body temperature. (Data from Monk, T.H. and Embrey, D.E., in *Night and Shiftwork: Biological and Social Aspects*, A. Reinberg, Ed., Pergamon Press, Oxford, U.K., 1981, pp. 473–480. With permission.)

the work itself maybe calling for maximum alertness. This is a powerful biological need (HSE, 2006) that requires careful consideration.

3. Another aspect of circadian rhythms, one that this time can be mildly beneficial, is that it has a slight tendency to run slow. In the absence of other cues, the circadian rhythm tends to extend forward in time, with people electing to go to bed later and later with a corresponding delay in waking up. Note, however, that this applies more to younger individuals whose circadian rhythm corresponds to an approximate 25-hour day (compared to older individuals, whose circadian rhythm gets close to 24 hours). This means that people show a tendency toward better adaptation to forward shifts in working arrangements (shifting to a later time schedule), than they do to backwards shifts (to an earlier time schedule).

9.4.2 Sleep

Sleep is an important restorative activity for shiftworkers, and its study has provided several powerful insights. First, sleep researchers have defined two types of sleep: rapid eye movement (REM) and non-rapid eye movement (NREM). REM sleep is distinct from NREM in that the patterns of brainwave activity are rather like those that occur when the individual is awake (Monk and Folkard, 1992). REM sleep is also regarded as dream sleep and the period of sleep in which most spontaneous waking occurs. NREM sleep is characterized by lower levels of measured brain-wave activity and is divided into four levels of sleep depth. According to the research, the deepest levels of NREM sleep (levels three and four) are thought to be the most restorative and give the worker a feeling of waking up refreshed. It is interesting to consider the normal pattern of sleep for a day worker before comparing it to that for a shift or night worker.

9.4.2.1 Sleep Patterns for the Day Worker

Approximately 15 to 20% of sleep is REM sleep. It occurs in a burst around 90 minutes after initially falling asleep, then around every 90 minutes thereafter. REM sleep has a tendency to increase in duration throughout the night. The amount of REM sleep is largely determined by circadian rhythms. As the episodes generally increase in length during the night, the chance of spontaneous waking also increases. NREM sleep accounts for approximately 50% of sleep. It varies in depth, where depth is largely determined by the length of wakefulness. This slow-wave sleep is normally the first to be recouped.

9.4.2.2 Sleep Patterns for the Shift or Night Worker

Slow-wave NREM sleep of various depths tends to be rapidly recouped, especially when sleep deprived. In general terms, it is REM sleep that causes most problems for shiftworkers: "Shift workers whose circadian systems are unadjusted are thus likely to have large amounts of REM sleep early on in their day sleeps and this may lead to premature awakenings" (Monk and Folkard, 1992). In general terms, the observation has been made that day sleep is usually lighter and shorter in duration and therefore less restorative than night sleep. It is more often disturbed because of warmer temperatures and daytime activity such as the phone ringing, noisy children, or domestic responsibilities. It is possible to resist normal sleep impulses to some extent, but this creates a situation of sleep debt, and, over time, sleep becomes increasingly difficult to resist.

9.4.2.3 Effects of Sleep Deprivation

Sleep deprivation is a difficult subjective experience to cope with, and shiftworkers can be particularly affected. At the subjective level, the effects of sleep deprivation can include:

- Poor mood
- Poor physiological activation and well-being
- Malaise and fatigue

Aside from the subjective experience of sleep debt and fatigue, the objective performance of the individual also suffers. The HSE (2006) noted that this applies particularly to tasks that require:

- Vigilance and monitoring
- Decision making
- Awareness
- Fast reaction time
- Tracking ability
- Memory

These are all relevant, to varying degrees, to the control room work setting, thus the aim of appropriately designed shiftwork arrangements and provisions is to specifically address these issues.

9.4.3 Social Challenges

Monk and Folkard (1992) rightly made a point of what they called the "strong social taboos" that protect the day worker's nighttime sleep compared to the complete absence of such strictures to protect the shift or night worker's daytime sleep. It is a truism that human society in general is designed around activity in the day and inactivity and sleep at night. Thus, the problems of circadian rhythms and REM sleep are compounded by the fact that sleeping during the day is rather more difficult, at a practical level, than attempting sleep at night.

Briefly stated, the dominant societal, neighborhood, and family norm is that of a day worker working 5 days per week. Shiftworking does not adhere to these norms:

- At the societal level, issues range from the apparently trivial (e.g., prime-time television is geared toward day workers) to the profound (a feeling of alienation from society at large).
- At the neighborhood level, similarly trivial-sounding issues range from events that are routinely scheduled for the day-working norm to the considerably more troublesome issue of noise.
- The familial level, out of all of them, is probably the most sensitive. Issues here frequently relate to the fact that partners and children do not have regular contact with the shiftworker and that the burden of family commitments, such as child care, can often default to the day worker.

All of these issues need to be recognized and carefully considered.

9.4.4 Health and Well-Being

While we have alluded to a number of immediate well-being issues connected with sleep deprivation, there remain a number of important longer term health issues to consider.

9.4.4.1 Heart Problems

Most researchers would agree that shiftwork, taken in isolation, does not cause heart problems directly; however, it is the case that shiftwork can create the conditions for certain unhealthy lifestyle choices, which in turn can lead indirectly to these problems. Examples include a relative lack of opportunity to exercise, poor dietary choices (due to availability or disrupted appetite), and stresses that arise from social challenges.

9.4.4.2 Mental Well-Being

"Put someone in a position where they are having very little sleep on a regular basis, their spouse is hostile, their children strangers, and they are permanently punch drunk from 'jet lag,' and it is not really surprising if they become a little nervous and irritable" (Monk and Folkard, 1992). Whether this translates into a risk of more serious mental disorder is as yet unresolved. What research does reveal (e.g., Gordon et al., 1986) is that feelings of nervousness and alcohol abuse are risk factors for those who work in shifts. Although complicated and highly interrelated to other social and lifestyle issues, the mental well-being of shiftworkers is doubtless a key issue and part of the health and safety duty of the employer.

9.4.4.3 Appetite and Digestion

By far the most common health-related complaint made by shiftworkers is disrupted appetite and digestion. It should perhaps come as no surprise that appetite is closely linked to circadian rhythm, thus disrupting one is likely to disrupt the other. Monk and Folkard (1992) stated: "Expert opinion is virtually unanimous in asserting that, apart from the sleep disorders, ulcers and other forms of gastrointestinal dysfunction are the most likely health consequences of shiftwork" (p. 36). Healthy eating education and the availability of healthy meal choices are often recommended as a two pronged approach to dealing with this problem.

From the above it is clear that shiftworkers often find themselves at potentially increased risk of health-related problems. It is also clear that these problems interact not only with each other but also within a wider context. Once again this highlights the need to adopt a systems perspective, as considering each aspect in isolation would not represent human factors best practice.

9.4.5 Handover

Poor-quality shift handovers mean that safety-critical information is not transferred adequately, that the incoming team's situational awareness is below par, and the potential for errors is increased. A number of major (and well-publicized) accidents have been attributed to poor shift handovers.

9.4.6 Summary

From the above, it should be clear that "maladjusted shiftworkers can become agents of risk to themselves and others" (Monk and Folkard, 1992, p. 38). There are three ways in which this can be understood:

1. The first, and most obvious, way in which shiftworkers can become agents of risk is by their being sleepy, which can lead to performance decrements that arise from trying to stay awake or actually dozing off while on duty. In both cases, the worker subjectively feels sleepy and can report as such.
2. The second way is also due to sleepiness but of the sort that is not subjectively felt by the shiftworker; thus, performance decrements can arise implicitly. It is entirely possible that, while shiftworkers may not feel subjectively tired, performance can nonetheless suffer. It becomes important, then, to realize that shiftworkers do not necessarily have the degree of meta-awareness necessary to know when this is happening, so the onus shifts to an organizational level and in the appropriate design and monitoring of shiftwork patterns.
3. Finally, the third way in which shiftworkers can become agents of risk does not relate directly to sleep at all. On the contrary, social and familial

problems (caused by shiftworking arrangements) can create operators that are not in a fit mood or disposition to operate the system safely. This occurs when they are otherwise awake and alert.

One final factor to consider is individual differences. Put simply, some individuals are better at coping with the demands of shiftworking than others. In fact, some authors (e.g., Monk and Folkard, 1992) suggest that the shiftworking population represents a form of "survivor population"; only those individuals who can cope with it elect to work shifts. Those who cannot, for whatever reason, tend to abandon their attempts. Interestingly, while it might seem safe to assume that older, more experienced shiftworkers would cope better, the more fragile nature of sleep for older persons means that the reverse is true. That said, lifestyle factors (such as better accommodations or quieter domestic situations) can ameliorate this. Another important individual difference is circadian rhythm type. There are two types: "morning larks," M-types who are early to bed and early to rise, and "night owls," E-types who are just the opposite. It is widely held that E-types are better at coping with shiftwork, as they are better able to stay awake at night and are possibly less susceptible to surrounding physical and social time cues (Monk and Folkard, 1992).

So, what we arrive at is a large collection of potentially adverse consequences of shiftwork affecting circadian rhythms and sleep, social interactions, and health and well-being, combined with a set of counterbalancing factors, such as individual differences, design of the shiftwork routine, and so on.

9.5 GUIDANCE AND RECOMMENDATIONS*

The Working Time Regulations are quite specific as to the minimum acceptable standards for shiftwork arrangements:

- A work week of 48 hours
- A right to 11 hours of rest per day
- A right to a day off each week
- A right to an in-work rest break if the shift is longer than 6 hours
- A right to 4 weeks of paid leave each year

In addition, extra rules apply to night workers:

- A maximum of 8-hour night shifts
- A right to free health assessments

As noted earlier, workers in the United Kingdom can elect to opt out of these provisions, but it is likely that the United Kingdom will be moving in the direction of implementing these minimum standards, which are, at the very least, likely to be a convenient benchmark to adopt. Compliance with minimum standards does require additional effort and insight in terms of defining the "average working time" and so on. To aid the data collection process for this more in-depth level of scrutiny, the following employer's checklist will be found useful.

* Adapted from Health and Safety Executive, *Managing Shift Work: Health and Safety Guidance*, HSE Books, London, 2006. With permission. Reproduced under the terms of the Click-Use Licence.

9.5.1 Employer's Checklist

1. What counts as working time?
2. How much time does each worker spend working?
3. If a worker is working more than an average of 48 hours a week, should we consider reducing his or her hours or ask if the worker wishes to sign an opt-out from the working time limit?
4. What records need to be kept?
5. Do we employ people who could be classified as night workers according to the definition of shiftwork?
6. How much working time do night workers normally work?
7. If night workers normally work more than 8 hours a day on average, can the hours of work be reduced or do any other exceptions or flexibilities apply?
8. Is a worker engaged in work that may be particularly hazardous?
9. How can or should a health assessment be conducted?
10. How often are health checks carried out?
11. Can any workers be transferred to day work?
12. How is the workers' working time arranged, and are they able to take rest breaks they are entitled to? Do any other exceptions or flexibilities apply?
13. Do we need to provide different rest break periods for young workers? Do different rules and entitlements apply?
14. Who is entitled to annual leave?
15. How much leave do workers currently receive? Is it enough?
16. Do workers receive a week's pay for each week of leave?

The purpose of this checklist is to gather together the information required to make a detailed assessment of compliance with the Working Time Regulations. In support of this endeavor the regulations also provide a sample health questionnaire for those about to embark on night work (or already engaged in it), which is reproduced below.

9.5.2 Health Questionnaire

The purpose of this questionnaire is to ensure that you are suited to working at night. All the information you provide will be kept confidential.

Type of work/duration of night work? ______________________

1. Surname ______________________
2. First and second names ______________________
3. Sex (M/F)
4. Date of birth ______________________
5. Permanent address ______________________
6. Job title ______________________
7. National insurance number ______________________
8. Department/clock number ______________________

Do you suffer from any of the following health conditions?

Diabetes (Yes/No)
Heart or circulatory disorders (Yes/No)
Stomach or intestinal disorders (Yes/No)
Any condition which causes difficulties sleeping (Yes/No)
Chronic chest disorders, especially if night-time symptoms are troublesome (Yes/No)
Any medical condition requiring medication to a strict timetable (Yes/No)
Any other health factors that might affect fitness at work (Yes/No)

If you have answered "yes" to any of the above, you may be asked to see a doctor or nurse for further assessment.

I, the undersigned, confirm that the above is correct to the best of my knowledge.

Signed ______________________________ Date __________________

Assessment
[This gives an indication of whether the worker is fit to work nights or should see a doctor or nurse for a medical examination]

Signed ______________________________ Date __________________

9.5.3 Shift Handovers

The Health and Safety Executive has developed a number of key principles for ensuring effective shift handovers:

1. Make shift handover a high priority.
2. Develop staff communication skills.
3. Provide procedures for shift handover.
4. Plan for safety-critical procedures to be completed within one shift.
5. Place greater reliance on written communication between handovers of 12-hour shifts.

Specifically, shift handover should be:

1. Conducted face to face
2. Two-way, with both participants taking joint responsibility
3. Done using both written and verbal communication
4. Based on an analysis of the information needs of incoming staff
5. Given as much time as necessary

These guidelines can be easily adapted to form a checklist when gathering data about current processes.

TABLE 9.1
Good Practice Guidelines for Shiftwork Schedule Design

Do:	Plan a workload that is appropriate to the length and timing of the shift.
	Rotate shifts either very quickly (every 2–3 days) or slowly (3–4 weeks).
	Limit night shifts to a maximum of 12 hours (including overtime) and consider the needs of vulnerable workers.
	Limit night shifts or shifts where work is demanding, monotonous, dangerous, or safety critical to 8 hours.
	Consider whether shifts of a variable length or flexible start and end times could offer a suitable compromise.
	Encourage and promote the benefit of regular breaks away from the workstation.
	Limit consecutive working days to a maximum of 5–7 days, and make sure rest time between successive shifts is adequate.
	Set a limit of 2–3 consecutive shifts where shifts are long (>8 hours), occur in the early morning, or take place at night.
	Allow workers a minimum of 2 nights of full sleep when switching from day to night shifts or *vice versa*.
	Build regular free weekends into the shift schedule.
Avoid:	Demanding, dangerous, monotonous, or safety-critical work during the night, early morning, toward the end of long shift, and during other periods of low alertness
	Placing workers on permanent night shifts
	Weekly or fortnightly rotating shifts schedules
	Split shifts, unless absolutely necessary to meet business needs
	Workers not using break time (and not taking appropriate breaks) to leave earlier
Where possible:	Schedule a variety of tasks to be completed during the shift, and allow workers some choice about the order in which they must be done.
	Offer workers a choice between permanent and rotating shift schedules.
	Adopt a forward-rotating schedule rather than a backward-rotating cycle.
	Try to avoid early morning starts before 7:00 a.m.
	Allow workers some discretion over when they take their breaks.

Source: Health and Safety Executive, *Managing Shift Work: Health and Safety Guidance*, HSE Books, London, 2006, p. 24. Reproduced under the terms of the Click-Use Licence.

9.6 ASSESSMENT AND EVALUATION*

Research into shiftworking now has a relatively long history, and human factors best practice today finds itself generally well embedded in the Health and Safety Executive's current guidance. A summary of this guidance is provided below in the next two sections, which highlight steps employers can take to demonstrate compliance with Health and Safety law and detail steps individuals can take in terms of their reciprocal duty of care. This guidance is supplemented with additional insights where appropriate.

9.6.1 Strategies for the Employer

As noted above, the Health and Safety at Work etc. Act (1974) requires employers to ensure the health, safety, and welfare of their employees, so far as is reasonably

* Adapted from Health and Safety Executive, *Managing Shift Work: Health and Safety Guidance*, HSE Books, London, 2006. With permission. Reproduced under the terms of the Click-Use Licence.

TABLE 9.2
Good Practice Guidelines for Improving the Shiftwork Environment

Do:	Provide facilities and opportunities for shiftworkers similar to those available for daytime workers (e.g., canteen, recreation).
	Ensure that workplace lighting is adequate and adjustable by workers.
	Ensure that the workplace temperature is adjustable and allows workers to carry out their tasks in reasonable comfort.
	Make sure that supervisors and team members with responsibility for shiftworking arrangements are aware of the risks associated with shiftwork and can recognize shiftwork-related problems.
	Control overtime and shift swapping by monitoring and recording hours worked and rest periods.
	Make provision in the work schedule to allow adequate rest for those workers carrying out standby/on-call duties or overtime.
	Provide training and information for workers, their families, and management on the risks associated with shiftwork and on coping strategies.
	Make provision to release staff for foreseeable training, development, and communication needs.
	Agree on and ensure that timing and procedures for transmitting information to the next shift team are followed at all times.
	Promote healthy living strategies such as increasing the amount of exercise and improving diet.
	Ensure that free health assessments are provided for night workers.
	Ensure that the workplace and its surroundings are well lit, safe, and secure.
Where possible:	Increase supervision during key periods of low alertness (e.g., during the night, early morning, toward the end of long shifts, and other periods of identified low alertness).
	Encourage interaction between workers and provide a means of communication for lone workers.
	Encourage workers to inform their doctors about their working arrangements (to help speed up diagnosis of shiftwork-related ailments).

Source: Health and Safety Executive, *Managing Shift Work: Health and Safety Guidance*, HSE Books, London, 2006, p. 25. Reproduced under the terms of the Click-Use Licence.

practical. Table 9.1 and Table 9.2 document some of the practical steps that can be taken to achieve this aim, all of which are based on the guidance published by the Health and Safety Executive (2006), albeit synthesized and tuned for the control room setting.

9.6.2 Strategies for the Employee

The employee has a reciprocal duty of care under health and safety legislation, and it is incumbent upon them to "take reasonable care of their own health and safety and that of other people who may be affected by their activities at work" (HSE, 2006, p 2). To this end, employees need to take positive steps in order to cope with the challenges of shiftwork. It is only possible to deal briefly with human factors best practice in this area, but some key advice follows.

9.6.2.1 Coping with the Shift and Managing Fatigue

- Develop self-awareness of known patterns of fatigue (for example, in the hours between 4:00 and 6:00 a.m.), as well as individual times when fatigue appears prevalent.
- Consider consuming small quantities of caffeine (coffee, tea, or cola drinks), but no later than 3 hours before the end of a shift (to prevent disrupted sleep).
- Take frequent breaks. If possible, increase the frequency of breaks at known fatigue times and leave the workstation.
- Take a proper lunch break halfway through the shift (as one would normally do during a day shift).
- In general, participate in social interaction, communication, and other such activities.
- Where opportunities exist to vary one's schedule, take advantage of these to avoid difficult or monotonous work between 4:00 and 6:00 a.m. (or other known fatigue-vulnerable times).

9.6.2.2 Traveling Home

Shiftwork not only increases fatigue generally but also puts the shiftworker on the road during potentially hazardous times. The general advice is to drive defensively or even to consider other modes of transport.

9.6.2.3 Sleep

Try to establish a suitable sleep schedule. Every individual is different, so it is sometimes useful to try going to bed at different times. When a successful pattern has been found, the advice is to try to stick to it. The sleep environment is important. A restful environment should be created; in addition to using heavy curtains, eye shades, earplugs, and so on, it is often advisable to have a decluttered bedroom that is not used for other activities (e.g., studying or storage). Following a normal bedtime routine (e.g., undressing, brushing teeth) can also be helpful in providing a set of cues that can prepare the individual for sleep. Other sleep-promoting techniques include engaging in relaxing activities prior to going to bed (a short walk, reading a book), avoiding caffeinated drinks, and not going to bed hungry.

9.6.2.4 Lifestyle

Dietary problems are a significant issue for shiftworkers. The general advice is to eat easily digestible foods (avoiding large heavy meals, highly spiced foods, and quick sugar fixes such as chocolate). Fruits, vegetables, and plenty of liquids provide a slower release of energy, and good hydration aids concentration. In general, adopting widely held current views about diet and eating is recommended. Exercise has two benefits for shiftworkers. Moderate exercise can increase alertness when performed before a shift (and before driving home). Light exercise can also aid relaxation and sleep prior to bedtime. On the social side, it is considered good advice, among other things, to (HSE, 2006, p. 30):

- Talk to friends and family about shiftwork. If they understand the problems you are facing, it will be easier for them to be supportive and considerate.
- Make your family and friends aware of your shift schedule so they can include you when planning social activities.

These simple steps can help to avoid isolation and conflict on a domestic level. They also constitute part of the advice and support that can be provided by the employer to the worker and to the worker's family.

In dealing with the strategies that employee and employer can deploy to mitigate risks and promote health and well-being, the reciprocal duty of care enshrined in the Health and Safety at Work etc. Act (1974) is given a practical manifestation. The procedures, processes, and materials developed to ensure that these practical steps are undertaken and managed are congruent with the Management of Health and Safety at Work Regulations (1999).

9.6.3 Additional Assessment Materials

Numerous tools and techniques are now available to diagnose various facets of shiftworking. Two fatigue-related indices, one objective and the other subjective, are briefly described below. A third technique is wider ranging and comprehensive and is also introduced.

9.6.3.1 Fatigue and Risk Index Tool

The fatigue index tool is designed to assess the level of fatigue experienced by shiftworkers based on a set of parameters that include commuting time, workload, breaks, and the shift rota. Apart from the subjective ratings of workload and attention, the fatigue index is based on objective data and thus allows different shiftworking arrangements and rotas to be easily compared. A worked example is shown in a following section. Figure 9.2 illustrates an easy-to-use, computer-based version of the tool freely available from the Health and Safety Executive (http://www.hse.gov.uk/research/rrhtm/rr446.htm). The shiftworking rota is entered into the appropriate columns, supported by a set of default data or shift-specific data (where available). It is then possible to switch between two modes. The risk index is a relative value where 1.0 maps onto the average risk of an accident or incident occurring on a day–day–night–night–rest–rest–rest schedule with 12-hour shifts. The fatigue index, on the other hand, relates to predicted values of sleepiness, with 100 being high levels of sleepiness, 0 low, and 20.7 corresponding to the shift pattern mentioned in the previous sentence. In Figure 9.2, the calculator is set to output the risk index merely for the sake of illustration. Both indices are calculated based on a number of objective and subjective fatigue and risk assumptions, for which the interface shown in Figure 9.3 is used. Detailed guidance on the use and interpretation of the indices is provided by the HSE, and it is strongly recommended to consult these sources when using the tool.

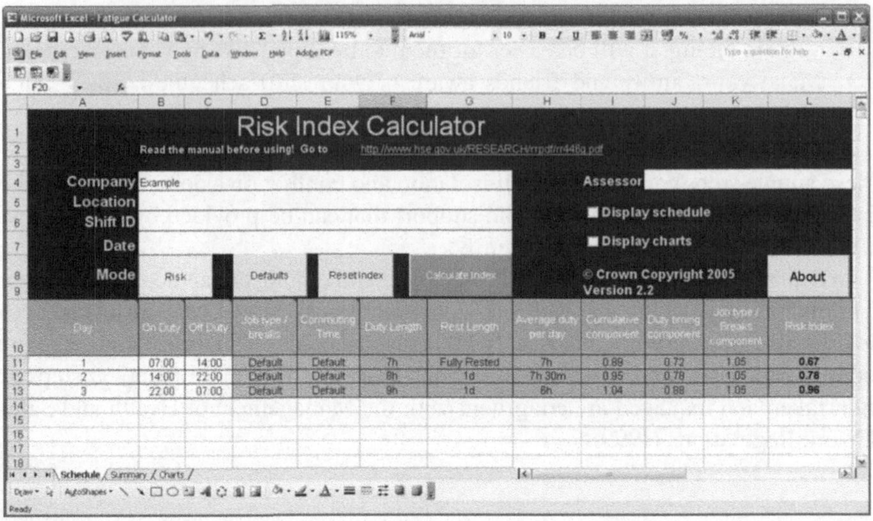

FIGURE 9.2 The fatigue index takes the form of an Excel® spreadsheet into which objective data about shift rotas can be added. (From Health and Safety Executive, *Fatigue Index Calculator: Version 2.2*, Health and Safety Executive, London, 2005. Reproduced under the terms of the Click-Use Licence.)

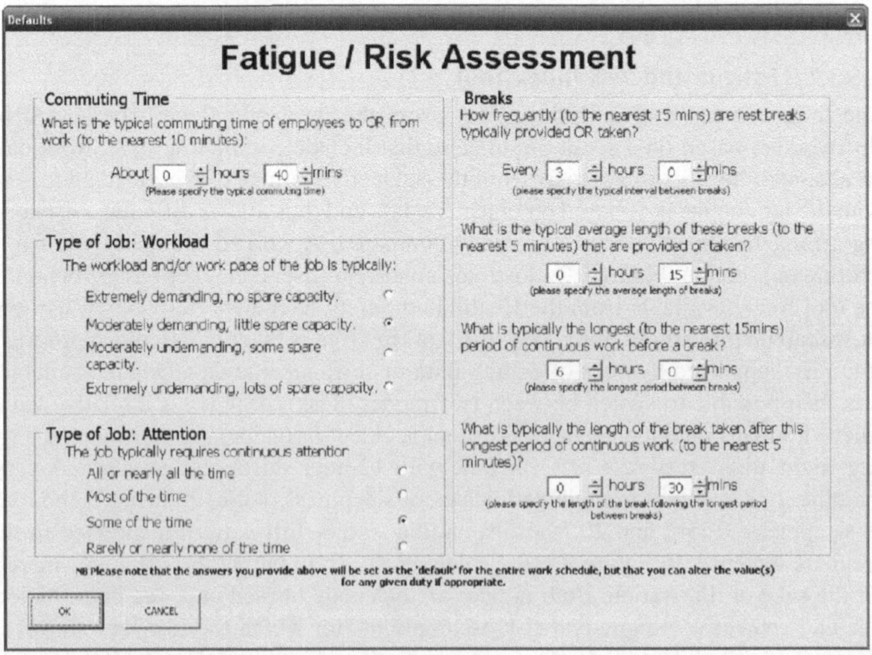

FIGURE 9.3 The fatigue index calculation is based on a range of objective and subjective values entered into an accompanying fatigue and risk assessment. (From Health and Safety Executive, *Fatigue Index Calculator: Version 2.2*, Health and Safety Executive, London, 2005. Reproduced under the terms of the Click-Use Licence.)

TABLE 9.3
Copy of the Epworth Sleepiness Scale

Rate how likely you are to doze off or fall asleep in the following situations.
Insert a 1 in the appropriate response column.

Example Activity	Would Never Doze	Slight Chance of Dozing	Moderate Chance of Dozing	High Chance of Dozing
Sitting and reading	___	___	___	___
Watching television	___	___	___	___
Sitting, inactive, in a public place (e.g., in a movie theater or a meeting)	___	___	___	___
Passenger in a car for an hour without a break	___	___	___	___
Lying down to rest in the afternoon (when circumstances permit)	___	___	___	___
Sitting and talking to someone	___	___	___	___
Sitting quietly after lunch (without alcohol)	___	___	___	___
In a car, while stopped for a few minutes in traffic	___	___	___	___
Raw score	___	___	___	___
Multiply the raw score by:	×0	×1	×2	×3
Column totals	___	___	___	___
Overall total (sum of column total)		___		

9.6.3.2 Epworth Sleepiness Scale

The subjective experience of fatigue can be further tested using the Epworth Sleepiness Scale. This is a simple self-report questionnaire that assesses shiftworkers' views and experiences of their shiftworking arrangements. A recommended strategy might be to use the fatigue index tool to narrow down a set of available shiftworking choices that can then be evaluated in practice and assessed with subjective tools such as this. A copy of the questionnaire is reproduced in Table 9.3.

9.6.3.3 The Standard Shiftwork Index

The Standard Shiftwork Index was developed in response to a perceived lack of standardization in the assessment of shiftwork. The solution is a set of scales that "reflect the most pertinent issues within shiftwork research … chosen on the basis of being both relatively short, easy to administer, and having good psychometric properties" (Barton et al., 1995). The scales are grouped into those that access data on outcomes (e.g., actual shiftwork problems), those that refer to modifiers (e.g., individual differences), and a set of general scales (e.g., shiftwork context).

9.6.3.4 Shift Handover

In terms of general guidance, it is usually the case that the sort of information to be passed between teams is well defined, and incoming and outgoing teams often undertake specific checking procedures. This is a minimum requirement. Human factors techniques such as the Critical Decision Method (CDM) (Klein and Armstrong, 2005) and hierarchical task analysis (HTA) (Annett and Stanton, 2000) are both good candidates for accessing such information requirements.

Aside from this explicit information exchange situation, there is also an implicit one to consider. In research performed by Stanton et al. (e.g., Stanton and Ashleigh, 2000; Stanton, Ashleigh, Roberts, and Xu, 2001) it was found that the overlapping shift handover period performed a valuable social and interactional function. Shift workers actually rather enjoyed aspects of the handover, and a great deal of continuous, redundant, recursive communication was found to take place above and beyond that required by set procedures. It is argued that this informal, implicit interaction and exchange play a similarly important role in developing team cohesion and situation awareness.

In terms of human factors best practice it can be broadly recommended that shift handovers should not necessarily default to an overtly streamlined act. The risk is that efficiencies gained in the explicit situation come at the cost of the equally valuable implicit situation. In general, human factors best practice would suggest ample opportunity for interaction and team overlap, actively supported to create the conditions for favorable implicit interaction and situational awareness.

9.7 EXAMPLE

A design and assessment can be provided based on the proposed shift rota for control room operators. Summer and winter shifts are proposed (see Table 9.4 and Table 9.5). The summer shift aims to provide the shiftworkers the maximum weekends off over the school holidays and enable staff to take holidays with their families. The proposed shift rota schedule for 12 weeks in the summer is comprised of 25 working days with 17 days of rest over 6 weeks. Over the 6 weeks of the shift cycle, the average working week is no more than 37 hours, including a 15-minute handover each shift. The proposed rota is thus nominally compliant with the Working Time Regulations in terms of a 48-hour work week maximum. The proposed winter shift rota schedule is comprised of 27 working days with 15 days of rest over 6 weeks. Over the 6 weeks of the shift cycle, the average working week is no more than 37 hours, including a 15-minute handover each shift. The proposed rota is thus nominally compliant with the Working Time Regulations in terms of a 48-hour work week maximum.

9.7.1 Best Practice Guidelines

A brief assessment can now be performed based on some of the more explicit guidance provided by the HSE (where these aspects are available to be assessed in the context of the present example) (see Table 9.6). A shift rota designed literally around these guidelines, ignoring any context-specific constraints/issues, might look like that shown in Table 9.7. It is a rapidly (forward) rotating schedule of 8-hour shifts,

TABLE 9.4
Proposed Shift Rota (Summer)

	Monday	Tuesday	Wednesday	Thursday	Friday	Saturday	Sunday
Week 1	Morning	Morning	Afternoon	Afternoon	Night	Night (12)	Night (12)
Week 2	Rest	Rest	Rest	Rest	Morning	Day (12)	Day (12)
Week 3	Afternoon	Afternoon	Night	Night	Rest	Rest	Rest
Week 4	Rest	Day	Morning	Morning	Afternoon	Rest	Rest
Week 5	Night	Night	Rest	Rest	Rest	Rest	Rest
Week 6	Day	Day	Day	Day	Day	Rest	Rest

Key: Morning = 0645–1400; afternoon = 1345–2100; night = 2045–0700; day = 0900–1700; night (12) = 1845–0700; day (12) = 0645–1900.

TABLE 9.5
Proposed Shift Rota (Winter)

	Monday	Tuesday	Wednesday	Thursday	Friday	Saturday	Sunday
Week 1	Morning	Morning	Afternoon	Afternoon	Night	Night	Night
Week 2	Rest	Rest	Rest	Rest	Morning	Morning	Morning
Week 3	Afternoon	Afternoon	Night	Night	Rest	Rest	Rest
Week 4	Rest	Day	Morning	Morning	Afternoon	Afternoon	Afternoon
Week 5	Night	Night	Rest	Rest	Rest	Rest	Rest
Week 6	Day	Day	Day	Day	Day	Rest	Rest

Key: Morning = 0645–1400; afternoon = 1345–2100; night = 2045–0700; day = 0900–1700.

TABLE 9.6
Compliance with Human Factors Best Practice for Shiftwork

Human Factors Best Practice	Compliance
Rotate shifts either very quickly (every 2–3 days) or slowly (3–4 weeks).	Yes
Limit night shifts to a maximum of 12 hours (including overtime) and consider the needs of vulnerable workers.	Yes (under 10 hours for winter; up to 12 hours in summer)
Limit night shifts or shifts where work is demanding, monotonous, dangerous, or safety critical to 8 hours.	Yes
Limit consecutive working days to a maximum of 5–7 days, and make sure rest time between successive shifts is adequate.	Yes
Set a limit of 2–3 consecutive shifts where shifts are long (>8 hours), occur in the early morning, or take place at night.	Yes
Allow workers a minimum of 2 nights of full sleep when switching from day to night shifts or *vice versa.*	Yes
Build regular free weekends into the shift schedule.	Yes
Avoid placing workers on permanent night shifts.	Yes
Avoid weekly or fortnightly rotating shift schedules.	Yes
Avoid split shifts unless absolutely necessary to meet business needs.	Yes
Adopt a forward-rotating schedule for rotating shifts rather than a backward-rotating cycle.	Yes
Try to avoid early morning starts before 7:00 a.m.	No (shift starts at 6:45 a.m.)

none of which begins before 7:00 in the morning; night shifts are followed by 2 days of recovery. Obviously, the guidelines could be interpreted in different ways to suit local conditions, but for present purposes this serves as a useful baseline for comparing the proposed rota.

9.7.2 Risk Assessment

An objective assessment of fatigue can now be undertaken using the HSE's fatigue and risk index tool. First, the following assumptions are made using the default values in the tool. The risk index results are shown in Figure 9.4 for the proposed summer and winter shift rotas together with the nominal ideal shift rota. Figure 9.4 presents the results of the risk index calculations carried out on the proposed rotas as well as the ideal rota (which interprets the guidance literally). The points on the lines correspond to the risk index value achieved on a given day of duty. The mean risk index value achieved was 0.91 for the summer rota and 0.91 for the winter rota, compared with 0.94 for the ideal rota, all of which are reassuringly close. At the level of individual duty days, it can be noted that the proposed rotas contain spikes that the ideal rota generally lacks. Having said that, even with a literal translation of the guidance there are points at which predicted risk rises above recommended levels. On the basis of this it would seem appropriate to further review the summer and winter rotas after a period of operation.

TABLE 9.7
Rota Based on Literal Interpretation of Guidance

	Monday	Tuesday	Wednesday	Thursday	Friday	Saturday	Sunday
Week 1	Morning	Morning	Afternoon	Afternoon	Rest	Night	Night
Week 2	Rest	Rest	Morning	Morning	Afternoon	Afternoon	Rest
Week 3	Night	Night	Rest	Rest	Morning	Morning	Afternoon
Week 4	Afternoon	Rest	Night	Night	Rest	Rest	Morning
Week 5	Morning	Afternoon	Afternoon	Rest	Night	Night	Rest
Week 6	Rest	Morning	Morning	Afternoon	Afternoon	Rest	Rest

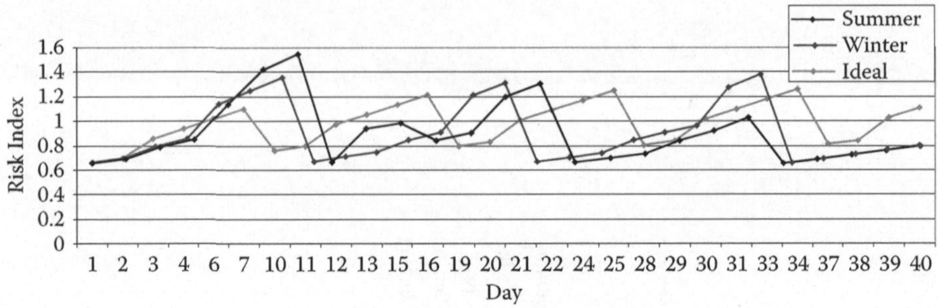

FIGURE 9.4 Graph of risk index values (*y*-axis) plotted over the 6-week period of the rota (*x*-axis).

9.7.3 Fatigue Assessment

Switching modes, it is now possible to model fatigue as shown in Figure 9.5. The mean fatigue index value achieved for the proposed rota is 14.7 for the summer rota and 13.5 for the winter rota compared with 12.86 for the ideal rota. Again, these means values are reassuringly close to each other. Differences do emerge at the level of individual days, however, and it can be seen that the proposed summer rota has one peak with a fatigue index in the region of 50 and two peaks that are in the region of a fatigue index value of 45. By comparison, the winter rota has three peaks with a fatigue index in the region of 45. The ideal rota, on the other hand, rarely exceeds 30. One general observation that can be made is that both of the proposed rotas are front loaded, with the more demanding shifts occurring near the beginning and more days of rest at the end. Although this arrangement may offer improved secondary benefits of shiftworking (e.g., more convenient holiday arrangements), it may increase risk and fatigue compared to more idealized complaint alternatives. It is a careful balance. The point of analyzing shiftwork in this way is to reach an optimal solution, maximizing secondary benefits without incurring unnecessary fatigue and risk problems while on duty. Human factors analyses such as these help to achieve this outcome.

9.7.4 Handover

The proposed shift rota allows 15 minutes for shift handover. Some sample data obtained indicate that during the summer months the shift handover requires in the region of 6 minutes (doubling to 12 minutes during the winter). Shiftworking has a two-pronged definition that relates to both out-of-hours working and the requirement for a shift handover. Control room operations, by their continuous nature, frequently rely on shiftwork. Shiftwork is, however, an unnatural way of working that without care can conflict with the normal cycle of wakefulness and rest to a dangerous degree for the individual and the system. Shiftworking arrangements come under the purview of the Working Time Regulations as well as more established Health and Safety Executive guidance. A comprehensive checklist approach allows current

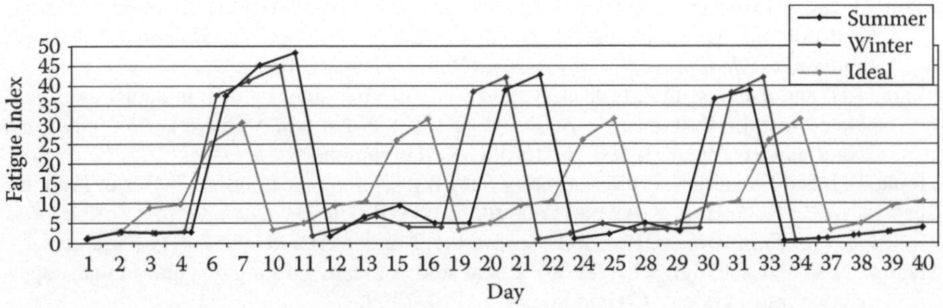

FIGURE 9.5 Graph of fatigue index values (*y*-axis) plotted over the 6-week period of the rota (*x*-axis).

arrangements to be assessed. Human factors best practice is well embedded in Health and Safety Executive guidance, and the use of a computerized risk and fatigue index tool serves as a way to model these factors prior to implementing a particular shift schedule that can increase insight and help to ensure compliance. Human factors best practice also reveals several more subjective techniques to complement this objective approach, both in terms of the time-period aspects of shiftworking and requirements for a safe and efficient handover. It is recommended that the health and well-being of shift staff be monitored and assessed after the new shift cycle has been operational for a period of time.

On the basis of the analysis undertaken on the shift pattern proposed by the organization, it was concluded that both the summer and winter shift cycles compared favorably with the notional ideal. In fact, the proposed shift pattern is more human-centered, as the summer and winter rotas offer four and three out of six weekends off, respectively, which is likely to be preferred by people with families. The time allocated for shift handover was assessed as satisfactory on the basis of observational data from a previous study.

BIBLIOGRAPHY

Annett, J. and Stanton, N.A. (2000). *Task Analysis*. London: Taylor & Francis.

Barton, J., Spelten, E., Totterdell, T., Smith, L., Folkard, S., and Giovanni, C. (1995). The standard shiftwork index: a battery of questionnaires for assessing shiftwork-related problems, *Work and Stress*, 9(1), 4–30.

Folkard, S., Dirkx, J., Harma, M., Knauth, P., Koller, M., Smith, P.A., and Wedderburn, A. (1993). Night and shiftwork. *Ergonomics*, 36(1–3), 1–319.

Gordon, N.P., Cleary, P.D., Parker, C.E., and Czeisler, C.A. (1986). The prevalence and health impact of shiftwork. *American Journal of Public Health*, 76, 1225–1228.

Health and Safety Executive (HSE). (2005). *Fatigue Index Calculator: Version 2.2*. London: Health and Safety Executive (http://www.hse.gov.uk/research/rrpdf/rr446cal.xls).

Health and Safety Executive (HSE). (2006). *Managing Shift Work: Health and Safety Guidance*. London: HSE Books.

Johns, M.W. (1991). A new method for measuring daytime sleepiness: the Epworth sleepiness scale. *Sleep*, 14(6), 540–545.

Klein, G. and Armstrong, A.A. (2005). Critical decision method. In N.A. Stanton, A. Hedge, K. Brookhuis, E. Salas, and H. Hendrick (Eds.), *Handbook of Human Factors and Ergonomics Methods* (58-1–58-6). Boca Raton, FL: CRC Press.

Monk, T.H. and Embrey, D.E. (1981). A field study of circadian rhythms in actual and interpolated task performance. In A. Reinberg (Ed.), *Night and Shiftwork: Biological and Social Aspects* (pp. 473–480). Oxford, U.K.: Pergamon.

Monk, T.H. and Folkard, S. (1992). *Making Shift Work Tolerable*. London: Taylor & Francis.

Spencer, M.B., Robertson, K.A., and Folkard, S. (2006). *The Development of a Fatigue/Risk Index for Shiftworkers*, Research Report 446. London: Health and Safety Executive.

Stanton, N.A. and Ashleigh, M.J. (2000) A field study of team working in a new human supervisory control system. *Ergonomics*, 43, 1190–1209.

Stanton, N.A., Ashleigh, M.J., Roberts, A.D., and Xu, F. (2001). Testing Hollnagel's contextual control model: assessing team behaviour in a human supervisory control task. *International Journal of Cognitive Ergonomics*, 5(1), 21–33.

10 Control Room Layout

10.1 OUTLINE

Control room layout is a significant factor in efficient system control. The job of the operator can be extremely demanding, and well-designed rooms and workstations can increase the physical comfort of the operator; reduce stress levels, and improve efficiency. Control rooms can be described as complex sociotechnical systems. The room represents a domain for significant levels of interaction, such as computer–plant, human–computer, and human–human interactions. Depending on the control room, certain levels of direct manipulation between humans and plant may also be possible. ISO 11064-3:2001 points out the differences between the terms *control room*, *control suite*, and *control center*:

- A control room is a core functional entity, along with its associated physical structure, where operators are stationed to carry out centralized control, monitoring, and administrative responsibilities (ISO 11064-3:1999, definition 3.4).
- A control suite is a group of functionally related rooms, colocated with the control room and including it, which houses the supporting functions to the control room, such as related offices, equipment rooms, rest areas, and training rooms (ISO 11064-3:1999, definition 3.6).
- A control center is a combination of control rooms, control suites, and local control stations which are functionally related and all on the same site (ISO 11064-3:1999, definition 3.1).

Essentially, a control center is comprised of one or more control suites which in turn are comprised of one or more control rooms which are comprised of one or more control workstations. This relationship can be seen graphically in Figure 10.2.

10.2 KEY ISSUES

The key issues involved in control room layout are summarized in Figure 10.1. Control room design requires a number of factors to be considered. At a high level, we can list some of these constraints shaping the layout of the room:

- Architectural considerations, such as the location of entrances and exits, windows (which can lead to glare), personal safety features such as guards and handrails, and areas considered for future expansion

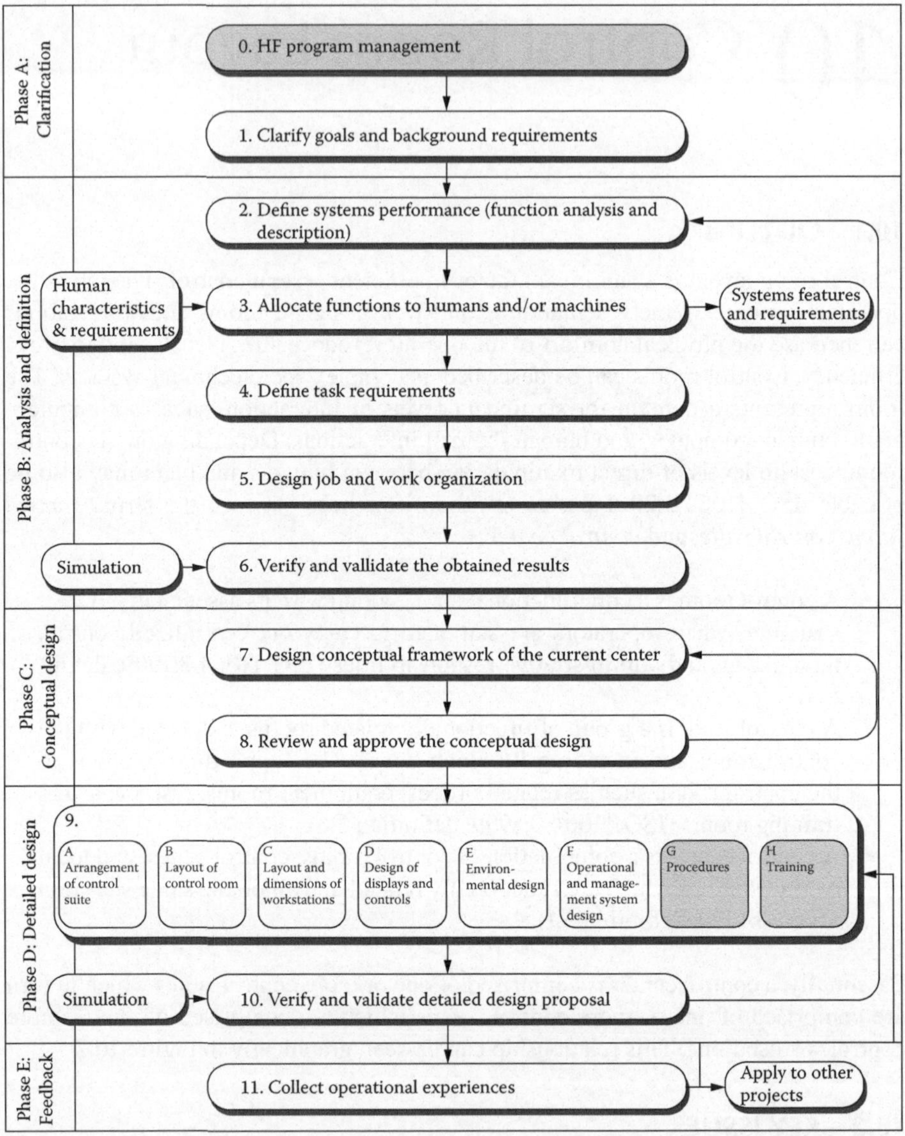

FIGURE 10.1 Human factors steps in control center design. (Adapted from ISO 11064-1:2001 and incorporating additions shown in gray from Green and Collier, 2001.)

- Operational considerations, such as the tasks that will be conducted within the room and how the layout impacts these, the required levels of team working and how they influence the layout, the need for supervision, the possible requirement of display sharing
- Anthropometrics and biometrics of the operator population

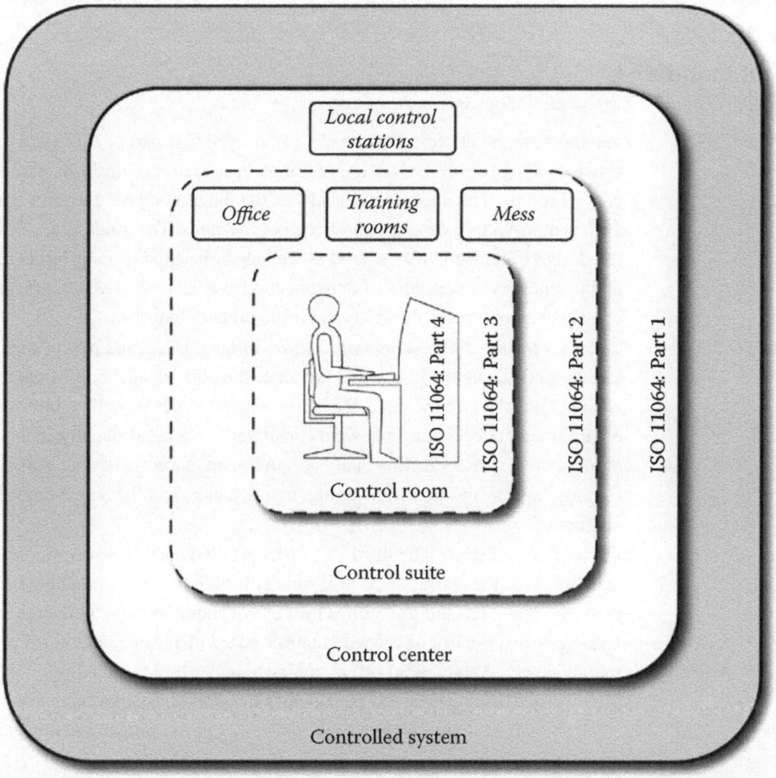

FIGURE 10.2 Diagram showing the scope of the relevant parts of ISO 11064.

Workstations within the room should be arranged so as to consider these factors. Also, access has to be provided for maintenance technicians and cleaning staff which can have significant implications in 24-hour operations. Particular attention also must be paid to the access of physically disadvantaged workers and visitors.

10.3 STANDARDS AND GUIDELINES

10.3.1 Relevant Standards

Many of the elements involved in control room design are captured over a range of ISO standards. Perhaps the most pertinent of these is ISO 11064. The standard is comprised of eight parts. The scope of the first four parts is shown in Figure 10.2. Beginning with Part 1, the standards address the design of control system in increasing levels of detail. Part 1 covers the entire control center made up of control rooms, control suites, and local control stations. Part 2 covers the design of a control suite made up of the control room and the supporting functions, such as related offices, equipment rooms, rest areas, and training rooms. Part 3 focuses on the actual control room and its associated physical structure, where operators are stationed to carry out

TABLE 10.1
Relevant Standards

ISO 11064	*Ergonomic Design of Control Centers*
ISO 11064-1:2001	*Principles for the Design of Control Centers.* The first part of ISO 11064 provides a high-level description of the process required for the design of control rooms. The document introduces the diagram reproduced in Figure 10.1 and provides a description of each of the steps. The guide is applicable to the design of new control centers, as well as in the expansion, refurbishment, and technological upgrades of existing control centers. It covers all types of control centers typically employed for the process industries.
ISO 11064-2:2001	*Principles for the Arrangement of Control Suites.* The second part of this standard considers the geospatial layout of the control suite and its composite rooms. The standard indicates that control suite layout should be informed by environmental conditions, physical constraints, task analysis, physical ergonomics, communication analysis, equipment access analysis, traffic and routing, ingress and egress, cleaning, maintenance, and the accommodation of visitors.
ISO 11064-3:1999	*Control Room Layout.* The third part of ISO 11064 establishes ergonomic principles for the layout of control rooms. It includes requirements, recommendations, and guidelines for control room layouts, workstation arrangements, the use of off-workstation visual displays, and control room maintenance. The standard offers guidance under topics including architectural and operational considerations, workstation arrangement, off-workstation shared displays, circulation of personnel, and maintenance access.
ISO 11064-4:2004	*Layout and Dimensions of Workstations.* The fourth part of ISO 11064 specifies ergonomic principles, recommendations, and requirements for the design of workstations found in control centers, with particular emphasis on layout and dimensions. This standard covers primarily seated, visual-display-based workstations, although sit/stand workstations are also addressed. The standard covers anthropometry, user requirements, posture, visual tasks, auditory tasks, and workstation layout.
ISO 11064-5	*Displays and Controls.* The fifth part of ISO 11064 is still under development.
ISO 11064-6:2005	*Environmental Requirements for Control Rooms.* The sixth part of ISO 11064 gives environmental requirements as well as recommendations for the ergonomic design, upgrading, or refurbishment of control rooms and other functional areas within the control suite. The following aspects are covered: thermal environment (temperate regions), air quality, lighting environment, acoustic environment, vibration, aesthetics, and interior design.

centralized control, monitoring, and administrative responsibilities. Part 4 addresses the design of the physical workstation. Parts 3 and 4 are the most relevant when designing a control room; however, the remaining parts of this standard also provide essential reading. Table 10.1 provides a list of some of the relevant standards available along with a short description of each.

TABLE 10.1 (continued)
Relevant Standards

ISO 11064-7:2006	*Principles for the Evaluation of Control Centers*. The seventh part of ISO 11064 establishes ergonomic requirements, recommendations, and guidelines for the evaluation of control centers. User requirements are a central theme of this part of ISO 11064, and the processes described are designed to take account of the needs of users at all stages.
ISO 11064-8	*Ergonomic Requirements for Specific Applications*. The eighth part of ISO 11064 is still under development
ISO 13407:1999	*Human-Centered Design Processes for Interactive Systems*. This standard covers human-centered design activities throughout the lifecycle of computer-based interactive systems. It is aimed at those managing design processes and provides guidance on sources of information and standards relevant to the human-centered approach. This standard is concerned with both hardware and software components of interactive systems.
ISO 13406-2:2002	*Ergonomic Requirements for Work with Visual Displays Based on Flat Panels*. This part of ISO 13406 establishes ergonomic image-quality requirements for the design and evaluation of flat-panel displays. It defines terms needed to address image quality on flat-panel displays and specifies methods of determining image. The standard establishes ergonomic principles for guiding these requirements.
ISO 9241-1:1997	*Ergonomic Requirements for Office Work with Visual Display Terminals (VDTs), Part 1: General Introduction*. This standard introduces a wider range of standards covering generic issues involved with VDTs.
ISO 9241-3:1992	*Ergonomic Requirements for Office Work with Visual Display Terminals (VDTs), Part 3: Visual Display Requirements*. This standard covers the requirements of the visual display. It covers the design of displays along with user performance tests.
ISO 9241-5:1998	*Ergonomic Requirements for Office Work with Visual Display Terminals (VDTs), Part 5: Workstation Layout and Postural Requirements*. This standard provides a detailed description of factors influencing workstation layout and operator posture. The standard offers mainly qualitative recommendations for design. A national annex is also included that covers some of the necessary quantitative values for the design of workstations.
BS/EN 1335-1:2000	*Office Furniture–Office Work Chair, Part 1: Dimensions—Determination of Dimensions*. This standard details the test requirement to determine the compliance of an office chair with the dimensions outlined. The standard provides quantitative dimensions for the office chair. Guidance is given on the level of adjustment required.

10.4 HUMAN FACTORS LITERATURE

As the previous section shows, the issue of control room design is already extremely well documented in international standards. A wide range of human factors research applies to this area both in the physical and the cognitive domains. The physical aspects of ergonomics are generally more focused on the layout of the workstation, ensuring

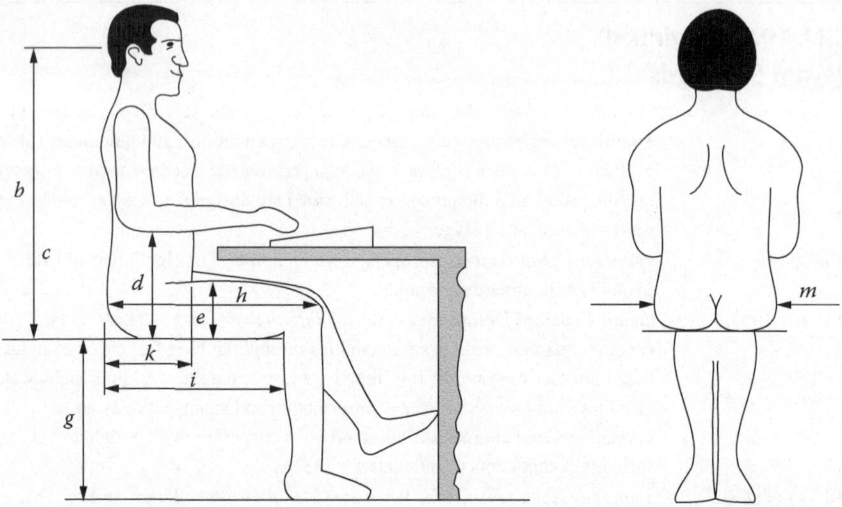

FIGURE 10.3 Dimensioning of a workstation. (Permission to reproduce figures from ISO 9241-5:1999 is granted by the British Standards Institution.)

correct levels of comfort, reach, and viewing angles (ISO 11064-3:1999; see also Corlett and Clark, 1995; Das, 2006; Lehto and Buck, 2008; Pheasant and Haslegrave, 2006). The cognitive aspects of human factors focus on the ability of the operator to acquire information, process it, and respond in a meaningful way when interacting with the system (primarily the display and controls within the workstation).

10.4.1 Dimensions of the Workstation

According to ISO-11064-4, the layout and dimensioning of control workstations should be governed by the anthropometric dimensions of the users and any requirements for movement to accomplish their tasks. Anthropometric data are usually given in terms of percentiles. ISO-11064-4 advocates accommodating the design to fit from the 5th to the 95th percentile of the population. In the available literature (Corlett and Clark, 1995; Pheasant and Haslegrave, 2006), percentile descriptions of the population are normally broken down by gender. This chapter will consider 5th percentile females to 95th percentile males. Data for the adult British population (ages 19 to 65) are taken from Pheasant and Haslegrave (2006). Care should be taken when extending this to other populations outside of the United Kingdom. Noticeable examples include Dutch men who are taller and Sri Lankan workers who are considerably smaller. By taking a range from the 5th to the 95th percentile, it should be considered that in any one body measurement 10% of the population are not actively designed for. Due to nonproportionality, this means that greater than 10% of the population may not be fully accommodated in a workstation design based around this range. Figure 10.3 illustrates a number of key dimensions required for the design of a workstation. The accompanying dimensions (for the adult British population) are included in Table 10.2.

TABLE 10.2
Key Dimensions of the Adult British Population

	Dimension	5th Percentile Female (mm)	95th Percentile Female (mm)	5th Percentile Male (mm)	95th Percentile Male (mm)
b	Eye height, sitting	685	795	735	845
c	Shoulder height, sitting	505	610	540	645
d	Elbow height, sitting	185	280	195	295
e	Thigh clearance, sitting	125	180	135	185
g	Popliteal height, sitting	355	445	395	490
h	Buttock knee length	520	620	540	645
i	Buttock popliteal length	435	530	440	550
k	Buttock abdomen depth, sitting	205	305	220	325
m	Hip breadth	310	435	310	405

Source: Pheasant, S. and Haslegrave, C.M., *Anthropometry, Ergonomics, and the Design of Work*, 3rd ed., Taylor & Francis, London, 2006. With permission.

Adjustability should be considered for those workstation-related dimensions which do not accommodate the 5th to 95th percentile users. This might be achieved by adjustable desk heights, seat heights, floor heights, viewing distances, or the orientation of displays. Great care should be taken when adopting recommendations from the literature on appropriate dimension for workstations. There is significant discrepancy between recommendations in the literature and ISO standards. The workstation and its components should be considered together so their combined adjustment range provides the greatest flexibility.

10.4.1.1 The Seat

The 1992 U.K. Health and Safety Regulations (OPSI, 1992) outline the minimum requirements for a chair used in conjunction with display screen equipment:

- The work chair shall be stable and allow the operator or user easy freedom of movement and a comfortable position.
- The seat shall be adjustable in height.
- The seat back shall be adjustable in both height and tilt.
- A footrest shall be made available to any operator or user who wishes one.

Figure 10.4 illustrates how measurements for office work chairs are to be taken, and the dimensions specified by EN 1335-1:2000 are shown in Table 10.3.

10.4.1.2 Seat Height

According to Pheasant and Haslegrave (2006), the optimum seat height is close to the user's popliteal height (measurement *g* in Figure 10.3) while wearing shoes (an additional 30 mm should be added to this value). Seat heights greater than the user's popliteal height will result in discomfort due to pressure on the underside of the thighs. Seat heights lower than the popliteal height will result in an acute angle between the thighs and trunk, resulting in a requirement for more leg room and the operator experiencing increased difficulty sitting down and getting up. Based on the British population (as specified in Table 10.2), to accommodate the 5th percentile female to the 95th percentile male, the chair height adjustment range should be 385 mm (355 + 30) to 520 mm (490 + 30). EN 1335-1:2000 specifies a range between 400 and 510 mm (accommodating from the 13th percentile female to the 91st percentile male). Users with shorter popliteal lengths, therefore, will require footrests to obtain an optimum seating position with EN 1335-1:2000-compliant chairs.

10.4.1.3 Seat Depth

If the seat depth (length) is increased beyond the buttock–popliteal length (measurement *i* in Figure 10.3), the user will be unable to engage with the backrest correctly. Seat length should always be compromised over appropriate back support. Deeper seats can also cause problems with standing up and sitting down. According to Pheasant and Haslegrave (2006), it is difficult to define a lower limit for seat depth; as little as 300 mm will still support the ischial tuberosities and may well be satisfactory in some circumstances. Based on the 5th percentile female buttock popliteal length (see Table 10.2), a seat depth should be provided that does not exceed 435

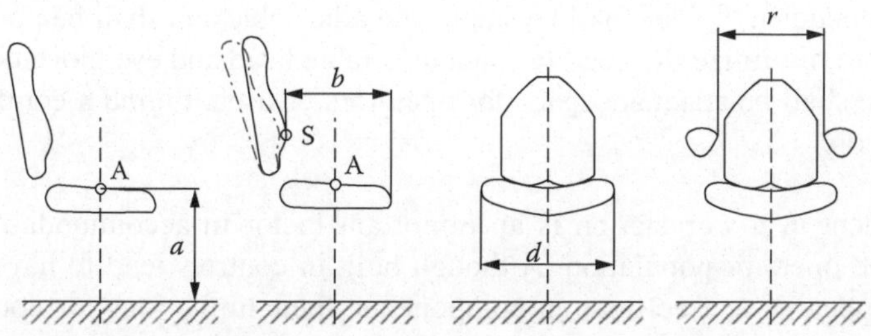

FIGURE 10.4 Measurement points for office chairs. (Permission to reproduce figure from EN 1335-1:2000 is granted by the British Standards Institution.)

TABLE 10.3
Dimensions for a Type A Office Chair

	Dimension	Minimum (mm)	Maximum (mm)
a	Seat height	400	510
b	Seat depth	400	420
d	Seat width	400	—
r	Clearance between armrests	460	510

Source: Permission to reproduce table from EN 1335-1:2000 is granted by the British Standards Institution.

mm. This recommendation is reflected in EN 1335-1:2000, which stipulates a range of 400 to 420 mm. It should be pointed out that, although adjustment is beneficial, it is not necessary from an accommodation perspective for seat depth.

10.4.1.4 Seat Width

According to Pheasant and Haslegrave (2006), in terms of support, a seat width that is 25 mm less on either side than the maximum hip breadth is all that is required (435 – 25 – 25 = 385 mm). This recommendation is compatible with EN 1335-1:2000, which recommends a minimum width of 400 mm

10.4.1.5 Lateral Clearance between Armrests

Pheasant and Haslegrave (2006) suggested that armrests should accommodate the largest hip width in the population (95th percentile females, 435 mm) with enough clearance for clothes and comfort. They offered a guidance value of 500 mm, which correlates with the specification in EN 1335-1:2000, ranging from 460 to 510 mm.

10.4.2 Workstation Desk Design

The 1992 U.K. Health and Safety Regulations outline the minimum requirements for a workstation used in conjunction with display screen equipment:

- The work desk or work surface shall have a sufficiently large, low-reflectance surface and allow a flexible arrangement of the screen, keyboard, documents, and related equipment.

- The document holder shall be stable and adjustable and shall be positioned so as to minimize the need for uncomfortable head and eye movements.
- There shall be adequate space for operators or users to find a comfortable position.

Adjustment in a workstation is an important factor in accommodating a wide range of the operator population. Although built-in controls tend to have a greater aesthetic appeal over free-floating components, their limitations for configuration mean that they are less adept at fitting the entire workforce. It is important that operators be able to adjust the mouse or trackerball for right- or left-hand use, and the position of the keyboard should be changeable, as well as the height, angle, and position of the screens. Fortunately, most freestanding table-top monitors come fitted with a wide range of adjustments. Monitor procurement decisions should be based not only on the screen size, quality, and resolution, but also on the level of adjustment within the monitor stand.

10.4.2.1 Desk Height

The operator is required to have access to three points when using the workstation: the floor, the chair, and the desk surface and keyboard. According to Pheasant and Haslegrave (2006), to accommodate the population two of these three components must be adjustable. Through height adjustment in the seat and through the addition of a footrest it is possible to manipulate the floor and sitting height. It may also be beneficial to adjust the height of the desk. For keyboard work, the home row of keys (30 to 50 mm above the desk surface) should be at sitting elbow height. For desks without adjustment, Pheasant and Haslegrave (2006) advocate a fixed desk height of 720 mm. This agrees with ISO 9241-5:1999, which recommends a height of 720 ± 15 mm. ISO 9241-5:1999 also points out that to accommodate smaller users the use of footrests and seats in high positions is necessary. For the required adjustment range, there is a general consensus for a range from 635 to 770 mm (see Table 10.4). For females with 5th percentile popliteal length of 355 mm, 5th percentile sitting elbow height of 185 mm, and a shoe height of 30 mm, the ideal desk height is 570 mm. In order to accommodate these users, it is essential that footrests be provided.

10.4.2.2 Leg Room

Pheasant and Haslegrave (2006) observed that when in the standard sitting position the distance between the underside of the elbows and the tops of thighs is on average only around 80 to 85 mm and may be a good deal less (see Figure 10.5). The thickness of the keypad is typically between 30 and 40 mm, so the desktop thickness should be kept to an absolute minimum. The adjustment range is based on the seat height allowing enough clearance for the thighs under the desk. Based on the male 95th percentile values for thigh thickness (185 mm), popliteal length (790 mm), and shoe thickness (30 mm), the minimum underside clearance for an optimal seating posture should be 705 mm; allowing for clearance to allow movement, this should be extended to 750 mm (underside). Figure 10.5 shows the recommended ISO 9241-5:1999 profile for minimum leg room. It should be noted that this profile does not allow taller elements of the population to assume the optimum seating position.

TABLE 10.4
Guidelines for Work Surface Height

Source	Minimum (mm)	Maximum (mm)
Pheasant and Haslegrave (2006)	650	750
ISO 9241-5:1999	660	770
Corlett and Clark (1995)	635	760

10.4.2.3 Size of the Work Surface

The work surfaces must be large enough to include the monitor as well as any input devices such as keyboards and mice. The work surface should be deep enough to allow for optimal eye-to-monitor distance plus the depth of the monitor. There seems to be consensus that the absolute minimum viewing distance is 500 mm; larger viewing distances improve depth of focus and minimize eye strain. For this reason, ISO 11064-4:2004 and Pheasant and Haslegrave (2006) have recommended distances of 700 and 750 mm, respectively. Pheasant and Haslegrave (2006) proposed that flexibility should be included to increase the distance to 1000 mm. ISO 9241-5:1999 states that the optimum viewing distance is 600 ± 150 mm.

The size of the desk is determined by the horizontal distance between the user and the display; this is a factor of the eye-to-monitor distance and the viewing angle. There is some controversy over the most applicable screen height and viewing angle.

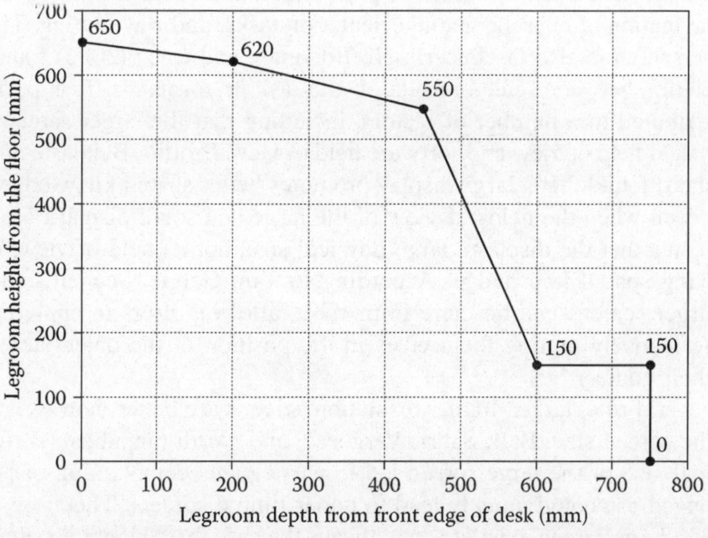

FIGURE 10.5 Legroom profile. (Permission to reproduce figure from ISO 9241-5:1999 is granted by the British Standards Institution.)

These values range from 0 to 40° below the horizontal; ISO 9241-5:1999 states that the viewing angle should not exceed 40°. To allow for this range it is imperative for the monitor to be adjustable. ISO 9241-5:1999 states that users should be able to angle, tilt, or swivel the display, and height adjustment is also preferred.

10.4.3 Display Screens

The 1992 U.K. Health and Safety Regulations outline the minimum requirements for display screen equipment:

- The characters on the screen shall be well defined and clearly formed, of adequate size, and with adequate spacing between the characters and lines.
- The image on the screen should be stable, with no flickering or other forms of instability.
- The brightness and the contrast between the characters and the background shall be easily adjustable by the operator or user and also be easily adjustable to ambient conditions.
- The screen must swivel and tilt easily and freely to suit the needs of the operator or user.
- It shall be possible to use a separate base for the screen or an adjustable table.
- The screen shall be free of reflective glare and reflections liable to cause discomfort to the operator or user.

The decision whether to use a small or large screen depends largely on the type of task being performed. Tasks that benefit from a larger display are those that also benefit from an egocentric approach or good situational awareness. Egocentric tasks include navigation of oneself, spatial orientation tasks, and way finding. The resolution of the screen is also of concern; Ni, Bowman, and Chen (2006) found a positive correlation between screen resolution and task performance. Task performance can be attributed to a number of factors, including that the larger screen allows a wider physical field of view and software field of view. Proffitt, Bakdash, and Banton (unpublished) found that a large display promotes better spatial knowledge of target locations even when the fields of view of the large and small displays are equated. This indicates that the display's large physical size, not its field of view, is critical to improving spatial knowledge. According to Tan, Gergle, Scupelli, and Pausch (2003), larger screens can be more immersive, allowing users to apply egocentric rotation (effectively putting themselves in the position of the object and mentally rotating their bodies).

As a general rule, larger, higher resolution screens are better; however, there is a limit to the screen size. Ball, Sabri, Varghese, and North (unpublished) found that the physical size of the large screen led to participants complaining of neck pain, and prolonged use could quickly lead to upper limb disorders. Therefore, although benefits of a large screen exist, it is not always the case that "bigger is better" or that the design of a smaller display can just be scaled up for use on a larger display. The situation, period of use, and ergonomic anthropometric needs must be considered carefully. Tasks that use an exocentric approach are not affected by smaller displays; reading comprehension tasks do not benefit from a larger display, and effort should be directed toward ensuring that the font height is correct for the distance the display is been viewed from.

10.4.4 Number of Displays

Research by Ball et al. (unpublished) looked at the difference between using a single 17-inch display, four 17-inch screens (2 × 2), and nine 17-inch screens (3 × 3) and the effect they had on winning a game. The game involved three players, one on each screen, battling against each other. The results showed that the reduction in navigation and the overview offered by the four- and nine-tiled displays increased the likelihood of winning a war game. EEMUA 201:2002 offers the following points for determining the number of screens required:

- How many overview displays are required?
- What alarm display philosophy is to be taken?
- How much simultaneous control action may have to be taken?
- How much general monitoring is required?
- What level of multiple windowing is to be allowed?
- What level of redundancy is considered adequate?
- Can screens be used for more than one process, or module?
- What future expansion may be required?

Factors affecting the arrangement of the screens and the layout of the room include:

- *Complexity of the plant*—This will have an impact on the amount of data that must be represented both for monitoring and for managing abnormal situation.
- *Manning levels*—The number of operators in the control room will have obvious implications on seating and room layout. The level of required communication between operators also has an impact on the way in which they should be arranged. If activities are shared, the room may have to be arranged to allow workers to share a single screen. The system must also protect against confusion about who is doing what.

Backup power supplies are often essential. It is important to also consider the implications of hardware or software failure. Redundancy should be built into any system. It is also prudent to consider a specific emergency shutdown device (ESD). No single failure of a component within the human–computer interaction (HCI) should cause the failure of the HCI.

10.4.5 Other Aspects Affecting Room Layout

10.4.5.1 Layout Analysis

Layout analysis has been proposed by Stanton and Young (1999) and Stanton, Salmon, Walker, Baber, and Jenkins (2005) as an approach to evaluate existing designs and suggest improvements based on functional groupings. The theory behind the approach is that the layout should mirror the user's structure of the task. The analysis starts with a schematic of the room and the components within it. These components are then clustered into functional groups. The functional groups are then arranged in the available space by importance, with the most important functional groups taking priority. The functional groups are then considered against sequence of use. The groups are finally arranged by frequency of use. Although link analysis serves as a good starting point for layout design, it assumes sequence is a more important consideration than frequency and that functionality is a more important consideration than sequence. A strong emphasis is placed on functional grouping, which may not always be the overriding consideration. Other factors such as environmental constraints (see Chapter 13), required supervision (see Chapter 8), social constraints, evacuation procedures, and maintenance may also have to be taken into consideration.

10.4.5.2 Social Organization

Communication among operators is often an essential part of a control room. Operators need to speak to each other as well as to the supervisor. The effect of avoidable distractions on performance also needs to be considered. During their observations of centralized control rooms, where each console had an alarm system to alert operators to significant process events, Cochran and Bullemer (1999) noted that in certain configurations operators were distracted by each other's alarm annunciations. In some cases, operators can even be confused as to whose alarm system is annunciating. The console operator can also be distracted by consultations going on at another console. Cochran and Bullemer (1999) discussed potential problems resulting from the distraction of other console users within the control room due to chatting to field operators or maintenance personnel. An effective balance needs to be struck between encouraging collaboration among operators and reducing the number of distractions. An indication of the required communication within the control room can be informed from task analysis during different activity scenarios.

10.4.5.3 Access

Consideration needs to be given to the different user groups of the room and their individual requirements (operators, maintenance staff, computer staff, managers, cleaners, and visitors). If the control room is in 24-hour use this problem is exacerbated, as maintenance and cleaning will have to be conducted while the operators continue to conduct their tasks.

10.4.5.4 Auxiliary Functions

Additional functions may also be required in or close to the control room, including meeting rooms, office facilities (e.g., PCs, printers, photocopiers), mess rooms, and restrooms. It may also be beneficial for safety- and performance-critical information and alarms to be displayed in some or all of these rooms.

10.4.5.5 Safety Considerations

Safety considerations influencing room layout include Health and Safety Executive regulations, fire prevention, fire control, and evacuation.

10.4.5.6 Flexibility

Flexibility is essential to ensure the long-term success of a control room. An approximation can be made of the optimum layout using task analysis; however, the way that a task is completed will be heavily influenced by the environment. This means that as a room is changed the behavior of the operators is also likely to change. This emergent behavior may result in subsequent iterative layout changes.

10.5 GUIDANCE AND RECOMMENDATIONS

10.5.1 Methods

10.5.1.1 Layout of the Room

Although the formal design of the control room should be conducted by an architect, it is possible to perform a rudimentary evaluation or feasibility study based on the required contents of the room along with an understanding of expected traffic flow. The design of the room should always be driven by the requirements of the operators rather than the physical constraints of the room. In most cases, the first step is to determine the required size of the operator's workstation (see Section 10.5.1.2) which is in turn informed by the Supervisory Control and Data Acquisition (SCADA) system housed on the workstation. Next, the number of required operators should be considered based on the workload (see Chapter 6). The arrangement of these workstations will be dependent on the required level of communication (see Chapter 4) as well as the required level of supervision (see Chapter 8). Other important factors to consider include the orientation of displays due to environmental considerations such as windows and the potential confusion that could arise from operators determining the ownership for audible alarms.

10.5.1.2 The Physical Layout of the Workstation

The physical layout of the workstation should be considered in its totality. The adjustment levels should be suitable such that any operator within the 5th to 95th percentile range can assume a correct posture with contact to the floor, seat, and work surface. Often the best approach is to start with a model of the extremities of the population, modeling their ideal position from the ground up. The seat height should be determined from the popliteal height (while wearing shoes). The keyboard height (desk height + 30 mm) can be determined by adding the sitting elbow height to the seat height. By modeling the extremities of the population being designed for, the adjustment range can be determined. With the addition of an adjustable footrest it is possible to adjust one of the three factors (seat height, work surface or keyboard height, or footrest or floor height). If possible, it is also worth adding additional adjustment to accommodate those outside of the 5th to 95th percentile range.

10.6 ASSESSMENT AND EVALUATION

Under the U.K.'s Health and Safety (Display Screen Equipment) Regulations of 1992, it is a legal requirement for the employer to perform a suitable and sufficient analysis of user workstations. An ergonomic evaluation will be based on an existing

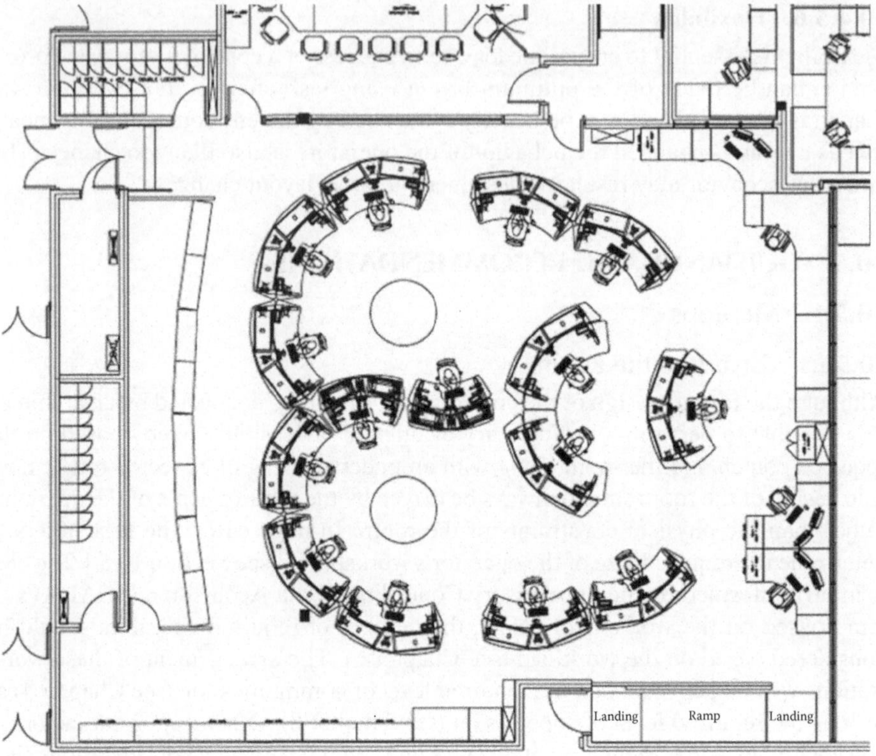

FIGURE 10.6 Layout of control room.

workstation shown schematically in Figure 10.6 and pictured in Figure 10.9. It is proposed that the room will contain five Shift Control Engineer terminals comprised of a desk, chair, and four LCD monitors. The Shift Control Manager has a similar workstation.

The Shift Control Engineer workstations are located to allow them to communicate with each other, communicate with the Shift Control Manager, and to leave the room to access the mess and washrooms. The Shift Control Manager workstation is located near the Shift Control Engineers and also allows access to the mess and washroom, as well as gate-keep entry to the control room.

Figure 10.7 shows the key traffic areas within the room. Based on a 95th percentile male's shoulder breadth of 510 mm, the required distance should be in excess of 510 mm for one person to pass and in excess of 1100 mm for two people to pass. All of the walkways are comfortably in excess of 510 mm, and those to and from the Network Manager and Principal Operations Engineers from the support room are in excess of 1100 mm.

From the initial evaluation, the room is large enough to allow personnel to move freely around the room. The Network Manager and Principal Operations Engineers have been strategically placed at the top of the room to enable them to control access to the room, ensuring that the operators are not exposed to unnecessary distractions

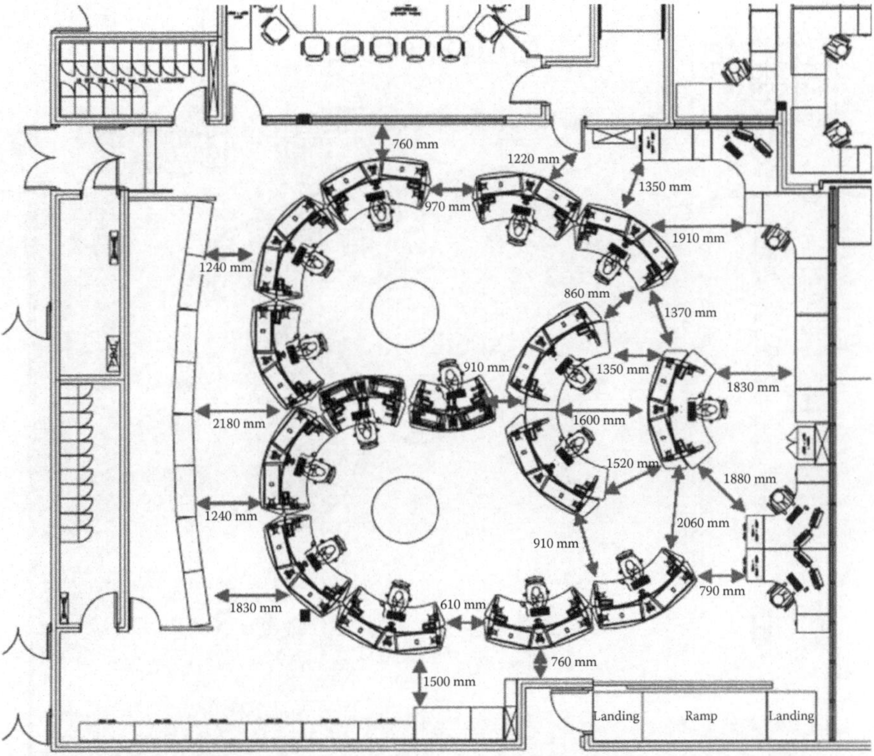

FIGURE 10.7 Diagram showing access around the desks (one person requires 510 mm, two people require 1100 mm).

and the room does not become overpopulated. The sources of natural light are controlled by blinds to eliminate any potential problems with glare. The room is also equipped with an additional desk containing terminals for people on the day shift.

Figure 10.8 shows a layout analysis of the control room. Functions have been allocated to operations engineers, supervision, hot desks, and the production and storage of documentation. The current layout shows that the components are grouped functionally. There are two meeting desks, one for each side of the room. Provision for day working is at one end of the room, facing away from the system overview display.

10.7 EXAMPLE

An example workstation is shown in Figure 10.9. The desk surface is divided in two with independent height controls. It is therefore possible to adjust the height of the desk surface as well as the base height of the monitors. Each of the monitors is freestanding, allowing them to be repositioned and manipulations to be made to their viewing angles and heights. The workstation also contains a repositionable mouse, keyboard, and chair. The desk of this workstation can be seen in Figure 10.10; it has two adjustable desk surfaces. The main part of desk at the front can be adjusted from 690 to 995

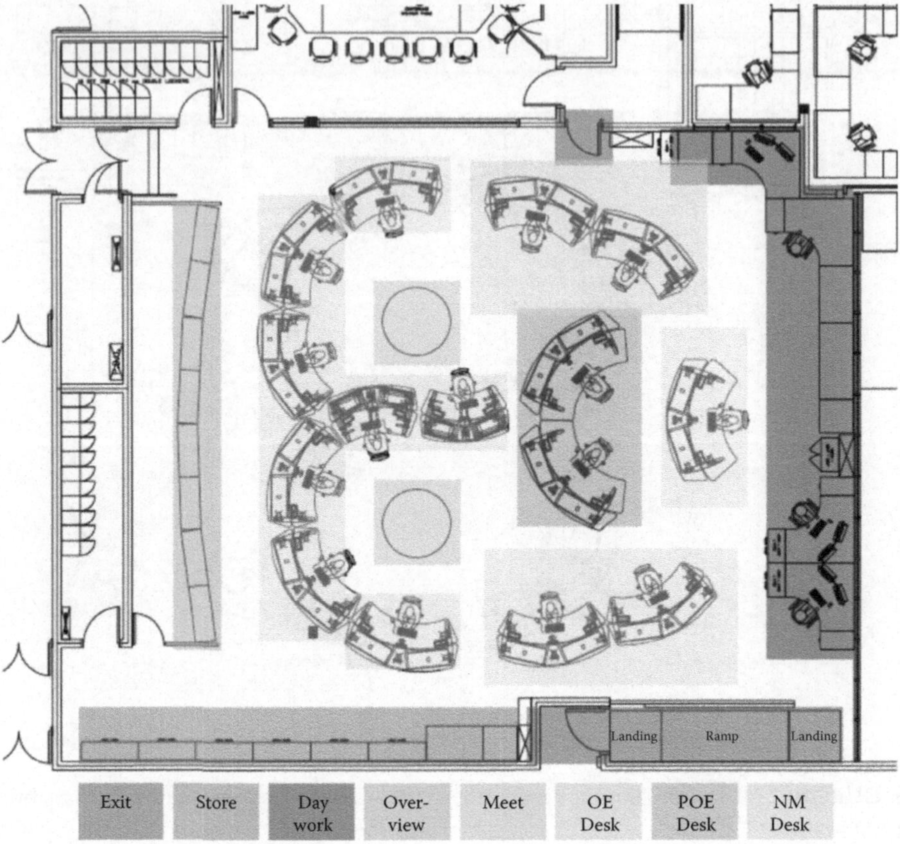

FIGURE 10.8 Layout analysis of control room.

mm above the floor. The smaller rear part of the desk can be adjusted from 695 to 1000 mm above the floor. The desk surface is of adequate size. As the figure shows, the desk can comfortably hold four screens with room for adjustment and repositioning. The keyboard and mouse shown in the figure can be readily repositioned. Assuming a monitor depth of 220 mm, the screens can be placed between 75 and 550 mm from the leading edge of the desk. This allows a good level of adjustment in line with the recommendations in ISO 9241-5:1999, which states that the optimum viewing distance is 600 ± 150 mm. The range of height adjustment on this desk is extremely good (305 mm); however, due to the height of the lower limit smaller operators will require a footrest to raise the floor level in order to achieve an optimal position.

10.1.2 The Chair

The chair shown in Figure 10.11 was developed for the use of control room operators 24 hours a day, 7 days a week. Its relevant dimensions are shown in Table 10.5. The chair is more than wide enough, at 533 mm (the recommendation for this is greater

FIGURE 10.9 Example workstation.

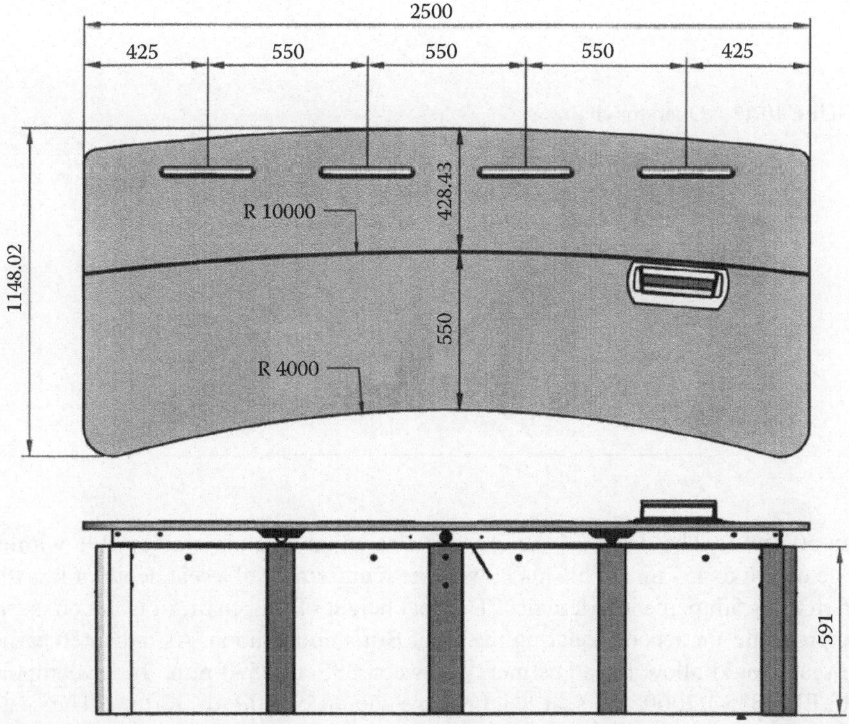

FIGURE 10.10 Adjustable desk.

FIGURE 10.11 Operator chair.

TABLE 10.5
Chair Dimensions

	Overall Dimension (mm)	Seat Squab (mm)	Seat Back (mm)
Width	723	533	533
Depth	686	406–457	—
Height	1270	457–558	711–787

than 500 mm). The depth of the chair is also suitable, as it is adjustable within a range of 406 to 457 mm. This fits in with the requirement of a seat depth of less than 435 mm for 5th percentile females. The seat height of this chair, in isolation, seems inappropriate for accommodating the adult British population. As indicated earlier, the seat should allow for adjustments between 385 and 520 mm. To be compliant with EN 1335-1:2000, the seat height range should be 400 to 510 mm. The seat is therefore too high for the lower limit and is also lacking in adjustment range, as the literature indicates an ideal range of 135 mm and EN 1335-1:2000 indicates a range

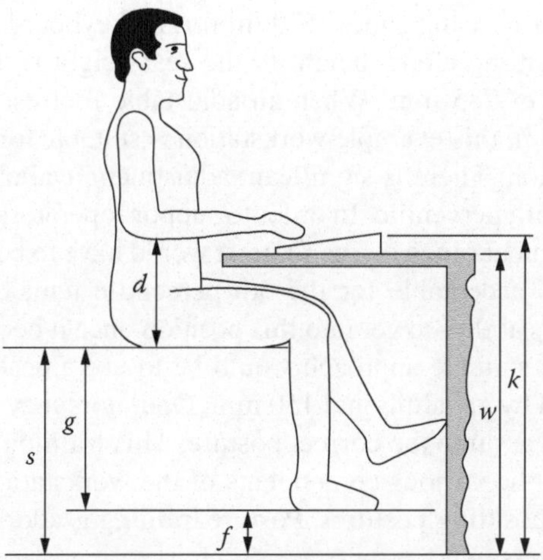

FIGURE 10.12 Dimensioning of workstation. (Permission to reproduce figure from ISO 9241-5:1999 is granted by the British Standards Institution.)

TABLE 10.6
Key Dimensions for Example Workstation

	Dimension	5th Percentile Female (mm)	95th Percentile Male (mm)
g	Popliteal height	355 (385 with shoes)	490 (520 with shoes)
d	Elbow height, sitting	185	295
s	Seat height (*k* – *d*, 720 – 185)	535	520
w	Desk height	690	785
k	Keyboard height	720	815
f	Footrest height (*s* – *g*, 535 – 385)	150	0

Source: Permission to reproduce table from ISO 9241-5:1999 is granted by the British Standards Institution.

of 110 mm. A footrest is therefore essential for use with this chair. The footrest should allow a height of 72 mm (457 – 385 mm).

10.7.1.1 Calculating Sitting Postures for Population Extremities

(See Figure 10.12 and Table 10.6.) For the 5th percentile female, the limiting factor is the desk, as this is higher than would ideally be recommended for a 5th percentile female. Starting with the minimum value (690 mm), the optimum seat height is calculated by subtracting the sitting elbow height from the keyboard height (720 – 185 = 535 mm); the result is then checked against the chair to ensure it is within the ideal range. The final stage is to calculate the footrest height by subtracting the popliteal height from the seat height (535 – 385 = 150 mm). For the 95th percentile male, it is possible to figure from the ground up. A seat height is required that matches the popliteal height when wearing shoes (520 mm). The keyboard height can be calculated by adding the sitting elbow height to the seat height (520 + 295 = 815 mm), giving a desk height of 785 mm. When an adjustable footrest is added allowing a height of up to 150 mm, this example workstation is suitable for accommodating the adult British population. There is significant adjustment available to support operators larger than the 95th percentile. In order to support operators smaller than the 5th percentile, the adjustment range on the footrest would have to be extended. Although the sitting posture is acceptable for the 5th percentile female, thought should be given to the operator's ability to get into this position once it becomes elevated above the floor. A far more suitable approach would be to use a desk allowing the work-surface to be lowered by an additional 150 mm. Operators may also require training to encourage them to assume the correct posture. This training should cover how to make adjustments to the various components of the workstation as well as provide information on correct sitting postures. Posture training is addressed in HSE display screen equipment assessments.

BIBLIOGRAPHY

Ahlstrom, V. and Longo, K. (2001). *Human Factors Design Standard (HFDS) for Acquisition of Commercial Off-the-Shelf Subsystems, Non-Developmental Items, and Developmental Systems*, Report No. DOT/FAA/CT-03/05 HF-STD-001. Washington, D.C.: Federal Aviation Administration (http://www.hf.faa.gov/).

Ball, R., Sabri, A., Varghese, M., and North, C. (unpublished). *More Pixels Wins the War*. http://people.cs.vt.edu/~rgb6/Professional/WarCraft.pdf.

Cochran, E. and Bullemer, P. (1999). Workplace Design in Effective Control Center Design for a Better Operating Environment. Paper presented at NPRA Computer Conference, Kansas City, MO, November 1999.

Corlett, A.N. and Clark, T.S. (1995). *The Ergonomics of Workspace and Machines: A Design Manual*. London: Taylor & Francis.

Das, B. (2006). Ergonomic workstation design. In W. Karwowski (Ed.), *International Encyclopedia of Ergonomics and Human Factors*, Vol. 2 (pp. 1598–1607). Boca Raton, FL: CRC Press.

Green, M. and Collier, S. (2001). Experience in Incorporating Human Factors into the Control Center Design Process. Paper presented at the People in Control: An International Conference on Human Interfaces in Control Rooms, Cockpits, and Command Centres, Manchester, U.K., June 2001.

Health and Safety Executive (HSE). (2006) *Working with VDUs*. London: Health and Safety Executive (http://www.hse.gov.uk/pubns/indg36.pdf).

Lehto, M.R. and Buck, J.R. (2008). *Introduction to Human Factors and Ergonomics for Engineers*. Boca Raton, FL: CRC Press.

Nachreiner, F., Nickel, P., and Meyer, I. (2006). Human factors in process control systems: the design of human–machine interfaces. *Safety Science*, 44, 5–26.

Ni, T., Bowman, D., and Chen, J. (2006). Increased display size and resolution improve task performance in information-rich virtual environments. In *Proceedings of Graphics Interface 2006*, Quebec, Canada, May 2000.

Office of Public Sector Information (OPSI). (1992). Health and Safety (Display Screen Equipment) Regulations 1992, http://www.opsi.gov.uk/si/si1992/Uksi_19922792_en_1.htm#end.

Pheasant, S. and Haslegrave, C.M. (2006). *Anthropometry, Ergonomics, and the Design of Work*, 3rd ed. London: Taylor & Francis.

Proffitt, D., Bakdash, J., and Banton, T. (unpublished). *When and for Whom Does Display Size Matter?* Charlottesville, VA: Department of Psychology, University of Virginia.

Stanton, N.A. and Young, M.S. (1999). *A Guide to Methodology in Ergonomics: Design for Human Use*. London: Taylor & Francis.

Stanton, N.A., Salmon, P.M., Walker, G.H., Baber, C., and Jenkins, D.P. (2005). *Human Factors Methods: A Practical Guide for Engineering and Design*. Aldershot: Ashgate.

Tan, D., Gergle, D., Scupelli, P.G., and Pausch. R. (2003). With similar visual angles, larger display improve spatial performance. In *Proceedings of Computer–Human Interaction 2003*, Ft. Lauderdale, FL, April 2003.

11 Supervisory Control and Data Acquisition (SCADA)

11.1 OUTLINE

The design of the Supervisory Control and Data Acquisition (SCADA) system is integral to the performance of the system, in terms of safety, efficiency, and effectiveness. SCADA systems essentially have two main roles:

1. Communicating information from physical sensors within the plant
2. Facilitating physical manipulations of components within the system

Descriptions of the controlled system can be presented at a number of levels of abstraction. These descriptions range from physical raw data from individual sensors up to higher order systems performance descriptions calculated through data fusion. SCADA has been around in one form or another for as long as the systems they represent and control have. Early SCADA systems were comprised of a series of readouts directly hardwired to the physical objects they represented. These hardwired display consoles were generally large, fixed, and difficult to modify. The advent of the microchip has allowed many of these systems to be digitized. It is now possible to remotely control these systems with mechanization, computers, and appropriate software.

11.2 KEY ISSUES

The introduction of these "soft desk" solutions also provides opportunities for automation, which affects the distribution of work within the system. Automation relieves the operator from the task of making physical manipulations to the plant equipment. In routine situations, the system is able to sense changes in a process, apply existing rules, and make physical component manipulations at a much faster rate than would be humanly possible. Automated systems are also able to cope with concurrent fault management, whereas human operators seem to favor serial fault management. Through automation, the responsibilities of the control room operator have migrated from performing physical control actions to higher level tasks such as system optimization. Physical control manipulations and fault diagnosis are skills still required by operators; however, these are normally only called upon where automation is unable to cope with situations that exist outside those defined in the software. Due to this redistribution of work within the system, the task of designing a soft desk for a SCADA system is far more complex than recreating a digital version of previous hard-desk solutions. SCADA systems are likely to be created to account for the uniqueness of each of the systems under its control. Although the way in which a SCADA system is developed and evaluated is heavily dependent on its purpose, it is possible to draw some broad guidelines by considering the SCADA system in generic terms.

11.3 STANDARDS AND GUIDELINES

See Table 11.1.

11.4 HUMAN FACTORS LITERATURE

In most SCADA systems, the data available to be displayed significantly exceed both screen real estate and the human operator's processing ability. It is therefore important to have a method for navigating through these data in a meaningful way. The choice of information displayed should be dependent on the system state and the task the operator is trying to perform. The role of the operator is heavily dependent on the current state of the system under control. As previously stated, the introduction of automation has relieved the operator of the role of physically manipulating the system under normal working conditions. In these situations, operators have available time and resources to turn their attention to system optimization. The Engineering Equipment and Materials Users' Association (EEMUA) standard EEMUA 201:2002 describes the control task as being comprised of four distinct situations or categories, listed in order of priority:

Category 1. Abnormal situation handling (includes startup and shutdown)
Category 2. Normal operation (most utilized role)
Category 3. Optimization
Category 4. General information retrieval (may be delegated to an information technology system)

The EEMUA is an industrial association of leading national and multinational organizations in the petroleum, oil, gas, chemical, and energy industries. The EEMUA shares resources and expertise with the aims of improving the effectiveness and efficiency of its member organizations. The EEMUA guides encapsulate best industrial practice and are therefore an excellent basis for a review of existing systems or a basis for design of new systems. The requirements of a SCADA system, which includes both the operator and the automation system, differ in each of the four identified conditions. The role of the operator will vary depending on the category of work. Typically, the operator responsibilities include plant operation, optimization of production, fault identification, and coordination of maintenance. Table 11.2 shows the situations in which typical activities are likely to be conducted.

TABLE 11.1
Visual Display Terminal Standards

Standard	Relevant Parts
ISO 9241:1997 *Ergonomic Requirements for Office Work with Visual Display Terminals (VDTs)*	Part 1. General introduction Part 2. Guidance on task requirements Part 3. Visual display requirements Part 4. Keyboard requirements Part 5. Workstation layout and postural requirements Part 6. Guidance on the work environment Part 7. Requirements for display with reflections Part 8. Requirements for displayed colors Part 9. Requirements for non-keyboard input devices Part 11. Guidance on usability Part 12. Presentation of information Part 13. User guidance Part 14. Menu dialogs Part 15. Command dialogs Part 16. Direct manipulation dialogs Part 17. Form filling dialogs Part 20. Accessibility guidelines for information/communication technology (ICT) equipment and services Part 110. Dialog principles Part 151. Guidance on World Wide Web user interfaces Part 171. Guidance on software accessibility Part 300. Introduction to requirements and measurement techniques for VDTs Part 302. Terminology for electronic visual displays Part 303. Requirements for electronic visual displays Part 304. User performance test methods Part 305. Optical laboratory test methods for electronic visual displays Part 306. Field assessment methods for electronic visual displays Part 307. Analysis and compliance test methods for electronic visual displays Part 400. Principles and requirements for physical input devices Part 410. Design criteria for products for physical input devices Part 308. Surface-conduction electron-emitter displays Part 411. Laboratory test and evaluation methods for the design of physical input devices Part 420. Selection procedures for physical input devices Part 421. Workplace test and evaluation methods for physical input devices Part 920. Guidance on haptic and tactile interactions
BS/EN/ISO 13406:2002 *Ergonomic Requirements for Work with Visual Displays Based on Flat Panels*	Part 2 of ISO 13406 establishes ergonomic image-quality requirements for the design and evaluation of flat-panel displays. It defines terms needed to address image quality on flat-panel displays and specifies methods of determining image. The standard establishes ergonomic principles for guiding these requirements.

TABLE 11.2
Taxonomy Showing Situations where Activities Are Typically Conducted

	Category 1 Abnormal Situation Handling	Category 2 Normal Operation	Category 3 Optimization	Category 4 General Information Retrieval
Plant operation	✓			
Optimization of production			✓	
Fault identification			✓	
Coordination of maintenance		✓		

Note: ✓ indicates that an activity is likely to be conducted.

11.4.1 Control Task Categories

11.4.1.1 Category 1. Abnormal Situation Handling

Abnormal situation handling requires operators to conduct decision making. Operators must be alerted to an abnormal situation, and they then need to observe data and identify the current system state. From this understanding, they must then evaluate the system to determine an actions goal. The operator then defines the task required, formulates a procedure, and executes it. This decision-making process can be modeled on Rasmussen's (1974) decision ladder (Figure 11.1), where the steps of decision making are broken down into data-processing activities and resultant states of knowledge. The system must provide the operator with some cue to indicate that the system should be observed (alert). The system also must provide the data required by the operator to diagnose the current system state. In hard-desk systems, this activation prompt would be provided by alarms or by abnormal readings on dials. The desk itself would provide all the available data all of the time. Soft desks also have the functionality to support this process. Due to their dynamic nature, they have the ability to reconfigure the display to raise the salience of certain pieces of data, thus drawing the operator's attention. In addition, they can more explicitly support the evaluation section of the decision-making process as well as the process of defining the task and formulating the procedure.

11.4.1.2 Category 2. Normal Operation

During normal operation, the system (the operator and SCADA system) observes the data and identifies the current system state, which then must be evaluated to decide if any action is required. This task is normally conducted by a level of automation. The operator is left to consider the system at a higher level of abstraction, based on performance-related metrics.

11.4.1.3 Category 3. Optimization

During optimization, the system under control is observed at a number of levels of abstraction. Changes to the current system configuration are considered to improve the overall performance of the system under control.

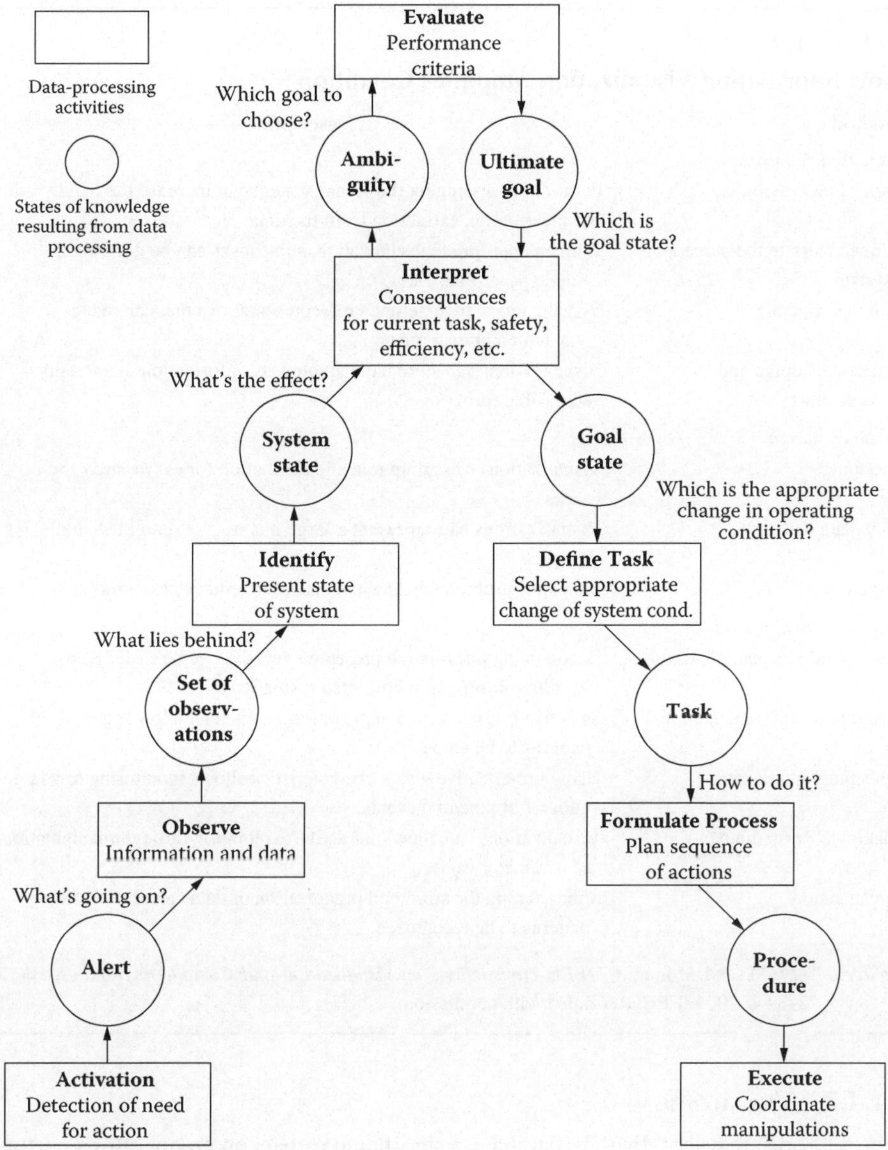

FIGURE 11.1 The decision ladder.

Each of these categories represents different activities, behaviors, and competencies. It is therefore not surprising that the display requirements for these categories of activity differ. Guidelines such as EEMUA 201:2002 advocate the use of multiple screens, such that read-only screens display only the current system picture and alarms status. These displays should be supplemented with additional displays for optimization and abnormal situation handling.

TABLE 11.3
How Information Visualization Amplifies Cognition

Method	Description
Increased resources	
Parallel processing	Parallel processing by the visual system can increase the bandwidth of information extraction from the data.
Offload work to the perceptual system	With an appropriate visualization, some tasks can be done using simple perceptual clues.
External memory	Visualizations are external data representations that can reduce demand on human memory.
Increased storage and accessibility	Visualizations can store large amounts of information in an easily accessible form.
Reduced search	
Grouping	Visualizations can group related information for easy search and access.
High data density	Visualizations can represent a large quantity of data in a small space.
Structure	Imposing structure on data and tasks can reduce task complexity.
Enhanced recognition	
Recognition instead of recall	Recognizing information presented visually can be easier than recalling information presented textually.
Abstraction and aggregation	Selective omissions and aggregation of data can allow higher level patterns to be recognized.
Perceptual monitoring	Using pre-attentive visual characteristics allows monitoring of a large number of potential events.
Manipulable medium	Visualizations can allow interactive exploration through manipulation of parameter values.
Organization	Manipulating the structural organization of data can allow different patterns to be recognized.

Source: Tory, M. and Möller, T., *IEEE Transactions on Visualization and Computer Graphics*, 10(1), 72–84 © 2004 IEEE. Reprinted with permission.

11.4.2 Visualization

The appropriate use of displays can have a significant impact on an operator's performance. According to Tory and Möller (2004), visualizations can provide cognitive support through a number of mechanisms. These mechanisms can exploit advantages of human perception, such as parallel visual processing, and they can compensate for cognitive deficiencies, such as limited working memory (see Table 11.3).

11.4.3 Display Design

A significant body of literature provides information on interface design. Although design is more of an art than a science, heuristics can be applied to determine the most efficient way of communicating data (Tufte, 1983, 1990, 1997). Many texts are available that inform interface design key principles (e.g., Bruce, Green, and Georgeson, 1996; Burns and Hajdukiewicz, 2004; Card, Mackinlay, and Shneiderman, 2004; Shneiderman, 1998; Spence, 2000; Tory and Möller, 2004; Ware, 2000). Usability assessment techniques are used to assess the usability associated with a device, system, or product. Typically, end users perform a series of tasks with the device, system, or product in question, and then a usability assessment technique of some sort is used to derive usability ratings from the end users. One of the simplest, yet most powerful, forms of usability assessment is a usability checklist. Checklists offer a quick, easy, and low-cost approach to interface assessment and typically involve analysts checking features associated with a product or interface against a checklist containing a predefined set of criteria. Checklist-style evaluations can occur throughout the lifecycle of a product or system, from paper drawings to the finished product. A number of specific human factors checklists exist, such as Ravden and Johnson's (1989) human–computer interaction (HCI) checklist, which is particularly suited to the assessment of software-based systems. The HCI checklist has ten sections of questions designed to assess the overall usability of a particular device or system. These questions also serve as important design considerations. The ten sections are briefly outlined below:

1. *Visual clarity*—The clarity with which the system displays information; information displayed on the screen should be clear, well organized, unambiguous, and easy to read.
2. *Consistency*—Consistency of the interface in terms of how it looks, the ways in which it presents information, and the ways in which users perform tasks; the way the system looks and works should be consistent at all times.
3. *Compatibility*—The system in question's compatibility with other related systems; the way the system looks and works should be compatible with user conventions and expectations.
4. *Informative feedback*—The level, clarity, and appropriateness of the feedback provided by the system; users should be given clear and informative feedback on where they are in the system, what actions they have taken, whether these actions have been successful, and what actions should be taken next.
5. *Explicitness*—The clarity with which the system transmits its functionality, structure, and capability; the way the system works and is structured should be clear to the user
6. *Appropriate functionality*—The level of appropriateness of the system's functionality in relation to the activities that it is used for; the system should meet the needs and requirements of users when carrying out tasks.
7. *Flexibility and control*—The flexibility of the system and the level of control that the user has over the system; the interface should be sufficiently flexible in structure, presentation, and functionality to suit the needs and requirements of all users and to allow them to feel in control of the system.
8. *Error prevention and correction*—The extent to which the system prevents user errors from either being made or impacting task performance (error

tolerance); systems should minimize the possibility of user error and have facilities for detecting and handling the errors that do occur, and users should be able to check inputs and correct errors or potential error situations before inputs are processed.
9. *User guide and support*—The level of guidance and support that the system provides to its end users; systems should be informative and easy to use and should provide relevant guidance and support, both on the system itself (e.g., help function) and in document form (e.g., user guide) to help users understand and use the system.
10. *System usability*—The overall usability of the system in question.

The U.K. Health and Safety Regulations (1992) outline the minimum requirements for software used in conjunction with display screen equipment. In designing, selecting, commissioning, and modifying software and in designing tasks using display screen equipment, the employer should take into account the following principles:

1. Software must be suitable for the task.
2. Software must be easy to use and, where appropriate, adaptable to the level of knowledge or experience of the operator or user; no quantitative or qualitative checking facility may be used without the knowledge of the operators or users.
3. Systems must provide feedback to operators or users on the performance of those systems.
4. Systems must display information in a format and at a pace that are adapted to operators or users.
5. The principles of software ergonomics must be applied, in particular to human data processing.

11.4.4 Elements

The design of data representation can be informed by the use of a number of heuristics. The data elements making up the display should be considered in the context within which they are required to be interpreted. By considering the type of data to be displayed, it is possible to narrow down the choices for display type. Typical considerations include:

- Is it binary (two states), multistate, or analog?
- Is it between limits?
- Are the limits fixed or variable?
- Is it critical to monitor?
- Does trend need to be tracked?
- Does rate of change need to be tracked?
- Is the value additive?
- Do values need to be considered in context to each other?

11.4.5 Grouping of Data Elements

The clustering of data can be organized in a number of ways. The approach taken for this clustering should be dependent on the system that is to be represented. Four of the most common approaches to display organization are described below:

- *Spatial importance*—Spatially arranged displays are used where the relative location of displays is important. An example of this would be a system that is geographically dispersed in a way that the operator is familiar with. This model can then be used to sort through the data.
- *Functional importance*—Functionally arranged displays may be used to monitor manufacturing processes. The display may be constructed to represent the flow of the product through the manufacturing process to supply information at each stage. This functional display may differ from the actual plant layout.
- *Hierarchical importance*—Hierarchically arranged displays can be used to show data in some kind of hierarchy with the data related to the most senior objects at the top and the data relating to the least senior objects at the bottom.
- *Critical importance*—Critically arranged displays represent data based on its importance. In this kind of arrangement, important data would take a prominent position and be likely to be larger than less critical data.

11.4.6 Information Density

According to Kandogan (1998), information density of an interface examines the compactness of an interface in terms of the amount of information. The information density metric is calculated as the quantity of visible node instances containing information divided by the screen space area of the whole interface in pixels. Although the argument here is not to encourage designs where the screen is always filled, efficient use of the screen space is crucial, as it has the effect of expanding the short-term memory of users. Screen space must be cleverly used for information related to the user's task and make maximum use of the screen real estate. EEMUA 201:2002 indicates that packing density should not be more than 50% and should be preferably around the 25 to 30% mark.

The information density gives us an indication of how much information is on any one screen, but it does not quantify the value of this information. Care should be taken in the interpretation of the results of this calculation. By cramming more information onto a single page, the need to navigate between pages is eliminated; however, the operator is still required to navigate within the page to find the desired information. If too much information is displayed, this task may become more difficult then navigating between better organized pages.

11.5 GUIDANCE AND RECOMMENDATIONS

A SCADA system can be evaluated in a number of ways, two of which are addressed below: Ravden and Johnson's HCI checklist and the EEMUA user's questionnaire.

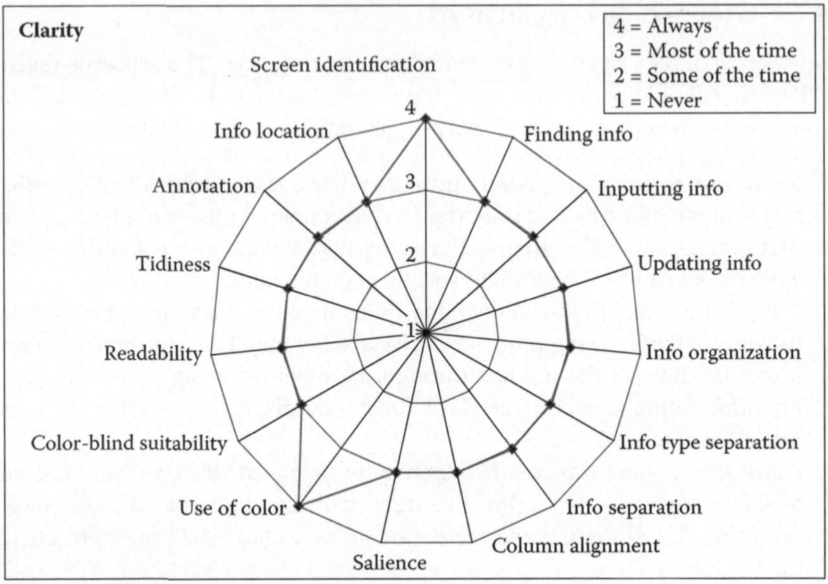

FIGURE 11.2 Median visual clarity results.

11.6 ASSESSMENT AND EVALUATION

To understand the user's subjective opinion of the SCADA system, the HCI checklist was distributed to all control room staff, and 33 Operations Engineers responded. The following sections discuss the findings. Radar plots were selected as the most appropriate representation for the results. Each plot presents an axis illustrating how the participants rated each factor. The farther away from the center the point is, the higher the factor was rated (and thus the higher in terms of usability the system was rated). The results of the graphs are described based on the results of the respondents. It should be noted that the conclusions drawn do not necessarily represent the opinions of the authors but represent an interpretation of the combined voices of the respondents.

11.6.1 HCI Checklist

11.6.1.1 Visual Clarity

The median visual clarity results are presented in Figure 11.2. The results presented indicate no significant problems with visual clarity. When compared against each of the parameters, the system seems to be clear the majority of the time. Participants found the screen identification (labeling and description of the screen) particularly clear. They also found the use of color particularly clear.

11.6.1.2 Consistency

The median consistency results are presented in Figure 11.3. The ratings for consistency in this software are high, indicating no significant problems; however, according to the participant feedback, smaller incremental improvements could be considered to increase the usability of the software.

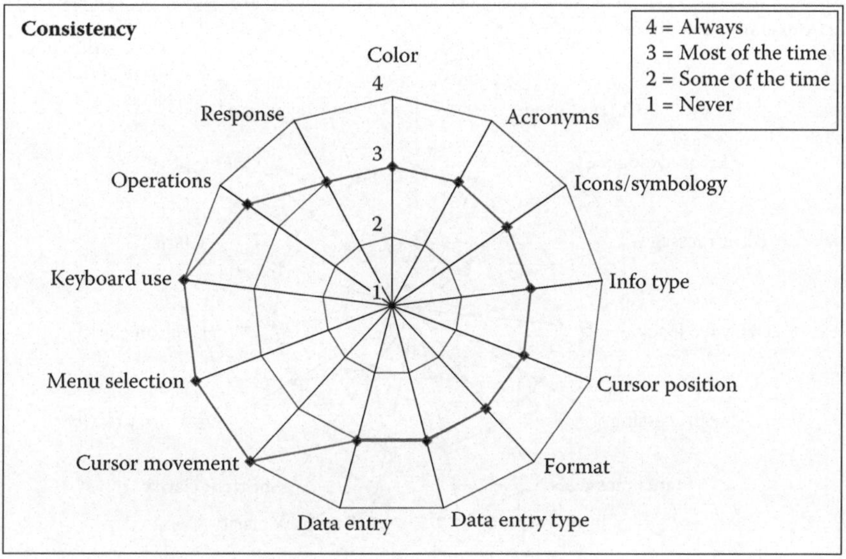

FIGURE 11.3 Median consistency results.

11.6.1.3 Compatibility

The median compatibility results are presented in Figure 11.4. Overall, the system scored well on its compatibility. The software performed exceptionally well for color, icons, and cursor movement. The results indicate that there are no significant issues.

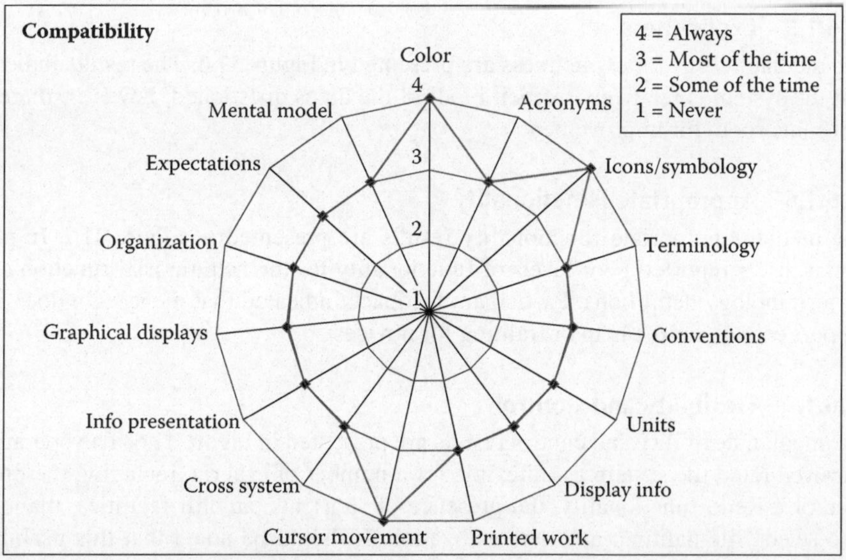

FIGURE 11.4 Median compatibility results.

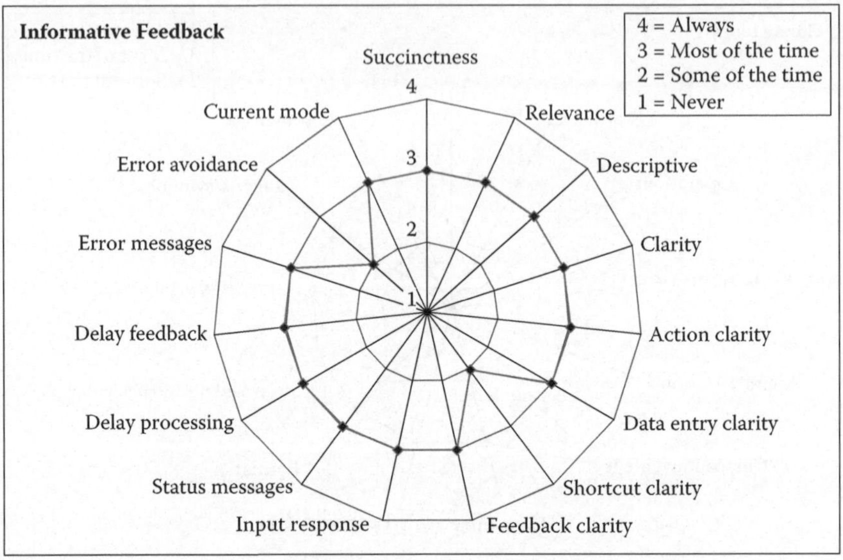

FIGURE 11.5 Median informative feedback results.

11.6.1.4 Informative Feedback

The median informative feedback results are presented in Figure 11.5. The system is providing adequate levels of feedback but performs poorly for error avoidance and shortcut clarity. Participant comments indicate that successful operation is more heavily dependent on operator knowledge than system feedback.

11.6.1.5 Explicitness

The median results for explicitness are presented in Figure 11.6. The results indicate that the system is relatively explicit in all of the areas questioned; however, there is still some room for improvement.

11.6.1.6 Appropriate Functionality

The median appropriate functionality results are presented in Figure 11.7. In particular, users reported poor levels of functionality for the help/tutorial function and for terminology definitions. Participant feedback indicated that the system does not support enough online help or training for novices.

11.6.1.7 Flexibility and Control

The median flexibility and control results are presented in Figure 11.8. The operators surveyed rated the system as inflexible for a number of criteria, including the presence of a Redo functionality, the presence of shortcuts, autofill facilities, manual/auto select, file naming, and personalization. It should be noted that this usability questionnaire is generic to all software. It is therefore highly reasonable that some of the functionality is correctly omitted from the SCADA system.

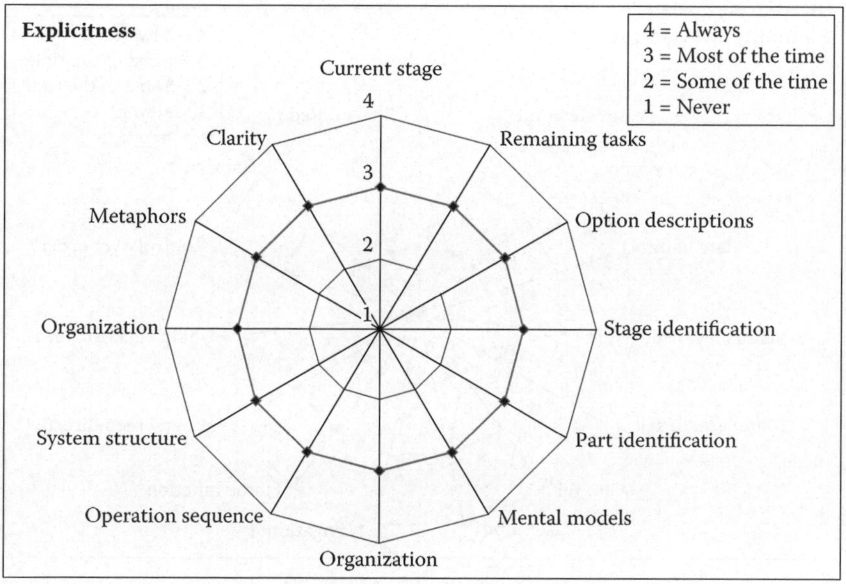

FIGURE 11.6 Median explicitness results.

11.6.1.8 Error Prevention and Correction

The median error prevention and correction results are presented in Figure 11.9. The error prevention results are particularly poor for Undo, diagnostics (when system errors occur, can the user access all necessary diagnostic information to resolve the

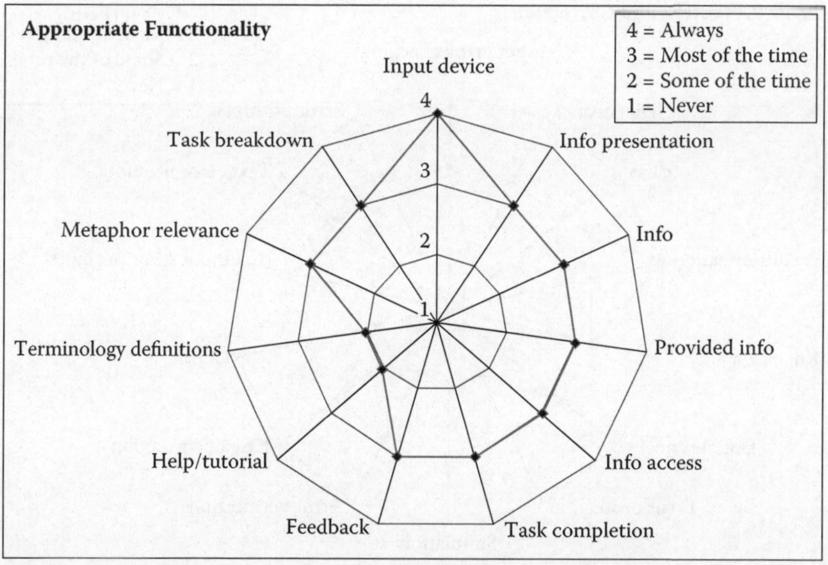

FIGURE 11.7 Median appropriate functionality results.

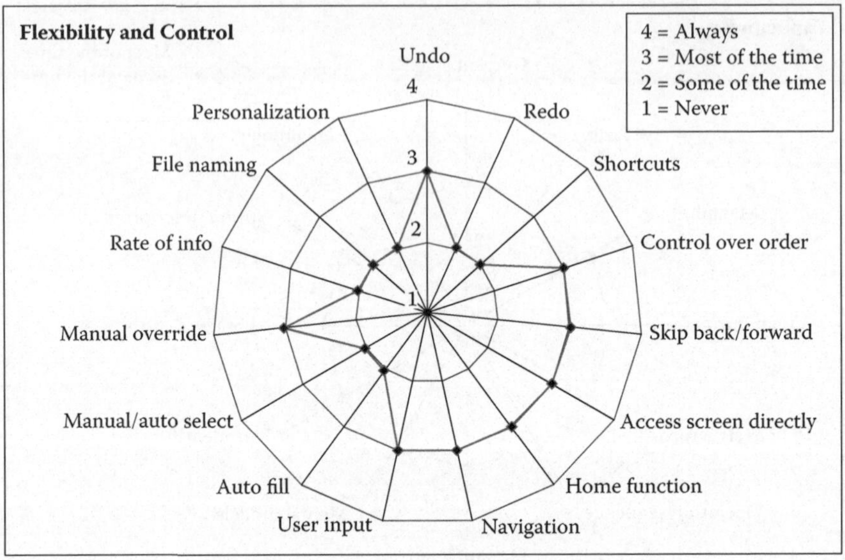

FIGURE 11.8 Median flexibility and control results.

problem, such as where and what the fault is and what is required to resolve it?), and simulation (can the user try out possible actions using a simulation facility without the system processing the input and causing problems?). Based on the operator feedback, it appears as though the system is not guiding the operators enough while they

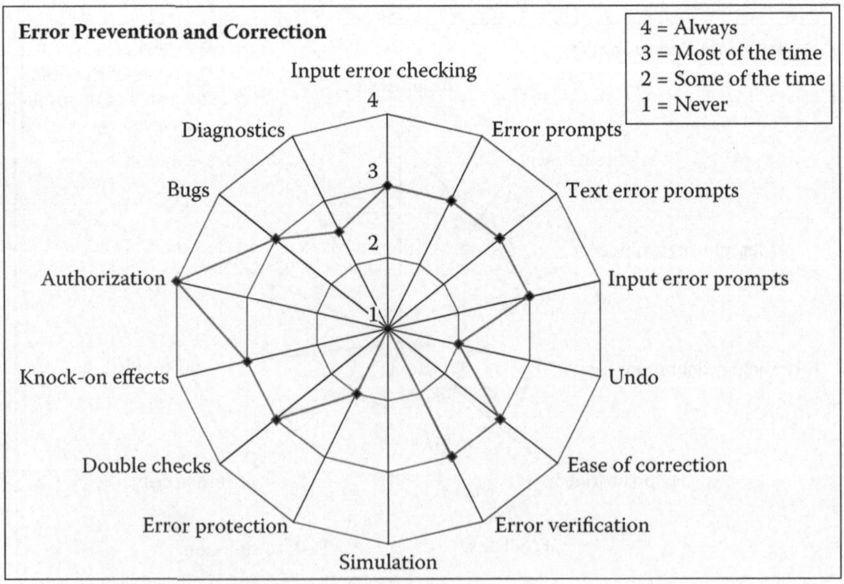

FIGURE 11.9 Median error prevention and correction.

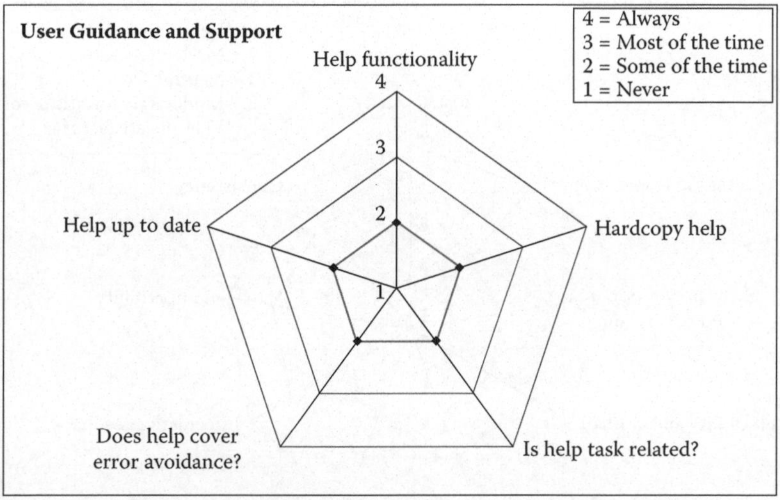

FIGURE 11.10 Median user guidance and support results.

are diagnosing problems within the system. Much of this diagnosis task must remain a human activity; however, the operators clearly feel that there is potential for further support in this process. Operators are not provided with a simulation environment within the software; therefore, they are unable to test hypotheses for fixing problems within the current system. This means that operators are required to calculate in their head any knock-on effects an action may have before making a decision. There may be potential for the software to support in this task.

11.6.1.9 User Guidance and Support

The median user guidance and support results are presented in Figure 11.10. The results show poor scores for all aspects of user guidance and support. There appears to be little support for operators confronted with an unfamiliar situation.

11.6.1.10 Overall System Usability

At the end of each section a question was asked requiring the participant to rate the overall system in relation to the section they had just answered; for example, at the end of the visual clarity section, users were asked, "Overall, how would you rate the system in terms of visual clarity?" These end-of-section questions are designed to elicit overall judgments for the ten checklist dimensions (visual clarity, consistency, compatibility, feedback, explicitness, functionality, flexibility and control, error prevention and correction, user guidance and support, and system usability problems). In this case, anything inside the red circle (rating of 3 or less) was considered unsatisfactory (see Figure 11.11). The subjective results collected indicate that the users generally held a positive view of the system; they rated user guide and support as neutral (possibly because there did not appear to be any).

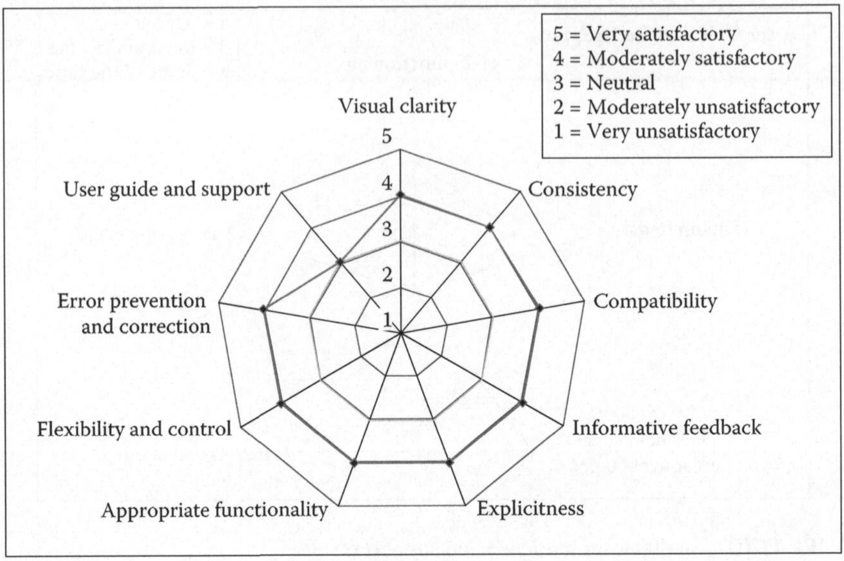

FIGURE 11.11 Comparison of summary results.

To summarize, the operators had a positive opinion of the SCADA system investigated. Opportunities exist to make small incremental improvements within the system. Perhaps the most interesting finding from this analysis is the lack of support provided for novices to the software system. There appears to be little in the way of help or guidance; arguably, the high scores for "appropriate functionality" and other parameters within the analysis could be linked to the fact that the majority of users could be considered to be expert due to the significant time spent using the system. Participant comments indicate that successful operation is more heavily dependent on operator knowledge than system feedback. One area for possible consideration would be the introduction of a simulation module.

11.6.2 EEMUA User's Questionnaire

The questionnaire taken from the appendix of EEMUA 201:2002 was distributed to all control room staff, and 31 responded. Participants were asked to provide subjective responses to the alarm system. As with all subjective assessments, care should be taken in the interpretation of this data. The users were taken from a variety of job titles; see Table 11.4. Operators had an average of 9.1 years (standard deviation, 7.4 years) of experience with the current system. The operators were asked to respond to a number of questions related to the SCADA system. The following represents a summation of their subjective opinion. As with all subjective opinions on system design and performance, care should be taken in the interpretation of these results. There is not always a direct correlation between user opinion and system performance.

TABLE 11.4
Breakdown of Participant Roles

Job Role	Number
Operations Engineer	22
Principal Operations Engineer	5
Trainee Operations Engineer	1
Shift Engineer	1
Network Manager	2

11.6.2.1 Physical Arrangement

When asked about the physical arrangement of their screens, 100% of participants responded that the layout was satisfactory (see Figure 11.12); 86% of the polled operators rated their overview screens as large enough, and 31% of respondents commented that having additional screens would help.

11.6.2.2 Access to Displayed Information

When asked if they could display all of the information needed in steady-state conditions, 100% of participants responded "yes" (see Figure 11.13). For start-up conditions 96% said "yes," and for abnormal situations 93% responded "yes." When asked about the amount of information displayed, 87% thought it was about right, 10% said it was too much, and 3% thought too little information was displayed.

11.6.2.3 Navigation through Information Hierarchy

When asked if they found it easy to navigate through the hierarchy, 89% of the polled operators responded "yes" (see Figure 11.14), 97% believed the navigation was quick enough, 43% of the participants admitted to getting lost in the display format hierarchy, and 57% of the participants believed that the navigation could be improved.

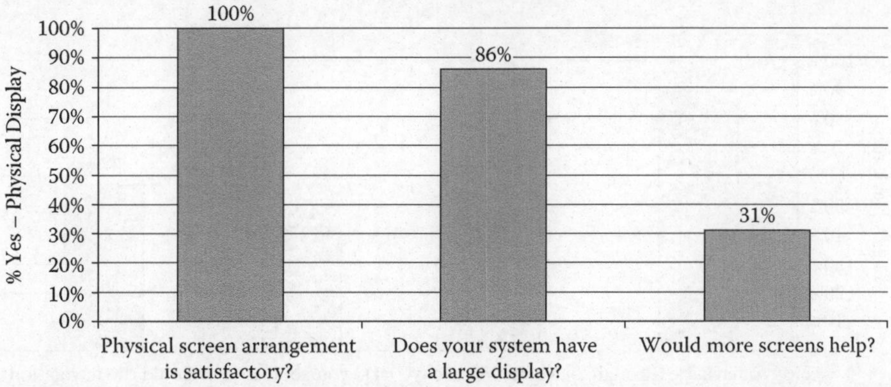

FIGURE 11.12 Percentage of operators responding "yes" to questions on the physical display.

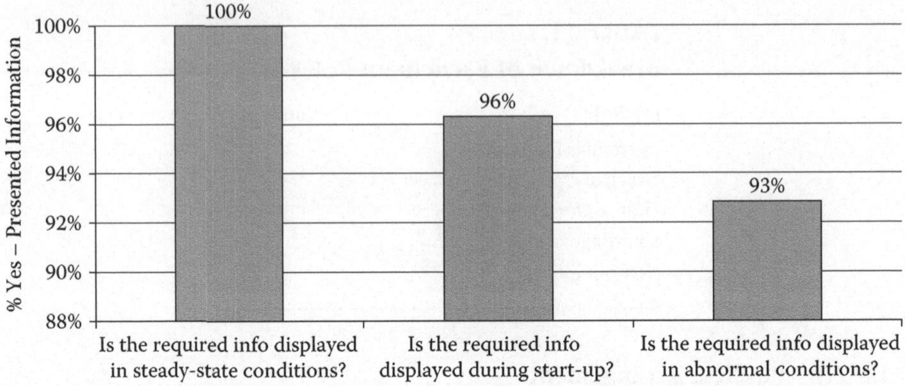

FIGURE 11.13 Percentage of operators responding "yes" to questions on the presentation of information.

11.7 EXAMPLE

The EEMUA 201 analysis showed that the SCADA human–computer interface was generally well regarded by the staff in the control room. Incremental improvements in system design were noted. Some navigational issues were identified, and the operators thought that this could be improved upon. The system was not thought of as error tolerant, and more system help and support could be provided. Finally, a system simulator would be appreciated for training and refresher training. All of these comments were in the context of a system that performed well on the EEMUA 201 criteria. The analysis of the system is summarized in Figure 11.15, which shows that the current system complies with 26 of the guidelines and partially complies with a further 7. Only 2 of the 35 guidelines are not currently complied with. The system also performed well on the SCADA and HCI usability questionnaires. As can be seen from the summary data, the human–computer interface was judged to

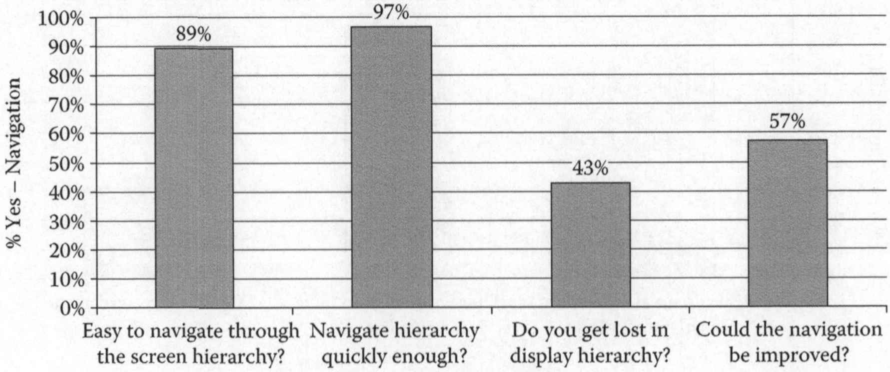

FIGURE 11.14 Percentage of operators responding "yes" to questions on navigation.

	Guidelines	Fully Compatible	Partially Compatible	Incompatible	EEMUA Page No.
1	The number of screens must allow for complete access to all necessary information under all operational circumstances.	✓			3
2	The design should allow for a permanently viewable overview display format.		✓		3
3	Continuous access to alarm information should be provided.	✓			3
4	The capability to expand the number of screens should be built into the design.	✓			3
5	The provision of screens should allow for start-up, steady-state monitoring, shutdown, and abnormal situations.	✓			7
6	The system should allow the user to resize windows easily.	✓			9
7	The system should enable ease of navigation up and down the hierarchy of screens.	✓			11
8	The system should have a simple hierarchical structure, with a total plant overview at the top and more detailed formats below.	✓			11
9	The system should allow ease of navigation across displays at the same level in the hierarchy.	✓			11
10	The system should permit users to switch between single and multiple window views.	✓			9
11	The system should limit the maximum number of windows to be opened, to prevent confusion.	✓			9
12	The system should allow users to save a default configuration of windows that suits their purposes.	✓			9
13	The system should allow important windows always to be on view.	✓			9
14	The system should allow users to configure their favourites and quick access buttons.		✓		9
15	The system should have the facility for shift handover notes.	✓			9
16	The system should allow the user to configure a list of displays that will cycle through automatically.		✓		9
17	The system should allow the user to capture a snapshot of the state of a variable, for discussion later (stored on a hard drive or printed).	✓			9
18	The system should have hotlinks to databases, such as operating procedures.			✓	9
19	The system should allow for the viewing of data in different formats (trend plots, site schematics).	✓			10
20	The system should allow quick and unambiguous selection of the required item.	✓			10
21	The system should allow different methods to interact with the interface (e.g., mouse, keyboard, touchscreen, trackerball).		✓		10
22	The system should give quick, consistent, and effective feedback to the user.	✓			10
23	There should be dedicated displays for emergency shutdown operations.		✓		12
24	There should be backup for loss of power (batteries and generators).	✓			15
25	There should be differentiation between display formats (e.g., to avoid confusion, site schematics should not be virtually identical).		✓		17
26	Access to schematics and detailed information must be quick.	✓			18
27	Displays density should be less than 50%, preferably less than 25%.		✓		18
28	Displays should not be cluttered.	✓			18
29	Spatial alignment on site schematics should give a good appearance.	✓			18
30	It should not be necessary to have more than one display on view to perform specific tasks.			✓	18
31	Concrete pictorial displays should be used where possible (abstract shapes are difficult to learn).	✓			20
32	A cool background color should be used.	✓			20
33	Garish colors should be avoided.	✓			20
34	Color coding should be used sparingly, and only where it adds value to the tasks.	✓			20
35	Animated graphics should be used sparingly; only use where it is really necessary to attract attention.	✓			20
	Total	26	7	2	

FIGURE 11.15 Summary data from the usability questionnaires.

be "moderately satisfactory" by the control room staff. Only the criterion "user guidance and support" was judged as "neutral." It is suggested that this area demands the most attention, while the other areas may simply need some fine tuning.

BIBLIOGRAPHY

Bruce, V., Green, P.R., and Georgeson, M.A. (1996). *Visual Perception: Physiology, Psychology, and Ecology*, 3rd ed. Exeter, U.K.: Psychology Press.

Burns, C.M. and Hajdukiewicz, J.R. (2004). *Ecological Interface Design*. Boca Raton, FL: CRC Press.

Card, S.K., Mackinlay, J.D., and Shneiderman, B. (1999). *Readings in Information Visualization: Using Vision to Think*. San Francisco, CA: Morgan Kaufmann.

Engineering Equipment and Materials Users' Association (EEMUA). (2001). *Process Plant Control Desks Utilising Human–Computer Interfaces: A Guide to Design, Operations, and Human Interface Issues*. London: Engineering Equipment and Materials Users' Association.

Kandogan, E. (1998). Hierarchical Multi-Window Management with Elastic Layout Dynamics. Ph.D. dissertation, University of Michigan (http://hcil.cs.umd.edu/trs/98-13/dissertation.html).

Office of Public Sector Information (OPSI). (1992). Health and Safety (Display Screen Equipment) Regulations 1992, http://www.opsi.gov.uk/si/si1992/Uksi_19922792_en_1.htm#end.

Rasmussen, J. (1974). *The Human Data Processor as a System Component: Bits and Pieces of a Model*, Report No. Risø-M-1722. Roskilde, Denmark: Danish Atomic Energy Commission.

Ravden, S.J. and Johnson, G.I. (1989). *Evaluating Usability of Human-Computer Interfaces: A Practical Method*. West Sussex, U.K.: Ellis Horwood.

Shneiderman, B. (1998). *Designing the User Interface*, 3rd ed. Reading, MA: Addison-Wesley.

Spence, R. (2000). *Information Visualization*. Reading, MA: Addison-Wesley.

Tory, M. and Möller, T. (2004). Human factors in visualization research. *IEEE Transactions on Visualization and Computer Graphics*, 10(1), 72–84.

Tufte, E.R. (1983). *The Visual Display of Quantitative Information*. Cheshire, CT: Graphics Press.

Tufte, E.R. (1990). *Envisioning Information*. Cheshire, CT: Graphics Press.

Tufte, E.R. (1997). *Visual Explanations: Images and Quantities, Evidence and Narrative*. Cheshire, CT: Graphics Press.

Ware, C. (2000). *Information Visualization: Perception for Design*. San Francisco, CA: Morgan Kaufmann.

12 Alarms

12.1 OUTLINE

Alarm systems are an essential part of the Supervisory Control and Data Acquisition (SCADA) control room system; they are required to alert the operator to a range of situations and system-state changes. Stanton (1994) pointed out that an alarm system is absolutely necessary in light of the sheer amount of information contained within the system; the operators could be monitoring up to 800 pages of information and up to 20,000 alarms. The design of the alarm system has a significant effect on the performance of the operator. A well-designed alarm system considers the operator's ability to recognize the situation, diagnose the fault, and develop a suitable course of action. A well-designed alarm system also considers the workload of the operator and the time available for action using this information to prioritize alarms. An alarm can be defined as a plant variable exceeding a setpoint threshold and resulting in some form of explicit communication to the Operations Engineer via the SCADA display. The communication may be auditory or visual and may utilize a text message, flashing icon, annunciator board, or some other means of attracting attention. The definition of the term *alarm* is often dependent on its purpose; Stanton (1994) offered the following definition: "An unexpected change in system state, a means of signaling state changes, a means of attracting attention, a means of arousing the operator, and a change in the operator's mental state."

12.2 KEY ISSUES

Much of the following description has been synthesized from the Engineering Equipment and Materials Users' Association (EEMUA) guide to alarm systems, EEMUA 191:2007 (EEMUA, 2007). The EEMUA guide provides a comprehensive description of the many factors influencing the design of an alarm system. EEMUA 191:2007 points out that alarm systems should:

- Be relevant to the operator's role at the time.
- Indicate clearly what response is required.
- Be presented at a rate that the operator can deal with.
- Be easy to understand.

Essentially, alarms exist to alert the operator to a change of condition in the system. Typically, if unattended to, these changes may result in either a safety-critical situation or a non-optimal system. The alarm system has two main roles:

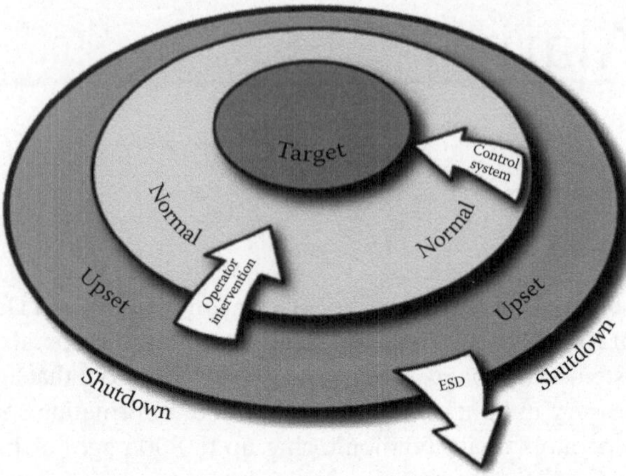

FIGURE 12.1 Diagram showing system states. (Reproduced with permission from EEMUA, *Alarm Systems: A Guide to Design, Management, and Procurement*, EEMUA 191:2007, Engineering Equipment and Materials Users' Association, London, 2007.)

1. Alert the operator that the system has moved to a non-normal state.
2. Provide data and assistance to aid the operator in maintaining or returning to the optimal system state.

When the system automation is unable to keep the process in its optimum state, operator intervention is required. EEMUA 191:2007 refers to this as an *upset state.* If the operator is unable to intervene, the system may then enter into its emergency shutdown device (ESD) state.

As illustrated in Figure 12.1, during normal operation the system automation should be working to optimize the system performance toward its target state. In this situation, the operator is not required to take action. At the point at which the automation is no longer able to cope, the system moves to a state of upset, and operator intervention is required. Effective alarms should mark the point at which the operator should take action. Alarms occurring that do not require attention become a nuisance. Long-term exposure to these false alarms can result in degraded operator system confidence, thus increasing the likelihood of operators ignoring the alarms. Alarms occurring too late do not allow operators enough time to correct the problem. Figure 12.2 shows a graphical representation of an effective and an ineffective alarm system. In the effective alarm system, the alarms mark the boundary at which operator intervention is required.

The role of the human operator in an alarmed SCADA system is often misconceived. Zwaga and Hoonhout (1994) questioned operators assuming a subservient role such that they only intervene when prompted; instead, they suggested that operators should actively follow the progress of the automated system, continually seeking evidence that everything is going according to expectations. EEMUA 191:2007 (p. 6) offers eight characteristics of a good alarm:

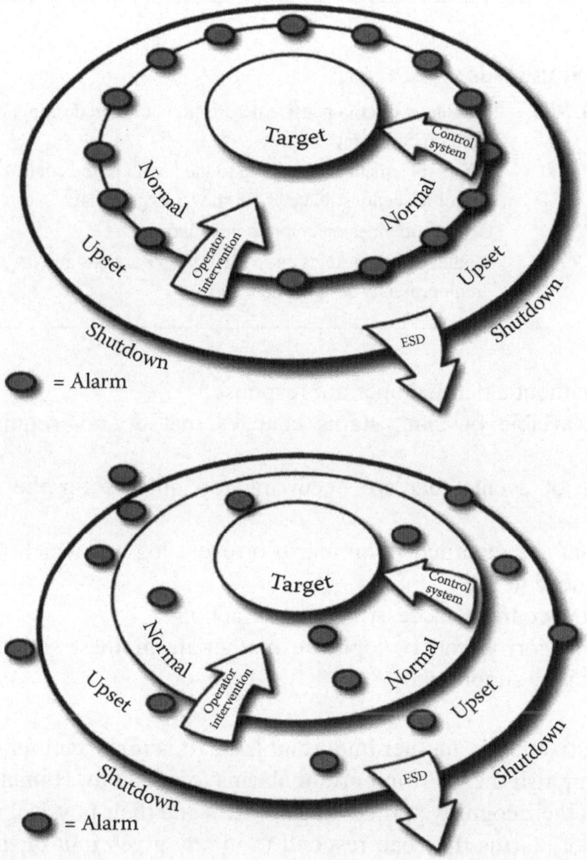

FIGURE 12.2 An effective (top) and ineffective (bottom) alarm system. (Reproduced with permission from EEMUA, *Alarm Systems: A Guide to Design, Management, and Procurement*, EEMUA 191:2007, Engineering Equipment and Materials Users' Association, London, 2007.)

- *Relevant*—Not spurious or of low operational value
- *Unique*—Not duplicating another alarm
- *Timely*—Not long before any response is needed or it is too late to do anything
- *Prioritized*—Indicative of the importance of the operator dealing with the problem
- *Understandable*—Clear and easy to understand
- *Diagnostic*—Identifies the problem that has occurred
- *Advisory*—Indicative of action to be taken
- *Focusing*—Draws attention to the most important issues

EEMUA 191:2007 (p. 18) provides examples of signals that should not be used as alarms. Although these alerts may have to be communicated to the operator, the medium for this should not be through the use of alarms:

TABLE 12.1
Relevant Standards

EEMUA 191:2007	Provides a comprehensive description of alarm design, management, and procurement
EEMUA 201:2002	Offers information pertaining to the layout of the control room; specific sections relate to alarm warning displays
BS EN 60073	Gives guidelines on color coding alarms
ISO 11428:1996	Relates to ergonomics and visual danger signals and their general requirements, design, and testing

- Signals without a defined operator response
- Process variable or plant status changes that do not require operator attention
- Warnings of events that are occurring too quickly for the operator to prevent
- Events that are recorded in an alarm or event log but which the operator does not need to see
- Signals that confirm successful operator actions
- Signals that corroborate or duplicate another alarm (these signals may need to be logically suppressed)

Alarm prioritization is another important issue. It is important for the operator to be able to distinguish the most important alarms to attend to. Human operators are limited by both their cognitive processing abilities and their physical response times to the number of alarms they can respond to in any given unit of time. Two main factors affect the prioritization of alarms: (1) the severity of the consequences that the operator could avoid by taking the corrective action, and (2) the time available compared with the time needed.

12.3 STANDARDS AND GUIDELINES

The EEMUA is an industrial association of leading national and multinational organizations in the petroleum, oil, gas, chemical, and energy industries. The EEMUA shares resources and expertise with the aims of improving effectiveness and efficiency among their member organizations. The EEMUA guides encapsulate best industrial practice and are therefore an excellent basis for a review of existing systems or a basis for design of new systems (see Table 12.1).

12.4 HUMAN FACTORS LITERATURE

The way operators respond to an alarm will depend on a number of factors. Operators responding to familiar alarms will react in a very different way to operators encountering an alarm for the first time. The response to alarms can be considered using

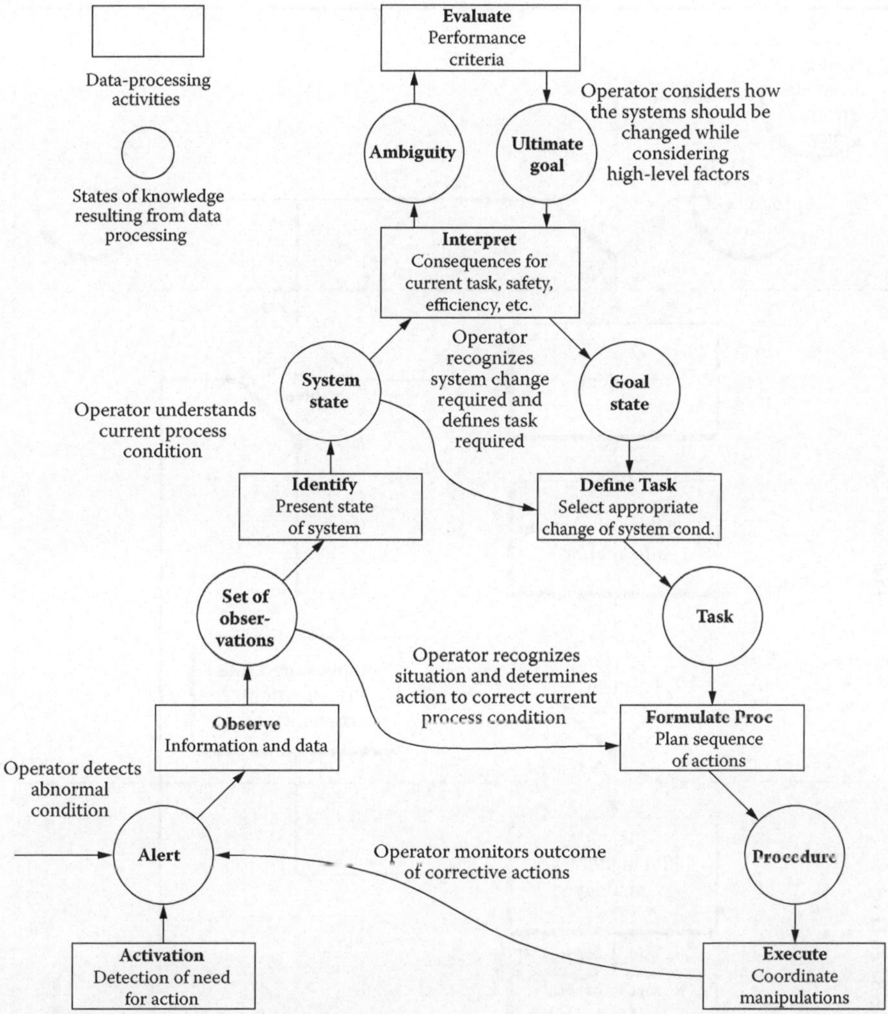

FIGURE 12.3 Decision ladder for operator.

decision-making theory (Rasmussen, 1974) (Figure 12.3). The operator is alerted to a situation and then observes the data. At this point, the operator may recognize the data from a similar familiar situation and then apply the familiar response (as shown in the middle link between the legs of the ladder). If the data are not familiar, the operator will use the information to identify the current system state. From this system state, the operator may be able to recognize the required system change and then move on to defining the task (shown by the highest link between the legs of the ladder). If the operator is unable to recognize the system change required, he must then consider the overall goals of the system, evaluating a number of options against their impact on performance and knock-on consequences. From this, the operator will develop a goal state, formulate a task and procedure, and execute the plan.

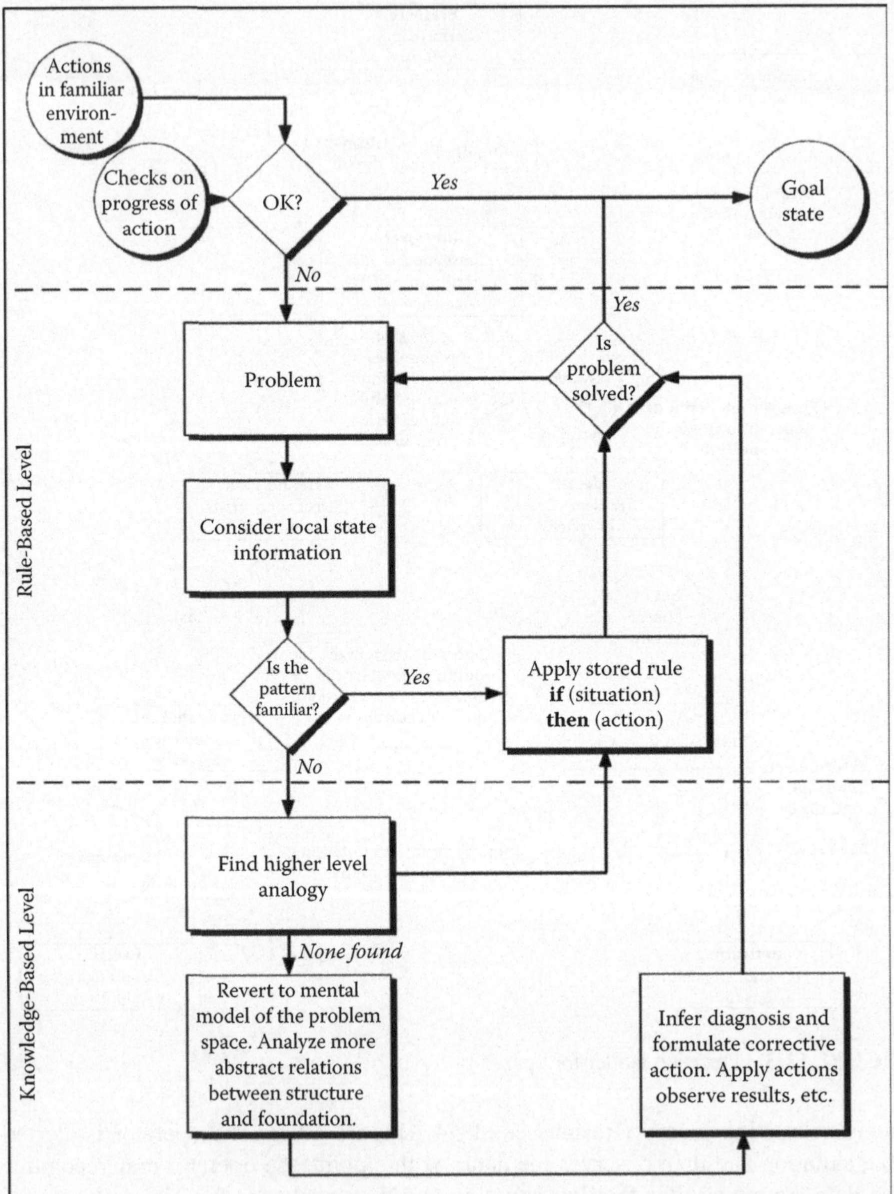

FIGURE 12.4 GEMS framework. (From Reason, J., *Human Error*, Cambridge University Press, New York, 1990. With permission.)

Another similar model of decision making was described by Reason (1990) (Figure 12.4). Reason used a flowchart to describe a process categorizing decision making into rule-based and knowledge-based behavioral levels. Rule-based decision making is fast, whereas knowledge-based decision making is slow and effortful. The time taken by operators to respond to an alarm will therefore be heavily influenced by their familiarity with the alert and their ability to associate this with a plan of action.

Alarm systems should be developed to support both rule-based and knowledge-based behavior. During rule-based decision making, operators should have the ability to short-cut the response process to allow them to act upon previous experience and mental models. To support the operator in knowledge-based behavior, the system should provide the operator with suggested causes for the situation, as well as information on higher order effects of any action taken to return the system to a desired state.

12.5 GUIDANCE AND RECOMMENDATIONS

An understanding of operator response times is essential in alarm design, both in terms of prioritization and in terms of how many alarms to present in any given unit of time. EEMUA 191:2007 suggests that there is little benefit in presenting a warning to the operator if the operator does not have time to respond to it. It is also not practical to provide operators with more alarms than they can possibly respond to without some level of prioritization. The time it takes an operator to respond to an alarm is dependent on the level of cognitive processing required. If an operator is primed to respond and the response is simply a matter of hearing an alarm and pressing a button to cancel the alarm, then one might expect performance (in emergency conditions) to be between 1 and 9 seconds (Stanton and Baber, 2008). This does not account for any analysis that must take place. From a review of the literature, Stanton and Baber (2008) found considerable variation in the reports of response times for human intervention in urgent alarm events, from 1 second to 90 seconds. In their paper, Stanton and Baber calculated approximations for operator response times based on experimental studies. The model of alarm-initiated activity (AIA) shows potential pathways for human activity involving alarms (Stanton, 1994; Stanton and Baber, 1995; Stanton and Edworthy, 1999). The term *activity* in this case refers to the ensuing behavior triggered by the occurrence of alarms. The main stages of AIAs are as follows:

- *Observation*, the initial detection of the alarm
- *Acceptance*, the act of acknowledging the alarm
- *Analysis*, the initial assessment and prioritization of the alarm
- *Investigation*, the activity directed toward determining the underlying cause for the alarm
- *Correction*, the stage at which the system controller implements a response to the alarm condition
- *Monitoring*, the assessment of success of the analysis, investigation, and correction activities
- *Resetting*, extinguishing the alarm and returning it to its inactive state

The alarm handling model in Figure 12.5 shows stages and pathways in alarm handling. It is recognized that not all alarms are dealt with in exactly the same way, as the activities are context dependent. After an alarm has been triggered, it may be

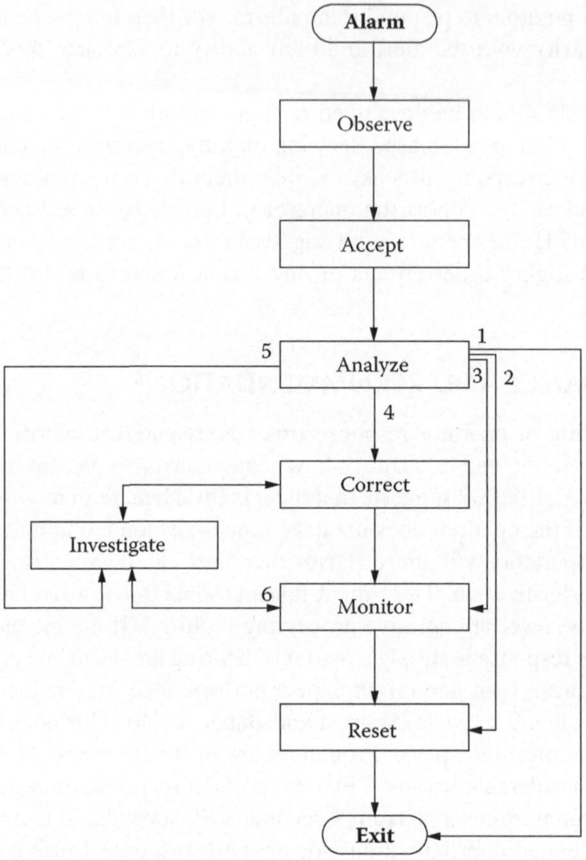

FIGURE 12.5 Alarm handling flowchart.

observed and accepted by the signaler, then some analysis of the likely trigger conditions is undertaken (which will be highly dependent on the situation awareness). Depending on the outcome of this analysis, one of five pathways could be followed, as indicated in Figure 12.5. These pathways include correcting the situation, monitoring the condition, resetting the alarm, investigating the condition, and exiting the analysis.

12.6 ASSESSMENT AND EVALUATION

By breaking down the alarm handling activity into a number of subprocesses and applying a time to each of these, it is possibly to develop an approximation of the operator response time. Stanton and Baber (2008) have assigned values that are generic to any domain. These times are based on underlying aspects of human performance that are independent of context. In this model, Stanton and Baber (2008) have taken the shortest time available in the literature in an attempt to define the minimum response time for an activity under consideration. It is not so easy to

TABLE 12.2
Estimates of Minimum and Maximum Response Times for Each Stage of Alarm-Initiated Activity in Simulated Emergencies

Alarm-Initiated Activity	Minimum Response Time (seconds)	Cumulative Minimum (seconds)	Maximum Response Time (seconds)	Cumulative Maximum (seconds)
Observation	1 (6%)	1	2 (2%)	2
Acceptance	1 (6%)	2	7 (8%)	9
Analysis	2 (12%)	4	6 (7%)	15
Investigation	6 (36%)	10	30 (33%)	45
Monitoring	Variable[a]	10	Variable[a]	45
Correction	7 (40%)	17	45 (50%)	90

[a] No figures are entered for the monitoring task, as it is difficult to estimate these from the data; it is assumed that monitoring is a continuous activity that can be performed in parallel with the other actions.

Note: Percentage of total response time is indicated in parentheses.

Source: Stanton, N.A. and Baber, C., *Ergonomics*, 51(4), 423–440, 2008. With permission.

specify an upper limit for the processing of the alarms due to individual differences in experience and cognitive abilities. By using a minimum value together with some safety margin, a suitable alarm system can be developed (see Table 12.2). The ability of an operator to respond to alarms is closely linked to the workload. Zwaga and Hoonhout (1994) pointed out that *cognitive overload* can occur due to too many alarms being presented. EEMUA 191:2007 offers more generalized heuristics for acceptable alarm rates. These are summarized in Table 12.3.

Standing alarms exist while the condition persists; these are sometimes referred to as *raised alarms.* Shelving is a facility that allows the operator to temporarily prevent an alarm from being displayed when it is becoming a nuisance. The EEMUA guide indicates that during normal operations alarm rates should be less than one alarm every 10 minutes. A survey of 37 unique operator consoles in the petrochemical industry revealed that one third of the surveyed consoles were able to achieve this alarm rate. An additional one quarter of the surveyed consoles were found to

TABLE 12.3
EEMUA Guidelines for Usability

Usability Metric	Benchmark Value
Average alarm rate in steady operation	Less than 1 per 10 minutes
Alarms in 10 minutes after plant upset	<10
Average number of standing alarms	<10
Average number of shelved alarms	<30

TABLE 12.4
EEMUA 191 Guidelines

Guidelines	Fully Compatible	Partially Compatible	Not Compatible	EEMUA Page No.
1. Each alarm should alert, inform, and guide.	✓			4
2. Every alarm presented should be useful and relevant to the operator.		✓		5
3. Every alarm should have a defined response.	✓			5
4. Adequate time should be allowed for the operator to carry out a defined response.		✓		5
5. The alarm system should be explicitly designed to account for human limitations.	✓			6
Each alarm should be:				
6. Relevant (not spurious or of low operational value)	✓			6
7. Unique (not duplicating any other alarm)	✓			6
8. Timely (not long before any response is needed, or too late to do anything)	✓			6
9. Prioritized (indicating the importance that the operator deals with the problem)			✓	6
10. Understandable (having a message that is clear and easy to understand)	✓			6
11. Diagnostic (identifying the problem that has occurred)	✓			6
12. Advisory (indicative of the action to be taken)		✓		6
13. Focusing (drawing attention to the most important issues)			✓	6
14. All alarm settings should be systematically determined and documented.	✓			17

be achieving the manageable rate of one or two alarms per 10-minute period. The surveyed consoles performed less well at achieving the EEMUA target of less than 10 alarms in a 10-minute period after an upset. Only 2 out of the 37 consoles even came close to achieving this.

12.7 EXAMPLE

The following analysis has been based upon assessing the alarm system's compliance with the guidelines given in EEMUA 191. In general terms, the analysis of the alarm system using EEMUA 191 as the benchmark showed that many of the criteria were met, but it fell short in some areas. Alarm prioritization is one of the areas where the current system has known and accepted shortcomings. The analysis is summarized in Table 12.4, which shows that the current system complies with 15 of the guidelines

TABLE 12.4 (continued)
EEMUA 191 Guidelines

Guidelines	Fully Compatible	Partially Compatible	Not Compatible	EEMUA Page No.
15. All changes to alarm settings should be documented with reasons.	✓			17
16. Alarm list displays should be designed such that repeating alarms do not cause them to be unstable.			✓	28
17. The alarm on the alarm list should have a quick navigation link to the appropriate operational display.		✓		28
18. Color coding should reflect the alarm priority and state.		✓		30
19. Alarm overload should be eliminated.			✓	21
20. An integrated design should be developed for all audible warnings in the control room.		✓		30
21. It should be possible to shelve individual alarms.			✓	144
22. It should be possible to automatically suppress alarms.		✓		144
23. The alarm status should be indicated on onsite schematics of the objects.		✓		145
24. It should be possible to accept alarms from site schematics.			✓	145
25. An alarm list should be provided.	✓			145
26. It should be possible to accept alarms from the alarm list without referring to the detailed plant display.	✓			145
27. It should be possible to accept alarms from the alarm list either individually or by the page with a single operator action.		✓		145
28. It should be possible for the operator to accept alarms when they are raised and when they are cleared.	✓			145
29. The alarm list should show alarm occurrences in strict chronological order according to their time tag.	✓			145
30. It should be possible for the operator to select a historical log of alarms in strict chronological order.	✓			145
31. It should be possible for the operator to select a filtered list that shows only alarms of a particular priority.		✓		145

(continued)

TABLE 12.4 (continued)
EEMUA 191 Guidelines

Guidelines	Fully Compatible	Partially Compatible	Not Compatible	EEMUA Page No.
32. It should be possible for the operator to select a filtered list which shows only alarms from particular plan areas.			✓	145
33. It should be possible for the operator to view a list of all standing alarms.			✓	145
34. The alarm response procedures should be informative.		✓		32
35. The alarm response procedures should be instructive.		✓		32
Total	15	12	8	

TABLE 12.5
EEMUA Classification of Alarm Systems

Performance Level	Characteristics	Typical Key Performance Indicators
1—Overloaded	Alarm system is difficult to use during normal operation and in practice is ignored during plant upset as it becomes unusable.	Average alarms/10 minutes, >100 Maximum alarms/10 minutes, >1000 Percent (%) hours when there were more than 30 alarms, >50%
2—Reactive	Alarm system is more stable and useful during normal operation but is often unusable in practice during plant upset.	Average alarms/10 minutes, <100 but >10 Maximum alarms/10 minutes, >1000 Percent (%) hours when there were more than 30 alarms, <50% but >25%
3—Stable	Alarm system is reliable during normal operation, providing early warning of impending plant upset, but is less useful during plant upset.	Average alarms/10 minutes, <10 but >1 Maximum alarms/10 minutes, <1000 but >100 Percent (%) hours when there were more than 30 alarms, <25% but >5%
4—Robust	Alarm system is reliable during all plant modes, including normal operation and plant upset.	Average alarms/10 minutes, <10 but >1 Maximum alarms/10 minutes, <100 but >10 Percent (%) hours when there were more than 30 alarms, <5% but >1%
5—Predictive	The alarm system is stable at all times and provides the operator with the right information at the right time to avoid process upset or to minimize the impact of any upset that does occur.	Average alarms/10 minutes, <1 Maximum alarms/10 minutes, <10 Percent (%) hours when there were more than 30 alarms, <1%

and partially complies with an additional 12. On the EEMUA 191 alarm questionnaire, most of the control room staff reported that most of the alarms were useful to them and that they understood the meaning of them. The rates of alarms presented to Operations Engineers are currently greater than ideal but are similar to the industry standard (i.e., 1 alarm every 1 to 2 minutes). In the event of a communications system failure, this rate can rise to 12 alarms per minute. Based on the data from the study, the alarm system has been classified into the EEMUA 191 performance level, as shown in Table 12.5. From the analysis of the alarm system with reference to the EEMUA 191 criteria, the current system can broadly be described as stable. Future work should be aimed at designing the system into the robust category.

BIBLIOGRAPHY

Baber, C. and Mellor, B. (2001). Using critical path analysis to model multimodal human–computer interaction. *International Journal of Human–Computer Studies*, 54, 613–636.

Bransby, M.L. (2001). Design of alarm systems. In J. Noyes and M.L. Bransby (Eds.), *People in Control: Human Factors in Control Room Design*. Piscataway, NJ: Institute of Electrical Engineers.

Bransby, M.L. and Jenkinson, J. (1998). *The Management of Alarm Systems*. London: HSE Books.

Card, S.K., Moran, T.P., and Newell, A. (1983). *The Psychology of Human–Computer Interaction*, Hillsdale, NJ: Lawrence Erlbaum.

Edworthy, J. (1994). Urgency mapping in auditory warning signals. In N.A. Stanton (Ed.), *Human Factors in Alarm Design* (pp. 15–30). London: Taylor & Francis.

Engineering Equipment and Materials Users' Association (EEMUA). (2007). *Alarm Systems: A Guide to Design, Management, and Procurement*. London: Engineering Equipment and Materials Users' Association.

Graham, R. (1999). Use of auditory icons as emergency warnings: evaluation within a vehicle collision avoidance application. *Ergonomics*, 42, 1233–1248.

Gray, W.D., John, B.E., and Atwood, M.E. (1993). Project Ernestine: validating a GOMS analysis for predicting and explaining real-world performance. *Human–Computer Interaction*, 8, 237–309.

Health and Safety Executive (HSE). (1996). *Guidance on the Health and Safety (Safety Signs and Signals) Regulations 1996*, Document L64. London: Health and Safety Executive.

Health and Safety Executive (HSE). (1999). *Reducing Error and Influencing Behaviour*. London: HSE Books.

Health and Safety Executive (HSE). (2007). *Better Alarm Handling*, Chemicals Sheet No 6. London: Health and Safety Executive (http://www.hse.gov.uk/pubns/chis6.pdf).

Hone, K.S. and Baber, C. (2001) Designing habitable dialogues for speech-based interaction with computers. *International Journal of Human–Computer Studies*, 54(4), 637–662.

John, B.A. and Newell, A. (1990). Toward an engineering model of stimulus–response compatibility. In R.W. Proctor and T.G. Reeve (Eds.), *Stimulus–Response Compatibility: An Integrated Perspective* (pp. 427–479). Amsterdam: North-Holland.

Olsen, J.R. and Olsen, G.M. (1990). The growth of cognitive modelling in human–computer interaction since GOMS. *Human–Computer Interaction*, 5, 221–265.

Rasmussen, J. (1974). *The Human Data Processor as a System Component: Bits and Pieces of a Model*, Report No. Risø-M-1722. Roskilde, Denmark: Danish Atomic Energy Commission.

Reason, J. (1990). *Human Error*. New York: Cambridge University Press.

Stanton, N.A. (1994). Key topics in alarm design. In N.A. Stanton (Ed.), *Human Factors in Alarm Design* (pp. 221–231). London: Taylor & Francis.

Stanton, N.A. and Baber, C. (1995). Alarm initiated activities: an analysis of alarm handling by operators using text-based alarm systems in supervisory control systems. *Ergonomics*, 38, 2414–2431.

Stanton, N.A. and Baber, C. (2008). Modelling of human alarm handling response times: a case study of the Ladbroke Grove rail accident in the UK. *Ergonomics*, 51(4), 423–440.

Stanton, N.A. and Edworthy, J. (1999). Auditory warning affordances. In N.A. Stanton and J. Edworthy (Eds.), *Human Factors in Auditory Warnings* (pp. 113–128). Aldershot: Ashgate.

Zwaga, H.J.G. and Hoonhout, H.C.M. (1994). Supervisory control behaviour and the implementation of alarms in control process. In N.A. Stanton (Ed.), *Human Factors in Alarm Design* (pp. 119–138). London: Taylor & Francis.

13 Environment

13.1 OUTLINE

The control room environment encompasses numerous basic physical conditions such as temperature, air quality, aesthetics, noise, and vibration. These physical conditions create subjective states in control room operators, states that have the potential to influence performance.

13.1.1 Accepted Definition

The broad intent of environmental ergonomics "is to create prevailing ambient conditions that are comfortable, acceptable and do not compromise work performance or worker health" (Hedge, 2005a). The relevant standard, BS/EN/ISO 11064-6:2005 (Ergonomic Design of Control Centers. Part 6. Environmental Requirements for Control Centers), states that factors such as lighting, humidity, temperature, vibration, noise, and aesthetics must be addressed, because poor environments can seriously affect operator performance.

13.1.2 Brief Description

It is fair to say that technology developments over the years have served to make the control room an increasingly sanitized and benign work setting in so far as extreme environmental conditions are concerned. In general terms, technology has brought with it a shift away from local manual control, whereby operators physically manipulate controls located on or around the plant itself and in doing so expose themselves to a wider array of potentially harmful environmental agents. Technological advances have progressively distanced the operator from the physical conditions extant at the item of plant being controlled and thereby limited the range and scope of environmental factors now relevant. From local automatic control through to remote and centralized control, the typical modern-day control room, in terms of its basic environmental conditions, is often allied more closely to an office setting than it is to an industrial one. Environmental ergonomics in control room design, therefore, focuses on comfort and amenities as well as safety.

13.2 KEY ISSUES

As an indoor activity, the range of environmental conditions in a contemporary control room setting is limited in extent and magnitude. In the majority of cases, the ergonomic design of these conditions is concerned with a range of more subtle and insidious phenomena such as, for example, the negative impacts of noise from information technology (IT) equipment and the so-called sick-building syndrome. In most cases, the resultant decrements to well-being and performance tend to accrue over a longer time period. Fortunately, by virtue of the indoor nature of modern control rooms, a significant degree of control over the physical environment can be exercised and a satisfactory set of conditions can, with proper design and guidance, be instantiated.

Safe, comfortable, and effective control room environments rely on the careful manipulation of the following key variables:

- Ambient temperature
- Ambient light
- Ambient noise
- Air quality
- Vibration
- Aesthetics

13.3 STANDARDS AND GUIDELINES

13.3.1 Relevant Standards

The principal standard covering environmental conditions in control rooms is BS/EN/ISO 11064-6:2005 (Ergonomic Design of Control Centers. Part 6. Environmental Requirements for Control Centers). This standard deals with the key areas of temperature, light, noise, air quality, vibration, and aesthetics mentioned above, as do several other companion standards. BS/EN/ISO 11064-6:2005 contains general principles for environmental design, requirements and recommendations for each key area, and more explicit informative guidance in terms of actual values to achieve. The general principles of environmental design, summarized in Table 13.1, convey the spirit of the standard. Figure 13.1 provides a high-level overview of how the general principles embodied in BS/EN/ISO 11064-6:2005 can be implemented in a more practical way. There are three key elements: a focus on operational goals as objectives, a hierarchical set of design steps that help to manage and document trade-offs, and specific requirements that take into account both users and the control room environment. In essence, the flowchart in Figure 13.1 represents a high-level overview of how to achieve joint optimization from a set of sometimes conflicting environmental demands. Requirements and recommendations from BS/EN/ISO 11064-6:2005 pertaining to thermal conditions, air quality, lighting, acoustics, vibration, interior design, and aesthetics are summarized in Table 13.2. Note also that the Health and Safety (Display Screen Equipment) Regulations (1992) also set out minimum requirements for several control room environment parameters.

13.4 HUMAN FACTORS LITERATURE

13.4.1 Objectively Measurable States vs. Subjectively Felt States

Environmental agents can be defined and measured in two ways:

TABLE 13.1
General Principles of Environmental Design

Principle	Description
Principle 1 Operator Focus	Operator task demands and comfort shall be the primary focus when designing control centre environments. ...The quality of the working environment shall be an integral part of the overall design process for control centers.
Principle 2 Environmental Adjustability	In order to optimize the operator's performance and comfort, levels of illumination as well as temperature shall be adjustable in accordance with the operator's needs.
Principle 3 Joint Optimization	Where conflicting demands exist between different environmental features (i.e., thermal conditions, air quality, lighting, acoustics, vibration, and interior design and aesthetics), a balance shall be sought which favors operational needs. To that end, the standard recommends an iterative and multidisciplinary approach to design.
Principle 4 External Factors	External factors providing operational information (e.g., security views, weather conditions) shall be taken into account when designing the control center.
Principle 5 Systems Perspective	Environmental factors work in combination and shall be taken into account in a holistic way; that is, the whole environmental entity needs to be taken into account (e.g., interaction between air-conditioning systems generating noise and the acoustic environment). In addition, consider that environmental design shall be used to mitigate the detrimental effects of shift work (e.g., raising ambient air temperature in the early morning).
Principle 6 Flexibility and Future Change	The design of environmental systems shall take account of future change (e.g., equipment, workstation layouts, and work organization).

Source: Permission to reproduce table from BS/EN/ISO 11064-6:2005 is granted by the British Standards Institution.

- In terms of their publicly observable physical state
- In terms of their subjectively felt state

Although the former enables the control room environment to be readily characterized in linear numeric terms (such as degrees centigrade, decibels, lux, parts per million, and so on), the latter state requires different units of measurement and is marked by considerable nonlinearity. *Loudness*, for example, is a subjective label that refers to the physical quantity properly referred to as sound *intensity*. The subjective experience of sound intensity (i.e., loudness) is not uniformly related to intensity, which means that sounds of identical intensity can be subjectively experienced as being of different loudness. To put it simply, the sense mechanisms by which control room operators experience the prevailing control room environment do not behave like the instruments used to measure them. The role of environmental ergonomics, therefore, is to help design the physical state of the environment to elicit the desired subjective experience—one of comfort, amenity, health, and performance.

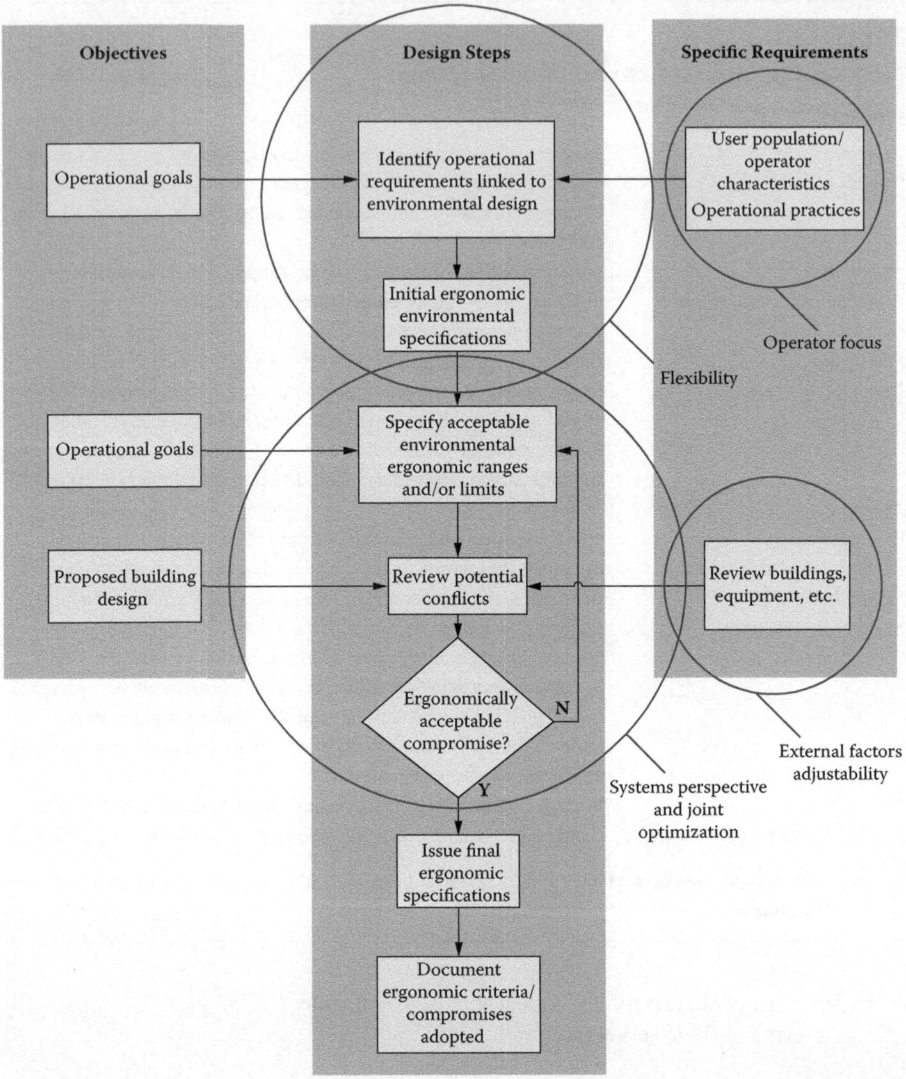

FIGURE 13.1 Process flowchart showing how general principles of BS/EN/ISO 11064-6:2005 can be turned into practice. (Permission to reproduce figure from BS/EN/ISO 11064-6:2005 is granted by the British Standards Institution.)

13.4.2 Regulatory Processes

The inherent nonlinearity between the physically measurable state of a control room environment and the comfortable subjectively felt state of it is ameliorated by the fact that most felt states exist not as points but as ranges. Because the human body is able to regulate itself, comfort, health, and performance can be sustained over a comparatively wide range of physical values. It is important to note that this regulatory characteristic is both physiological (e.g., pupils dilate in darker conditions,

TABLE 13.2
Environmental Requirements and Recommendation

Section	Description
Section 5.1. Ambient Temperature	The goal is to achieve a comfortable thermal environment, devoid of localized temperature-related anomalies, one over which the operators have some degree of control. Concomitantly, HVAC equipment should be fit for the task and able to maintain a stable temperature despite normal variations in external conditions. The thermal environment is not to be specified by air temperature alone but also by radiant temperature, air velocity, and humidity.
Section 5.2. Air Quality	An appropriate cycling rate of fresh outside air into the control room should be provided but not in such a way as to introduce drafts nor to admit pollutants, particulates, and/or stimulate the growth of bacteria. Positive air pressure (or even air tightness) can be beneficial in terms of helping to prevent ingress of pollution, especially in high hazard situations. The HVAC equipment itself should not introduce vibration, while grills and vents should be of a type, and installed in such a way, as to facilitate cleaning, draft-free airflow, and good air circulation.
Section 5.3. Lighting	Lighting conditions are to be designed so they are appropriate to the visual demands of operators while reducing at all times artifacts such as glare. A reference to natural daylight is to be provided, and the lighting scheme should incorporate diversity, with a mixture of artificial and natural light (both of which should be operator adjustable). Levels of illumination should, in addition, refer to the "maintained level" over the lifespan of the lighting device.
Section 5.4. Acoustics	An appropriate acoustic environment should be provided given the operators' task, communication, and alarm handling needs. To that end, a control room acoustically isolated from noise-generating external elements, and one with low reverberation times, should be provided. Consideration should be given to further acoustic isolation of noise-generating equipment found within the control room environment itself.
Section 5.5. Vibration	Vibration is, as far as practicable, to be eliminated from the control room environment. This means constructing the control room away from vibration-generating equipment or else isolating the equipment and/or the control room itself. Note that many of the measures for acoustic isolation will be applicable to vibration isolation as well.
Section 5.6. Interior Design and Aesthetics	Although often a matter of taste and corporate brand identity, some interior design and aesthetics guidelines are offered. Generally, the use of low ceilings (or otherwise poorly proportioned rooms), sharp color contrasts, heavily textured surfaces, and a brutalist, overly mechanical look should be avoided. Instead, it is recommended that pale colors be used (lighter ceilings, darker floors) along with appropriate furniture and plants in order to "humanize" the space. Consideration needs to be given to wear and tear and the implications of 24-hour usage. It is also recommended that operators participate in the design of the control room and in the selection of wall and floor coverings, etc.

perspiration occurs in hot conditions, respiratory changes occur in dry conditions) but also behavioral (e.g., individuals will move in closer to objects and squint in poor light, individuals will move more in cold conditions). What needs to be recognized is that sustained exploitation of these regulatory process comes at both a physiological and behavioral cost to the individual and, ultimately, to the performance of the system being controlled. In broad terms, physiological consequences tend to be related to those longer term and insidious health effects such as eye strain, sick-building syndrome, and other low-level issues of well-being. Behavioral consequences tend to relate more overtly to the performance of the system in the form of maladaptive behaviors developed over longer or shorter time spans.

13.4.3 Systems Perspective

Environmental agents (and their effects) do not occur in isolation. The subjectively felt state often arises as a product of individual environmental agents rather than their mere summation. This can mean that the subjectively felt state of an environment is likely to be disproportionately more (or indeed less) than its component parts. A good part of the effort involved in creating an optimum control room environment is in managing tradeoffs between apparently conflicting goals; for example, an optimum temperature may only be achievable with an increase in noise from ventilation equipment. Likewise, the corollary of sound emission is vibration; the physical attenuation of one can exacerbate the other. Adopting a systems perspective (and in doing so recognizing the interrelatedness of environmental factors) is an appropriate conceptual response, one that is enshrined in BS/EN/ISO 11064-6:2005.

13.4.4 Latest Thinking

Comfort, amenity, health, and well-being, combined with the environmental parameters that can be adjusted to achieve this outcome, create the impression of a rather bland work environment. The natural extension of the systems perspective, however, is to consider the ambient environment as somewhat more dynamic. This arises because:

- Human operators will adapt themselves to the system, and the system to them, by making whatever small or large adjustments that are permissible or possible in the context of their work.
- Distributed cognition occurs when the control room environment itself is informational; for example, the ambient temperature may inadvertently provide information about the time of day (perhaps it is warmer at midday than it is in the afternoon). Although trivial sounding, a collection of these environmental cues can have a meaningful impact on how operators perform their task.

A specific example can be given from air-traffic control. In this setting, a considerable component of ambient noise is the relentless clicking sound of plastic flight strip holders being placed and moved around in their slots. Controllers report that, far from being distracting, this noise serves as an instantaneous clue as to the presence of new aircraft in their sectors (e.g., Walker et al., in press). The latest thinking in the human factors field suggests, then, that care must be taken that environmental design does not become an end in itself and that valuable implicit cues useful for task performance are not inadvertently designed out. An ergonomically optimized control room environment may not be the bland, warm, softly lit, silent place that it might at first be assumed.

13.5 GUIDANCE AND RECOMMENDATIONS

Annex A of BS/EN/ISO 11064-6:2005 contains practical recommendations for creating safe and comfortable control room environments by specifying values for each environmental factor covered. These can be regarded as minimum standards. They serve as useful measurement criteria and thresholds in the context of a rapid ergonomic assessment. They do not, however, always provide a particularly fine-grained level of analysis in terms of mapping objective measurements to subjective states. The values are also based on the assumption that control room work is ostensibly sedentary in nature, which may or may not be the case. Reference to the human factors literature identifies what can be regarded as current best practice. Techniques under this heading address the key issues mentioned above in that they encourage a systems perspective, provide a much finer level of resolution, and more successfully map subjective states onto objective environmental conditions. Both minimum requirements and best practice are presented in the sections below. Although minimum requirements must be met, best practice offers enhanced environmental ergonomics with often quicker, more efficacious methods to ensure compliance.

13.6 ASSESSMENT AND EVALUATION

13.6.1 Ambient Temperature

13.6.1.1 Minimum Requirements

For the purposes of compliance with BS/EN/ISO 11064-6:2005, temperature is expressed in degrees centigrade (°C). Ignoring the cooling effect of wind, 10°C equates to a seasonal average in London for a typical day in March; likewise, 20°C

TABLE 13.3
Control Room Conditions

	Sedentary Activity	
Factor	**Winter Conditions**	**Summer Conditions**
Temperature	20–24°C	23–26°C
Vertical air temperature difference	<3°C	<3°C
Surface temperature of the floor	19–26°C	19–26°C
Mean air velocity	0.15 m/sec	0.15 m/sec
Radiant temperature asymmetry	<10°C	<10°C
Relative humidity	30–70%	30–70%

Source: Permission to reproduce table from BS/EN/ISO 11064-6:2005 is granted by the British Standards Institution.

equates to the same in mid-July. Comfortable control room conditions relate more to the subjective experience of the latter (see Table 13.3). In addition to the above, the standard points out that "control room operators should be able to increase control room temperatures by 1°C to 2°C in the early hours of the morning to compensate for diurnal rhythms."

13.6.1.2 Human Factors Best Practice

The predicted mean vote (PMV) index combined with the predicted percentage dissatisfied (PPD) index represents established ergonomic best practice in terms of assessing thermal comfort (Toftum, 2005). The PMV index predicts the mean value of the overall thermal sensation of a large group based on estimates of metabolic rate and clothing insulation and estimates or actual measurements of air temperature, mean radiant temperature, relative air velocity, and humidity. Based on the measurements or estimates of these six variables, the PMV value predicts how a large group of people will rate the thermal environment according to the following scale:

+3 = Hot
+2 = Warm
+1 = Slightly warm
0 = Neutral
–1 = Slightly cool
–2 = Cool
–3 = Cold

The PMV index is able to undertake this prediction by being based on a model derived from actual large-scale data. The PMV index is thus an effective way of translating objective environmental measures into subjectively felt states. The goal, generally, is to achieve a thermally neutral situation whereby the heat generated by activity and metabolism (as affected by clothing insulation) is in balance with the heat loss due to the temperature differential of the room.

It is not possible to suit everyone, so, although the PMV index provides a good overall measure of sensation, the PPD index provides a measure of dissatisfaction. The graph in Figure 13.2, used to plot the obtained PMV value, makes it possible to obtain an estimate of the percentage of people likely to be thermally dissatisfied. Generally, the goal is to achieve a PPD value of less than 10%. Values over 10% are normally taken as a cue to consider further interventions.

For a control room with a mean air velocity of 0.15 m/sec (as per BS/EN/ISO 11064-6:2005), a relative humidity of 50%, and operators wearing light clothing (0.5 clo) and engaged in sedentary activity seated at desks (with a metabolic rate of 70 W/m^2), an air temperature of 18°C would yield a PMV value of –2.17, which would be a subjectively felt rating of cool. The corresponding PPD value, according to the graph in Figure 13.2, would be <80%, meaning that more than 80% of people would find this thermally dissatisfying. Raising the temperature to 26°C yields a PMV value of 0.26, which is very close to thermally neutral. Under these

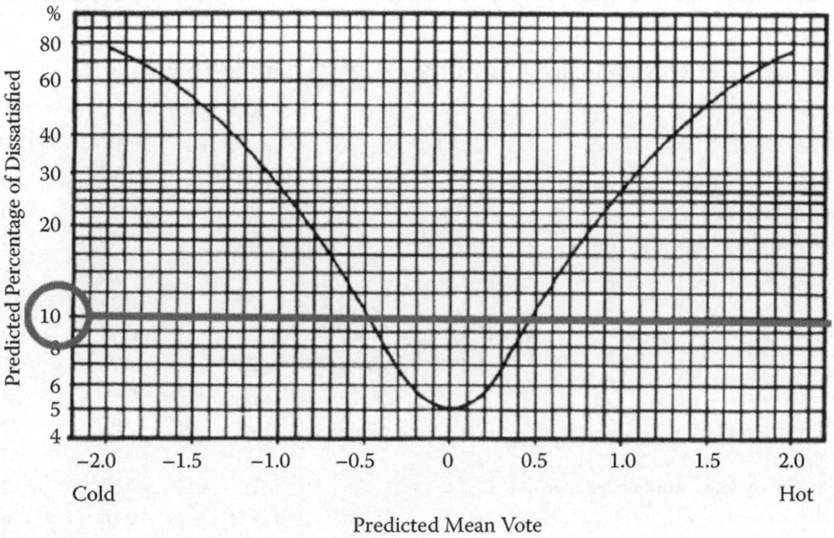

FIGURE 13.2 Graph showing how predicted mean vote (PMV) values map onto the predicted percentage of people thermally dissatisfied; a value of 10% or less is desired. (Permission to reproduce figure from BS/EN/ISO 7730:1995 is granted by the British Standards Institution.)

conditions, the corresponding PPD rating is just 5%. Note that a similar effect can be achieved with heavier clothing (raising the value in the calculation from 0.5 clo to 1.0 clo).

13.6.1.3 Ambient Temperature Example

Table 13.4 has been created from the information and estimates contained in BS/EN/ISO 7730:1995 (Moderate Thermal Environments—Determination of the PMV and PPD Indices and Specification of the Conditions for Thermal Comfort). It represents a subset of values most likely to be useful and relevant to environmental design in control rooms. Note that some of the air temperature values are higher than those set in BS/EN/ISO 11064-6:2005 as minimum requirements. This illustrates the importance of a more fine-grained analysis linked to the subjective experience of the thermal environment.

An environmental assessment was undertaken at the control room. A full thermal assessment requires specialist equipment (in particular, for measuring air velocity and radiant temperature) that is beyond the scope of this assessment. To that end, measurements took place at six workstation positions, an example of which is shown below in Figure 13.3. The control room thermostat was set at 20°C on the day of assessment. The external temperature was 12°C, and it was overcast with clouds. The windows were closed, and the blinds were drawn at the windows. Table 13.5 presents the outcomes of the analysis in more detail and indicates where compliance was achieved.

TABLE 13.4
Summary Table of Relevant Control Room PMV and PPD Values

Activity Level	Clothing	Air Temperature	PMV	PPD (%)
Sedentary (e.g., desk-based work undertaken while seated) Metabolic rate ≈ 70 W/m²	Light (e.g., light trousers and short sleeves)	20	−1.58	60
		22	−0.97	26
		24	−0.36	8
		26[a]	0.26	6[a]
		28	0.88	18
	Heavy (e.g., shirt, trousers, jacket)	20	−0.45	9
		22[a]	0.00	5[a]
		24	0.46	9
		26	0.91	22
		28	1.37	45
Light activity (e.g., work undertaken whilst standing) Metabolic rate ≈ 90 W/m²	Light	20	−0.62	12
		22[a]	−0.16	6[a]
		24	0.30	7
		26	0.76	16
		28	1.23	36
	Heavy	20[a]	0.19	6[a]
		22	0.53	10
		24	0.87	22
		26	1.21	38
		28	n/a	>80
Medium activity (e.g., work undertaken while moving and walking) Metabolic rate ≈ 116 W/m²	Light	20	−0.04	5
		22	0.35	8
		24	0.75	18
		26	1.15	35
		28	1.55	52
	Heavy	20[b]	0.61[b]	13[b]
		22	0.91	22
		24	n/a	>80
		26	n/a	>80
		28	n/a	>80

[a] Compliance has been achieved.
[b] A temperature lower than 20°C is required to yield further improvements in PMV and PPD.

The analysis showed very small differences in air temperature around the control room. One work station measured a little cooler than the others, and it was determined that the air-conditioning equipment in the direct vicinity sometimes created a slight cooling draft. Likewise, the workstation in front of it was a little warmer. The differences are very small, however, and the air conditioning system installed in this example has over 20 possible air-flow patterns and a wide range of adjustment. The small differences noted appear to fall within the domain of fine tuning.

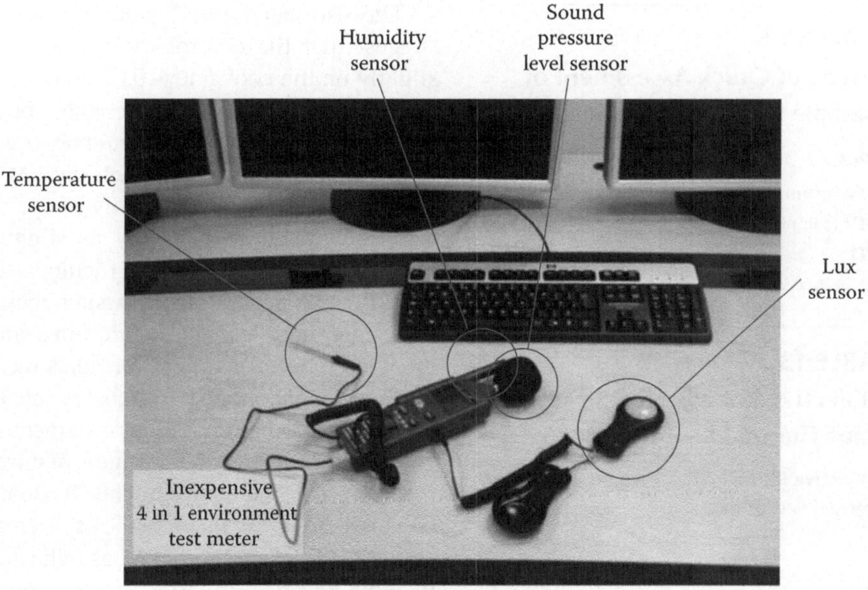

FIGURE 13.3 Test equipment used to take environmental measurements at workstations.

A visual assessment of the clothing worn by the control room operators suggested that it is light in nature (0.5 clo) and serves as a reasonable basis for calculating the PMV value. A similar visual assessment of physical activity was also undertaken. During the observation period, operators mostly performed their duties while seated at their desks with occasional walking and standing. BS/EN/ISO 7730:1995 suggests that a metabolic rate of approximately 90 W/m^2 is appropriate. The resultant PMV and PPD values are shown in Table 13.6 in relation to desired values. In addition, a sample of 22 operators rated the thermal comfort of the environment directly onto the PMV scale, and these results are also shown.

TABLE 13.5
Actual vs. Obtained Values Pertaining to the Thermal Environment

Variable	Acceptable Values	Obtained Values
Air temperature	20–24°C	20.1–22°C (mean, 21°C)[a]
Air velocity	0.15 m/sec	Not measured
Relative humidity	30–70%	30.6–35.7% (mean, 33.8%)[a]
Radiant temperature asymmetry	<10°C	2°C[a,b]
Surface temperature of floor	19–26°C	19°C[a]
Adjustment range	1–2°C	2°C within control room, a wider range from a central control point[a]

[a] Compliance has been achieved.
[b] Indicates air temperature asymmetry.

TABLE 13.6
Results of Quick Assessment of Example Control Room Setting

Index	Desired	Actual
PMV (estimated)	≈0	–0.62
PMV (sampled)	≈0	0.82
PPD	<10%	12%

TABLE 13.7
Subjective Ratings on a Seven-Point Thermal Sensation Scale

Subjective Ratings of Thermal Sensation	Percent Agreement (n = 22)
Hot (+3)	18 (n = 4)
Warm (+2)	27 (n = 6)
Slightly warm (+1)	9 (n = 2)
Neutral (0)	23 (n = 5)
Slightly cool (–1)	9 (n = 2)
Cool (–2)	5 (n = 1)
Cold (–3)	9 (n = 2)

The estimated PMV value seems to suggest that the control room might be slightly on the cool side (–0.62), meaning that the percentage of people who would feel thermally dissatisfied would be in the region of 12%. However, when the PMV scale was administered directly to operators, they rated it thermally as slightly warm (0.82). Given the difficulties in quickly and accurately assessing metabolic rate, heat loss due to clothing, and the wide range of temperature adjustment available to operators, it is probably safe to conclude that the thermal environment is compliant in terms of minimum requirements for thermal environments. It should be noted, however, that a range of thermal comfort ratings was reported, as indicated in Table 13.7; 54% reported they are warm to hot, 41% reported they are slightly cool to slightly warm, and 14% reported they are cool to cold. The data appear to be bimodal, with two peaks at neutral and warm. Four respondents reported that they were hot, and two respondents reported that they were cold. Handwritten comments on the questionnaire suggested that the control room is warmer during the day and cooler at night.

13.6.2 Air Quality

13.6.2.1 Minimum Requirements

The overriding air quality requirement is that the control room should be supplied with outdoor air in sufficient quantities to dilute internally generated pollutants. Air quality does not have such a fixed subjective experience (compared to light or temperature, for example), but air quality issues can result in odors and in a general feeling of stuffiness. Problems with air quality have also been highlighted as giving rise to longer term diffuse problems related to sick-building syndrome and a deleterious feeling of malaise and lethargy. The objective values and thresholds below which such sensations can be avoided are shown in Table 13.8. Note that these measurements apply to a fully occupied control room. Note also that the per-person fresh air supply can be calculated based

TABLE 13.8
Objective Values for Air Quality

Per-person fresh air supply	29 m^3/hr
Maximum CO_2 concentration	910 ppm

Source: Permission to reproduce table from BS/EN/ISO 11064-6:2005 is granted by the British Standards Institution.

on the volumetric capacity of any relevant heating, ventilating, and air-conditioning (HVAC) equipment in relation to the measured volume of the room. Carbon dioxide concentrations require specialist metering equipment.

13.6.2.2 Human Factors Best Practice

A viable strategy for an ergonomic assessment of air quality is shown in Figure 13.4 (Hedge, 2005b). The approach adopted involves an initial walkthrough, comprised of a visual inspection and interviews with staff members. This is followed by objective measurement of the environment, should issues be raised, followed by control interventions and follow-ups, if required. In general, precise air quality measurements in terms of CO_2 concentrations, particulates, and pollution require specialist test equipment. For our purposes, the initial walkthrough and assessment phases (prior to this stage) are offered as best practice. Best practice, then, is a rational approach used to determine whether such measurement and control interventions are required. The checklist shown in Figure 13.5 was completed in the context of a walkthrough type of analysis. Having performed the initial walkthrough assessment, the analyst can then proceed to poll individual control operators on their subjective experience of air quality. A number of air quality assessment questionnaires exist, and the Cornell Office Environment Survey (Hedge and Erickson, 1988) has been suggested as a candidate suitable for control rooms (Figure 13.6). These procedures and the supporting tools, then, serve as the rational basis for assessing whether objective measurement should proceed and, furthermore, what control measures might have to be taken.

13.6.2.3 Air Quality Example

An air quality assessment was carried out at the control room. In terms of minimum requirements, 15 permanent workstations are located in the control room, each requiring a per-person air supply of 29 m^3/hr (or a total of 435 m^3/hr). The volume of the control room is 790 m^3; therefore, the air must be completely changed around once every 2 hours. This is a minimum. In the example, a Biddle Metair system air conditioner has been installed (Figure 13.7) that is able to supply 132 m^3/min of a mixture of recycled and fresh air (or 7920 m^3/hr). The fresh air supply is 54 m^3/min (or 3240 m^3/hr), which means that the air is replaced 4 times per hour. Standards laying down minimum requirements are, therefore, easily achieved.

Carbon dioxide measurements require specialist test equipment and were not carried out in the context of this quick evaluation. The office survey did not detect any subjective phenomena to suggest a problem with CO_2 concentrations. The results of the subjective evaluations are presented in Table 13.9 and Table 13.10. Table 13.9 shows that respondents reported that the control room was both too cold and too hot, in addition to too little air movement and the air being too dry. Measurement of temperature (sampled range was between 20.1 and 22°C with a mean of 21°C) and relative humidity (sampled range was between 30.6 and 35.7% with a mean of 33.8%) showed that they were within acceptable limits. Table 13.10 shows the reports of symptoms experienced by people in the control room that may be associated with the environmental conditions. Over half of the respondents reported sore throats and congested noses, and approximately half reported sore eyes, mental fatigue, headache, and tiredness.

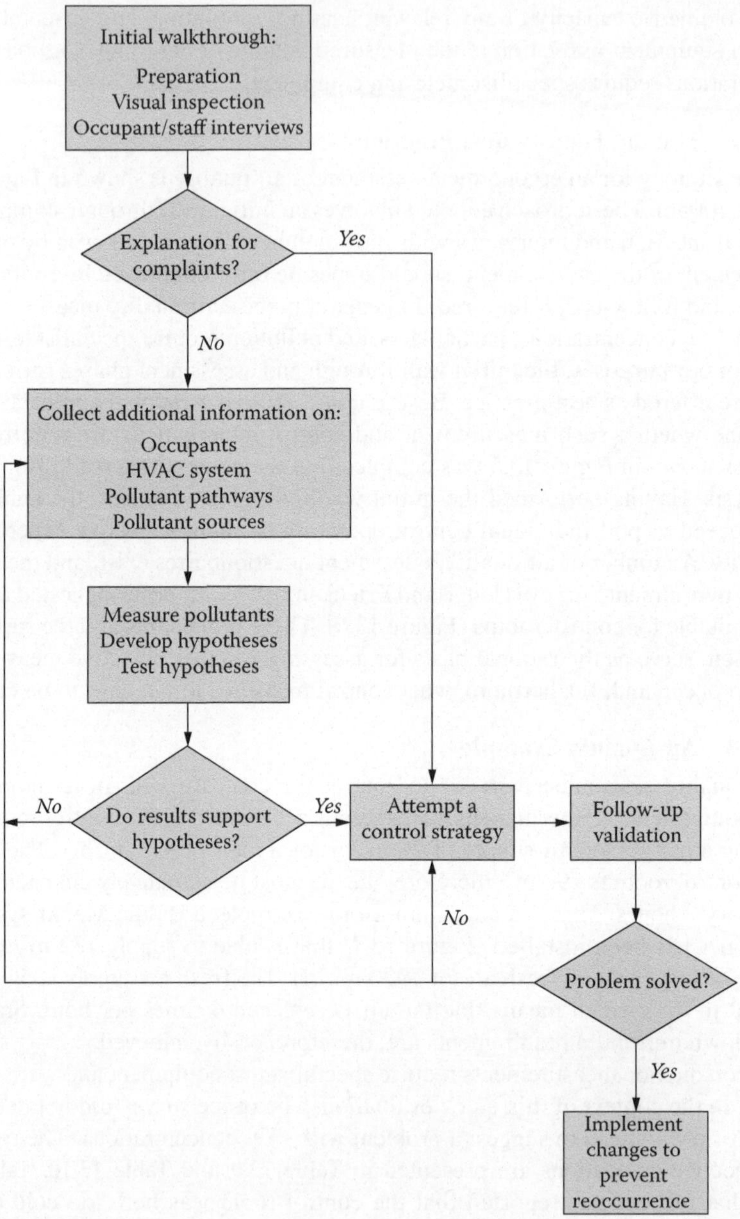

FIGURE 13.4 Flowchart for assessing indoor air quality. (From Hedge, A., in *Handbook of Human Factors and Ergonomics Methods*, N.A. Stanton et al., Eds., CRC Press, Boca Raton, FL, 2005, pp. 604–617. With permission.)

The results of our small sample of air quality walkthrough checklists and office environment surveys suggest that further investigation into air quality is required. There is clearly a discrepancy between the objective and subjective analyses. This

Are there any noticeable drafts? Y ☐ N ☒

Are any air intakes located too close to outlets? Y ☐ N ☒

Is there any marked vibration or noise? Y ☐ N ☒

Is the air supply filtered? Y ☒ N ☐

Are there any nearby external air pollution sources? Y ☐ N ☒

Are measures in place to prevent the malicious introduction of material? Y ☐ N ☒

Are there safety procedures and PPE available/provided in cases of air quality failure? Y ☒ N ☐

FIGURE 13.5 Air quality checklist.

Office Environment Survey

Please answer the following questions about environment comfort condition and health symptoms that you may have experienced in the office during the past month (4 weeks).

What is your gender? Male ☒ Female ☐

Please indicate whether you have experienced any of the following environmental conditions in the workplace over the past month:

	Yes	No
Air temperature too cold	☐	☒
Air temperature too warm	☐	☒
Too little air movement	☐	☒
Air too dry	☐	☒
Unpleasant odor in air	☐	☒
Air too stale	☐	☒
Air too dusty	☐	☒

Please indicate whether you have experienced any of the following symptoms at least once over the past month and whether they got better when you were away from the office (e.g., in the evenings or over the weekend):

	Yes	No
Irritated, sore eyes	☐	☒
Sore, irritated throat	☐	☒
Hoarseness	☐	☒
Stuffy, congested nose	☐	☒
Excessive mental fatigue	☐	☒
Headache across forehead	☐	☒
Unusual tiredness, lethargy	☐	☒

FIGURE 13.6 The Cornell Office Environment Survey. (From Hedge, A., in *Handbook of Human Factors and Ergonomics Methods*, N.A. Stanton et al., Eds., CRC Press, Boca Raton, FL, 2005, pp. 604–617. With permission.)

FIGURE 13.7 Illustration of the heating and ventilation system showing air introduction in floor (top) and extraction in ceiling (bottom).

anomaly could be related to the perceived dryness of the air (despite the measured humidity values being within acceptable limits, albeit the lower end of those limits), and addressing the humidity may alleviate some of the symptoms.

TABLE 13.9
Subjective Assessment of Control Room Environment

Control Room Environment Conditions	Percent Agreement (n = 25)
Air temperature too cold	88 (n = 22)
Air temperature too warm	96 (n = 24)
Too little air movement	68 (n = 17)
Air too dry	68 (n = 17)
Unpleasant odor in air	48 (n = 12)
Air too stale	52 (n = 13)
Air too dusty	28 (n = 7)

TABLE 13.10
Subjective Assessment of Symptoms Experienced

Symptoms Experienced in the Control Room Environment	Percent Agreement (n = 25)
Irritate, sore eyes	44 (n = 11)
Sore, irritated throat	60 (n = 15)
Hoarseness	16 (n = 4)
Stuffy, congested nose	64 (n = 16)
Excessive mental fatigue	48 (n = 12)
Headache across forehead	44 (n = 11)
Unusual tiredness, lethargy	48 (n = 12)

13.6.3 Lighting

13.6.3.1 Minimum Requirements

Lighting requirements adopt the unit lux (lx), which is a measure of illuminance or of the intensity of light. It is rather similar to the way sound is measured in that the measure corrects for the effects of light wavelength, and a measure of brightness that approximates more closely the way the human visual system perceives light is thus obtained. In everyday terms, a typical living room (softly lit) measures approximately 50 lx, whereas a television studio (very brightly lit) measures approximately 1000 lx. BS/EN/ISO 110640-6:2005 states that control rooms, in comparison, should be lit to a maximum level of 500 lx. Minimum requirements for control room lighting are shown in Table 13.11.

BS/EN/ISO 11064-6:2005 states that electric lighting should achieve a glare index (UGR) of 19 or less for all work positions. UGR stands for Unified Glare Index and is calculated based on a number of formulas whose inputs include foreground and background luminance, the size of the lit area, and the number of luminaries. BS/EN/ISO 11064-6:2005 also states that lighting should possess a color rendering index (CRI) of over 80. The perceived color of illuminated objects depends on the light shining on them. Sodium street lighting, for example, is considered monochromatic because everything will appear as varying shades of orange; as such, its color rendition is very poor (a CRI of 0 would be typical of this light source). An incandescent bulb,

TABLE 13.11
Minimum Requirements for Control Room Lighting

Maximum illuminance levels for video display units (VDUs)	500 lux
Maximum illuminance levels (e.g., non-VDU, paperwork)	750 lux
Maintained illuminance levels (e.g., non-VDU, paperwork)	500 lux
Maintained minimum illuminance levels	200 lux

Source: Permission to reproduce table from BS/EN/ISO 11064-6:2005 is granted by the British Standards Institution.

TABLE 13.12
Five Steps of an Office Lighting Survey

Step	Procedure	Additional Requirements
1.	Administer the office lighting survey to control room operators.	—
2.	Measure the illuminance at the working area.	A representative sample of work stations can be selected if there are more than ten. If there are fewer than ten workstations, then it is good practice to take measurements at each one. Likewise, best practice is to take illuminance measures at various points in the normal working area and average them. Be sure that the lights have been on for at least a half hour.
3.	Assess reflections.	Place the laminated sheet of paper on the illuminated work area and check for the presence of bright reflected images. If present, replace the laminated sheet with the mirror and identify the source of the reflection.
4.	Assess lighting sources.	Identify lighting equipment in terms of manufacturer and type and in sufficient detail to cross-check specifications and obtain color rendering index (CRI) values.
5.	Assess discomfort glare.	Glare indexes are often difficult to calculate, but the flowchart in Figure 13.8 may be helpful in rapidly identifying particular glare problem areas.

on the other hand, facilitates excellent color rendition and has a CRI of 100. Typical office and control room lighting is provided by fluorescent lights, which have CRI values ranging from approximately 60 to 90. The standard, therefore, is advocating the careful selection of such lighting, with triphosphor and other varieties being favored for their relatively high CRI values. While on the topic of fluorescent lighting, high-frequency systems are also recommended to reduce flicker effects.

13.6.3.2 Human Factors Best Practice

An extremely tractable method of evaluating the subjective conditions of control room lighting is briefly outlined in Table 13.12 (Boyce, 2005). The procedure requires an office lighting survey, an illuminance meter, a sheet of white paper laminated in clear plastic, and a mirror. Five steps are required.

13.6.3.3 Lighting Example

The control room was the subject of a lighting survey based on the methods just described. At the time of the assessment, the lighting in the control room was switched to its maximum setting and the blinds were closed. Figure 13.9 shows the overall lighting environment and some of the fixtures and fittings used. In addition, the use of high-frequency light sources with high CRI values and dimmer controls can also be noted, as can the presence of windows, natural daylight, and blinds. In

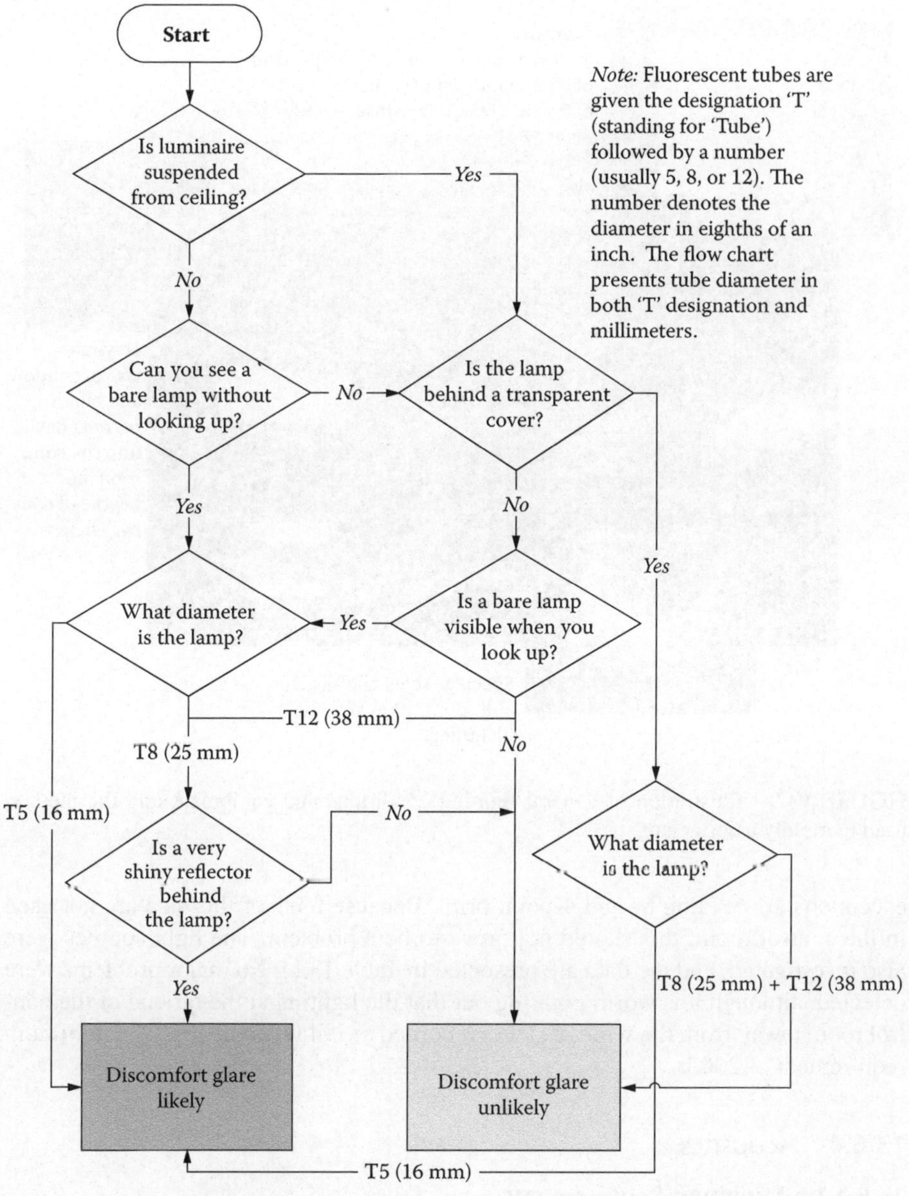

FIGURE 13.8 Discomfort glare flowchart.

this example, the results are presented in two parts. Table 13.13 presents the results of the office lighting survey carried out at this site compared to the data cohort upon which the questionnaire is based. The data in Table 13.13 show that the majority of people (i.e., 87%) are satisfied with the lighting in the control room. Generally, the data from the control room study compare well with the general population data. The

FIGURE 13.9 Illustration of general lighting conditions and equipment and the method used to rapidly assess glare.

exceptions are reading 6- and 4-point print. Because fonts of this size are not used in the control room, this should not prove to be a problem. The light sources were also investigated, and the data are presented in Table 13.14. No major problems were detected, although it is worth pointing out that the lighting at the far end of the control room (away from the windows) was recorded at 194 lx, compared to a minimum requirement of 200 lx.

13.6.4 Acoustics

13.6.4.1 Minimum Requirements

The primary measurement unit for the acoustic environment is decibels (dB). Decibels are an objective measure of sound intensity which is related, albeit nonlinearly, to the subjective impression of loudness. A subjective experience of very loud (e.g., a stereo at high volume) might be in the range of 95 to 100 dB. A very quiet domestic setting, on the other hand, would typically measure around 30 dB.

It is important to point out that the subjective sensation of loudness is frequency dependent. Sounds of the same intensity will generally be perceived to be less loud at lower and higher frequencies. As a result, different corrections and weightings are

TABLE 13.13
Office Lighting Survey

Survey Question		Percent Agreeing with Statement: Population Data (n = 1259)	Sample Data (n = 24)
Overall, the lighting is comfortable.		69	87
The lighting is uncomfortably bright for the tasks I perform.		16	8
The lighting is uncomfortably dim for the tasks I perform.		14	8
The lighting is poorly distributed here.		25	24
The lighting causes deep shadows.		15	8
Reflections from the light fixtures hinder my work.		19	16
The light fixtures are too bright.		14	8
My skin is an unnatural tone under the lighting.		9	8
The lights flicker throughout the day.		4	4
How does the lighting compare to similar workplaces in other buildings?	Worse	19	4
	About the same	60	84
	Better	22	12
Able to read 8-point print and larger.		99	100
Able to read 6-point print and larger.		94	84
Able to read 4-point print and larger.		76	37

Source: Eklund, N. and Boyce, P., *Journal of the Illuminating Engineering Society*, 25, 25–40, 1996. With permission.)

often applied to the dB measure so it reflects the subjective experience of loudness better; dB(A) and dB $L_{Aeq,y}$ are examples. In addition, the decibel measure is logarithmic. In practice, this means that an increase of 3 dB actually means an approximate

TABLE 13.14
Results of Tests Carried Out on Illuminance, Reflection, Light Source, and Discomfort

Factor	Minimum Requirement	Assessment Results
Illuminance of working area	200–500 lux	194–338 lux
Reflections	Situations in which glare is present should be eliminated or ameliorated; a glare index of less than 19 is a minimum requirement.	No glare detected in this measurement location
Light sources	Color rendering index (CRI) > 80	Four Sylvabia Activa 172 (full-spectrum 6500 K color output) fitted inside 600 × 600-mm diffusers; CRI = 98
Discomfort glare results	Discomfort glare unlikely	Discomfort glare unlikely

TABLE 13.15
Everyday Sounds, Their Levels of Loudness, and Associated Subjective Responses

Range	dB(A)	Source	Subjective Response
Safe (for average populations)	0	No sound	Hearing threshold
	10	Rustle of leaf	
	20	Buzzing insect	
	30	Quiet whisper	
	40	Quiet office	
	50	Window air conditioner	
	60	Conversation	
	70	Freight train	
	80	Computer print room	
Risk	90	Heavy vehicle	Limit exposure to 30 minutes at 90 dB(A) or provide protection
	100	Subway station	
	110	Rock drill	
Harmful	120	Propeller airplane	Pain threshold; protection required
	130	Riveting hammer	
	140	Jet airplane at 30 m	

doubling of loudness (all things being equal). Examples of everyday sounds and their associated loudness are presented in Table 13.15. The minimum requirements for the acoustic environment of a control room are given in Table 13.16.

Although more sophisticated acoustical test equipment is capable of deriving d(B) $L_{Aeq,y}$ directly, it is far more common for test equipment to have merely A and C weighting. A and C weighting both reduce the contribution that low frequencies

TABLE 13.16
Minimum Objective Requirements for the Acoustical Environment of a Control Room

Maximum ambient noise	45 dB $L_{Aeq,y}$
Maximum background noise	35 dB $L_{Aeq,y}$
Minimum background noise[a]	30 dB $L_{Aeq,y}$
Maximum ratio of alarm noise to background noise	>15 dB
Mid-frequency reverberation times[b]	0.4 sec

[a] A minimum level of background noise is specified to maintain privacy; if it is too quiet individual conversations stand out.

[b] Reverberation times are a significant factor in the intelligibility of communications. Lower reverberation times help to improve this situation.

Source: Permission to reproduce table from BS/EN/ISO 11064-6:2005 is granted by the British Standards Institution.

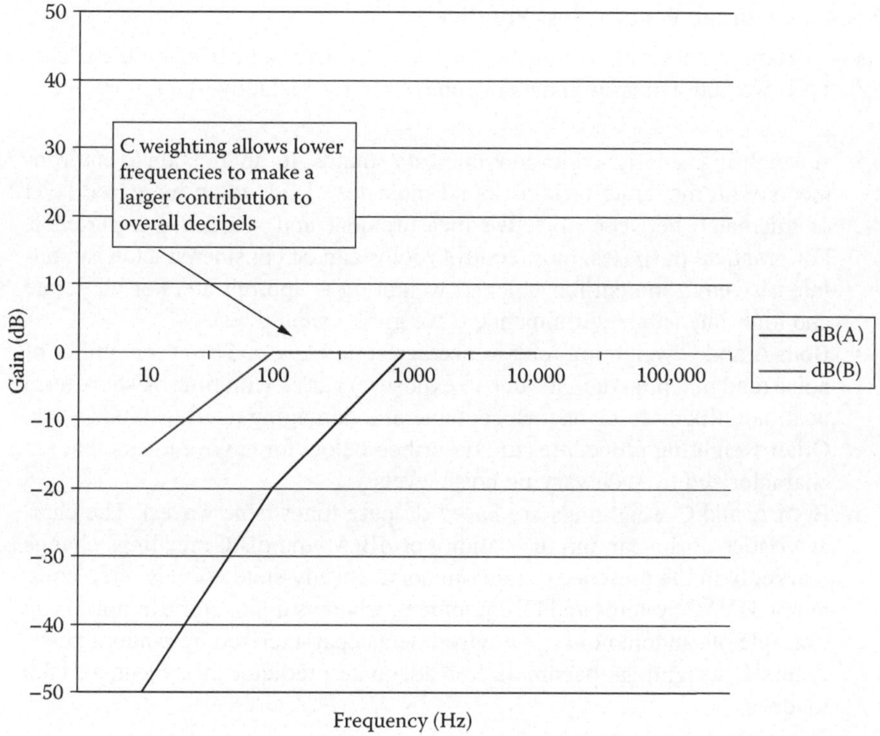

FIGURE 13.10 A and C weighting curves for taking measures of overall decibel levels.

make to a decibel reading. They do this to approximate the way the human ear perceives sound. A weighting approximates the subjective experience of loudness at lower volumes, such as those found in control rooms, and therefore deemphasizes low frequencies quite sharply. C weighting is a better approximation for noisier environments, where the ear tends to have a more uniform, flatter frequency response. Here the ear's sensitivity to low-frequency sounds is relatively greater. Figure 13.10 shows A and C weighting curves in relation to frequency. Note that human hearing is able to detect the vibration of air molecules from 20 Hz (deep bass) to 20,000 Hz (high treble), although this is fairly arbitrary. A more reasonable upper limit for adult human hearing is probably in the region of 15,000 Hz.

ISO 532 (Acoustics—Method for Calculating Loudness Level) contains standards that pertain to the measurement of loudness; however, these are equipment and resource intensive, and the minimum standards expressed by BS/EN/ISO 11064-6:2005 can be assessed using much simpler methods. In addition, very specific procedures for laboratory and free field assessment of IT equipment is dealt with in BS/EN/ISO 7779:2001. For the purposes of assessing control room environments, however, the minimum standards tabulated above, combined with the best practice described below, represent a much simplified strategy for assessment, certainly one that is able to identify problems that require more specialist treatment.

13.6.4.2 Human Factors Best Practice

It is important to deal with a number of caveats that have a bearing on the adequacy of A and C weighted measurements in control rooms, including the following:

- A weighting is designed for low-intensity sounds. Its application to environments with moderate or high sound intensities leads to an increased level of mismatch between objective measurement and subjective experience. For practical purposes, most control rooms can be considered a low sound-intensity environment for which A weighting is appropriate. For moderate and high-intensity environments, C weighting should be used.
- Both A and C weighting refer to average sound levels. The acceptability of noise (and potential damage due to exposure) varies with time. A short-term peak is unlikely to be as unacceptable and damaging as a sustained peak. Other weighting procedures are described below for environments that are characterized by such varying noise levels.
- Both A and C weightings are based on pure tones (sine waves). The characteristics of the ear and the validity of dB(A) and dB(C) readings change markedly in the presence of random noise. Steady-state noise is often emitted by HVAC systems and IT equipment, whereas dialog, for example, is an example of random noise. In environments characterized by random noise, A and C weightings become a less adequate predictor of subjectively felt loudness.

In summary, if the control room environment is of low sound intensity and if the sounds are of a more or less continuous nature (such as the noise from IT equipment, steady-state background noise, and so forth), then the A weighting procedure is likely to provide a good approximation of the subjective experience of loudness in that context. In noisier but still ostensibly steady-state acoustic environments, the C weighting procedure provides a better approximation. If the acoustic environment is characterized by fluctuating and random noise (e.g., a great deal of changing background noise), then additional weighted measurements are appropriate:

- dB $L_{Aeq,y}$ is based on the A weighting curve but uses longer term averaging and maps better onto the subjective experience of loudness as a function of time exposure. dB $L_{Aeq,y}$, therefore, will tend to ignore the effects of short-term, high-intensity noise (which may attract attention but be of little annoyance) compared to long-term higher intensities (with greater annoyance potential).
- ITU-R 468 is a weighting curve designed for environments characterized by random noise for which the sensation of loudness is considerably different compared to steady-state noise. ITU-R 468 is a counterpoint to A weighting, the limitations of which are the subject of an amended standard ISO 226:2003.

In cases where the use of these weightings is deemed appropriate, equipment somewhat beyond the technical capability of most inexpensive sound meters is required. Where the facility does exist, however, then BS/EN/ISO 11064-6:2005 would recommend the use of dB $L_{Aeq,y}$ in preference to dB(A). Likewise, when encountering control rooms characterized by significant random noise elements, additional measurements using the ITU-R 468 weighting are recommended. In terms of best practice, a good strategy to employ, one that requires only basic equipment but goes beyond minimum requirements, is the rapid sound quality assessment technique (Torres, 2005). This method has three steps and requires the use of a sound level meter with A and C weighting curves (a common feature). The three steps are as follows:

1. Configure the sound level meter to read in dB(A) and take a number of measurements at individual workstations. Take the average of the readings at each workstation.
2. Configure the sound level meter to read in dB(C) and take a number of measurements at individual workstations. Take the average of the readings at each workstation.
3. Subtract the averaged dB(C) reading from the averaged dB(A) reading.

The difference value contains useful information concerning the character of the ambient sound in the control room, where:

- A level of 10 dB or greater relates to the ambient noise having a "hissy" character. These mid- to higher frequency components can potentially affect the intelligibility of dialog and may even have a subtle masking effect on alarm sounds if they are of similar spectral content.
- A level of approximately 15 dB is subjectively neutral, the ambient noise being biased in neither high- nor low-frequency directions.
- A level of 20 dB or higher indicates that the ambient sound has a low-frequency rumbling character. This is a clue that excessive low-frequency content may be present which, in itself, has been associated with poor cognitive performance. In terms of auditory perception, low-frequency sound is omnidirectional, so subjectively it can be difficult to take action to alleviate the sensation (i.e., moving around the room or even localizing the source is difficult or has no effect). Furthermore, high levels of low-frequency content may suggest the presence of vibration in the control room and should be a cue for further analysis and intervention. It is also a clue that infrasonic phenomena (Pelmear and Benton, 2003) may be present which again can have subliminal effects on control room operators and be a contributor to sick-building syndrome.

In cases where noise pollution may be present, techniques exist to survey individuals as to its effects and causes. Various questionnaires exist for this purpose (e.g., Soames Job, 2005).

TABLE 13.17
Acoustical Environment Results Compared to BS/EN/ISO 11064-6:2005

Factor	Standard	Measured
Maximum safe noise	80 dB(A)	69.4 dB(A)
Maximum ambient noise	45 dB $L_{Aeq,y}$	49.7 dB(A) (marginal)
Maximum ratio of alarm noise to background noise	15 dB	No measurable difference, but the alarm sound was distinctive
Mid-frequency reverberation times	0.4 sec	Not measured, but none noticed in subjective assessment
dB(C) – dB(A)	15 dB (neutral/acceptable)	13 dB (neutral/acceptable)

13.6.4.3 Acoustics Example

The acoustical environment of the control room in our example was measured in accordance with the procedures just described and the results are shown in Table 13.17. $L_{Aeq,y}$ is based on the dB(A) measure but is averaged over time so they are not directly comparable. What does draw note is the fact that the difference between minimum and maximum sound levels is comparatively wide. This quick assessment has served its purpose in identifying, at little cost (time or financial), that there was no need for further investigation. Note, however, that the character of the background or ambient sounds is acceptable in terms of human factors best practice.

13.6.5 Vibration

13.6.5.1 Minimum Requirements

BS/EN/ISO 11064-6:2005 does not specify minimum criteria for objectively measurable vibration. Many standards do indeed refer to vibration but this typically takes the form of whole or part body vibration of the sort caused by operating machinery or vehicles. Generally speaking, it is not relevant to control room design. In terms of minimum requirements, compliance revolves around:

- Positioning the control room the maximum distance possible from vibration-generating sources (such as generators and compressors)
- Insulating the control room from vibratory sources using, if necessary, vibration absorbers in the floor, ceilings, and walls

13.6.5.2 Human Factors Best Practice

Reference to Section 13.6.4, particularly the rapid sound quality assessment, provides a quick assessment route to detecting the presence of problematic building-borne vibration. If these initial checks reveal that such stimuli exist in the control room, then further investigation to identify the source is required along with accompanying intervention. This sort of specialist intervention is beyond the scope of this chapter.

TABLE 13.18
Control Room Aesthetics and Surface Reflectivity

Control Room Element	Reflectance Value
Floor finishes	0.2–0.3
Wall finishes	0.5–0.6
Partitions, etc.	0.5–0.6

Source: Permission to reproduce table from BS/EN/ISO 11064-6:2005 is granted by the British Standards Institution.

13.6.5.3 Vibration Example

The control room was not sited near any vibration-generating equipment and the acoustical (and other checks) revealed no issues present. No further analysis is necessary.

13.6.6 Interior Design and Aesthetics

13.6.6.1 Minimum Requirements

Minimum aesthetic and interior design requirements can be quickly assessed in the context of a walkthrough type of analysis covering the following items:

- A calming backdrop has been installed to control activities.
- Excessive use of dark or light finishes is avoided.
- Pale, untextured wall finishes are in use.
- Excessively strong patterns are avoided.
- Heavy contract-grade carpets with small random pattern are used.
- Ceilings are lighter than walls, and walls are lighter than floors.
- Humanization of the environment (e.g., plants, furnishings) is apparent.
- In general, the selection of finishes, furnishings, and design is suitable for 24-hour operation.

BS/EN/ISO 11064-6:2005 is quite specific in terms of reflectance values. For the purposes of the standard, *reflectance* is defined as the "ratio of the luminous flux reflected from a surface … to the luminous flux incident on it." Note that reflectance also depends on the direction from which light strikes an object and its spectral content. For practical purposes, if all of the incident light is reflected off of a surface then the reflectance value (ρ) would be 1, and the surface would be something akin to a mirror. Somewhat less reflectivity is specified in the standard as shown in Table 13.18. These general guidelines are adjusted in the specific case of indirect lighting, where the standard specifies matte white ceilings and a minimum surface reflectance of 0.8.

13.6.6.2 Human Factors Best Practice

The procedure for checking glare can be easily adopted for other surfaces to provide a quick check requiring minimal equipment. The analysts can survey walls, floors, and other potentially specular surfaces for glare and high reflectance. Because the reflectance can readily be detected by the naked eye, this suggests the presence of

FIGURE 13.11 Overview of the control room environment with aesthetic and interior design features highlighted.

reflectance above ρ = 0.3/0.6. In these cases, a mirror can be placed over the top of the noted glare point and the source of the glare identified. Appropriate intervention can then take place. Note also that the Cornell Office Environment Survey dealt with above (Figure 13.6) also provides valuable information and prompting in terms of interior design and aesthetic problems.

13.6.6.3 Aesthetics Example

Figure 13.11 shows the general appearance of the control room environment, pinpointing some of the materials and finishes used. The aesthetic and interior design guidance offered above can be used in a checklist format as shown in Table 13.19. In summary, as far as interior design and aesthetics are concerned, the example control room is broadly compliant with many beneficial features.

13.7 SUMMARY

The guidance provided in the sections above represent a pragmatic way to ensure compliance with BS/EN/ISO 11064-6:2005. In general, the minimum standards posit easily measurable objective criteria for control room environments, and the human factors best practice goes further to offer not just quicker, more efficient ways of measuring compliance but also ways that map onto the subjectively felt experience of the control room environment itself. As the flowchart presented earlier suggests, having undertaken a first pass and perhaps manipulated some environmental variables, the opportunity should be taken to reevaluate the situation. This iterative cycle helps to ensure that the appropriate systems perspective is adopted (per BS/EN/ISO 11064-6:2005) and that optimum trade-offs are applied.

TABLE 13.19
Interior Design and Aesthetics Checklist

Checklist Item	Checklist Response
Calming backdrop has been installed to control activities.	Yes
Excessive use of dark or light finishes is avoided.	Yes
Pale, untextured wall finishes are in use.	Yes
Excessively strong patterns are avoided.	Yes
Heavy contract-grade carpets with small random pattern are used.	Yes
Ceilings are lighter than walls, and walls are lighter than floors.	Yes, except the two-tone walls, which have green above white
Humanization of the environment (e.g., plants, furnishings) is apparent.	To a limited degree; natural wood finishes used for work surfaces
In general, the selection of finishes, furnishings, and design is suitable for 24-hour operation.	Yes

In the environmental analysis, measurements were taken of ambient temperature, air quality, lighting, acoustics, vibration, and aesthetics. Most of these measurements were within acceptable limits. Although the objective measures of air temperature and air quality were within acceptable limits, the subjective ratings suggest that people consider the room to be either too hot or too cold, and they reported symptoms of stuffy noses and sore throats. It is recommended that the concept of creating local climates be investigated to see if the environment can be tailored to individuals. In addition, although the humidity was within limits, it was at the lower end of the scale. It is recommended that measures be investigated to see if humidity can be raised to 50%. The subjective ratings of the lighting suggest that it is satisfactory, although the objective measures identified a darker area of the control room registering 194 lux against a minimum standard of 200 lux. It is recommended that the lighting be increased in this area. Finally, it was noted that the dark green part of the control room wall went against the human factors guidelines, and it is recommended that the wall be lightened in accordance with the standard.

BIBLIOGRAPHY

Boyce, P.R. (2005). Evaluating office lighting. In N.A. Stanton, A. Hedge, K. Brookhuis, E. Salas, and H. Hendrick (Eds.), *Handbook of Human Factors and Ergonomics Methods*. Boca Raton, FL: CRC Press.

Eklund, N. and Boyce, P.R. (1996). The development of a reliable, valid and simple office lighting survey. *Journal of the Illuminating Engineering Society*, 25, 25–40.

Hedge, A. (2005a). Environmental methods. In N.A. Stanton, A. Hedge, K. Brookhuis, E. Salas, and H. Hendrick (Eds.), *Handbook of Human Factors and Ergonomics Methods* (pp. 15–18). Boca Raton, FL: CRC Press.

Hedge, A. (2005b). Indoor air quality: chemical exposures. In N.A. Stanton, A. Hedge, K. Brookhuis, E. Salas, and H. Hendrick. (Eds.), *Handbook of Human Factors and Ergonomics Methods* (pp. 604–617). Boca Raton, FL: CRC Press.

Hedge, A. and Erickson, W.A. (1998). Indoor environment and sick building syndrome complaints in air conditioned offices: benchmarks for facility performance? *International Journal of Facilities Management*, 1, 1–8.

Klein, G. and Armstrong, A.A. (2005). Critical decision method. In N.A. Stanton, A. Hedge, K. Brookhuis, E. Salas, and H. Hendrick. (Eds.), *Handbook of Human Factors and Ergonomics Methods* (pp. 347–356). Boca Raton, FL: CRC Press.

Pelmear, P. and Benton, S. (2003). *A Review of Published Research on Low Frequency Noise and Its Effects*. London: Defra Publications.

Soames Job, R.F. (2005). Noise reaction indices and assessment. In N.A. Stanton, A. Hedge, K. Brookhuis, E. Salas, and H. Hendrick. (Eds.), *Handbook of Human Factors and Ergonomics Methods* (pp. 667–682). Boca Raton, FL: CRC Press.

Toftum, J. (2005). Thermal comfort indices. In N.A. Stanton, A. Hedge, K. Brookhuis, E. Salas, and H. Hendrick. (Eds.), *Handbook of Human Factors and Ergonomics Methods* (pp. 590–603). Boca Raton, FL: CRC Press.

Torres, R.R. (2005). Rapid sound-quality assessment of background noise. In N.A. Stanton, A. Hedge, K. Brookhuis, E. Salas, and H. Hendrick. (Eds.), *Handbook of Human Factors and Ergonomics Methods* (pp. 670–676). Boca Raton, FL: CRC Press.

Walker, G.H., Stanton, N.A., Wells, L., Gibson, H., Young, M.S., and Jenkins, D.P. (in press). Beyond ethnography: A new approach to "doing" distributed cognition. *Theoretical Issues in Ergonomic Science*.

14 Human Error

14.1 OUTLINE

At its simplest, human error can be defined as the performance of an incorrect or inappropriate action or a failure to perform a particular action. Senders and Moray (1991) observed that an error is something that has been done that:

- Was not intended by the actor
- Was not desired by a set of rules or an external observer
- Led the task or system outside of its acceptable limits

Hollnagel (1993) labeled errors as "erroneous actions" and defined an *error* as "an action which fails to produce the expected result and which therefore leads to an unwanted consequence." Probably the most widely recognized definition of human error was offered by Reason (1990, p. 9), who formally defined *human error* as "a generic term to encompass all those occasions in which a planned sequence of mental or physical activities fails to achieve its intended outcome, and when these failures cannot be attributed to the intervention of some chance agency."

14.2 KEY ISSUES

Human error remains the main cause of accidents in complex systems (Nyssen and Blavier, 2006). Depending on where you look within the literature, human error is typically implicated as the main causal factor in around 70% of all accidents and incidents occurring in most safety-critical systems. Due in part to the explosion of research into the concept in the 1980s and 1990s, safety practitioners now have a range of approaches designed to combat the problem of human error. Known collectively as human error management approaches (Reason, 1997), these techniques can be used to inform the development of measures designed to reduce the occurrence of human error, mitigate the consequences of human error, or provide recovery strategies designed to deal with errors once they are made. One such category of approaches includes human error identification (HEI) techniques, which are used to predict the errors that are likely to be made during a particular task or procedure. This knowledge of potential errors is then used to design remedial measures for the errors identified to either eradicate their potential occurrence or to mitigate the consequences once they have been made.

At the most basic level of error classification, a distinction between errors of omission and errors of commission is proposed. Errors of omission are those instances where an actor fails to act at all, such as failing or forgetting to perform a particular action. Errors of commission are those instances where an actor performs an action either incorrectly or at the wrong time, such as pressing the wrong button or performing an action too early or too late.

Payne and Altman (1962) proposed a simplistic error classification scheme based on information-processing theory:

- *Input errors*, which occur during the input sensory and perceptual processes (e.g., visual perception and auditory errors)
- *Mediation errors*, which occur or are associated with the cognitive processes employed between the perception and action stages
- *Output errors*, which occur during the selection and execution of physical responses

The most commonly referred to error classification within the literature, however, is the slips, lapses, mistakes, and violations classification proposed by Reason (1990), an overview of which is presented below.

14.2.1 Slips and Lapses

The most common form of human error is the slip-based error. Slips are categorized as those errors in which the intention or plan was correct but the execution of the required action was incorrect. Within a process control context, an example of slip-based errors would be when a control room operator unintentionally presses the wrong button on a control panel. In this case, the intention (i.e., to press the appropriate button on the control panel) was correct, but the physical execution of the action was incorrect (i.e., pressing the wrong button on the control panel). Slips are therefore considered to be actions with the appropriate intention followed by the incorrect execution and are labeled action execution failures (Reason, 1990).

Whereas slip-based errors are observable errors involving an incorrect execution of a correct plan, lapse-based errors refer to more covert error forms that involve a failure of memory that may not manifest itself in actual behavior. Lapses typically involve a failure to perform an intended action or forgetting the next action required in a particular sequence. Examples of lapses within the process control domain would include an operator failing to perform a required task, such as communicating a key piece of information during a shift handover or recording a fault in the alarm log. Although slips occur at the action execution stage, lapses occur at the storage stage, whereby intended actions are formulated prior to the execution stage of performing them. Reason proposed the following definition of slips and lapses: "Slips and lapses are errors which result from some failure in the execution and/or storage stage of an action sequence, regardless of whether or not the plan which guided them was adequate to achieve its objective" (Reason, 1990, p. 9).

14.2.2 Mistakes

Although slips reside in the observable actions made by operators, mistakes reside in unobservable plans and intentions. A mistake is therefore categorized as an inappropriate intention or wrong decision followed by the correct execution of the required

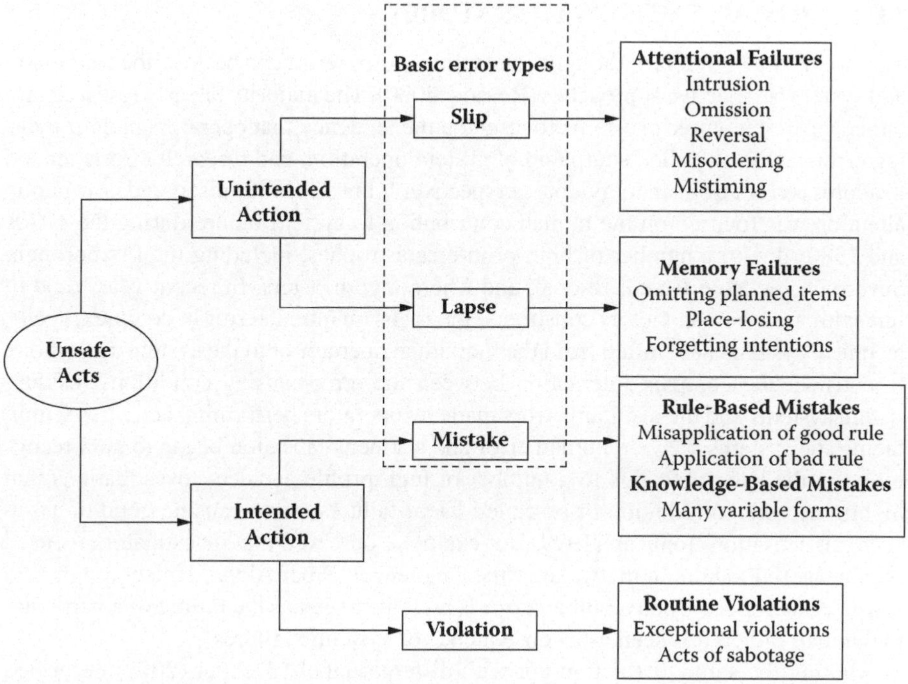

FIGURE 14.1 Unsafe acts taxonomy. (Adapted from Reason, J., *Human Error*, Cambridge University Press, Cambridge, U.K., 1990.)

action. A mistake occurs when an actor intentionally performs a wrong action; therefore, mistakes originate at the planning level, rather than the execution level, and can also be termed planning failures. Mistakes involve a mismatch between the prior intention and the intended consequences and are likely to be more subtle, more complex, less well understood, and more difficult to detect than slips. Reason defined mistakes as "deficiencies or failures in the judgmental and/or inferential processes involved in the selection of an objective or in the specification of the means to achieve it, irrespective of whether or not the actions directed by this decision-scheme run according to plan" (Reason, 1990, p. 9).

14.2.3 Violations

Another more complex category of error is violations. Violations are categorized as any behavior that deviates from accepted procedures, standards, and rules. Violations can be either deliberate or erroneous (Reason, 1997). Deliberate violations occur when an actor deliberately deviates from a set rules or procedures

In addition to the simplistic slips, lapses, mistakes, and violations classification described above, further error types have been specified within each category; for example, Reason (1990) proposed a taxonomy of unsafe acts that identifies a number of different error types within each of these four error categories. The taxonomy of unsafe acts is presented in Figure 14.1.

14.3 HUMAN FACTORS LITERATURE

The two distinct schools of thought concerning human error can be described as person and system perspective approaches (Reason, 2000). The majority of early research into human error in complex systems focused on the tendency that operators had for making errors at the so-called sharp end of system operation, and research efforts tended to address errors from an individual perspective. Johnson (1999) discussed how public attention was focused on the human contribution to system failure during the 1970s and 1980s due to a number of high-profile catastrophes, including the Flixborough, Seveso, Three Mile Island, Bhopal, and Chernobyl disasters. In recent years (and in line with most human factors constructs) the focus on human error in complex, sociotechnical systems has shifted from the individual operator onto the system as a whole, to consider the complex interaction between the error-causing conditions residing within systems and the resultant errors made by operators performing activities within them. The systems view on human error and accident causation began to gain recognition in the late 1980s due to a number of high-profile accident investigations that highlighted the contribution of so-called latent failures (error-causing conditions) in accident causation. Johnson (1999), for example, observed that investigators focused on managerial factors in the wake of the Challenger, Piper Alpha, Hillsborough, and Narita catastrophes. Today, human error is no longer seen as the failure of a particular individual; rather, it is seen as a consequence of systemic failures.

Making the same distinction but with different labels, Dekker (2002) described the old view and new view on human error. In the old view, human error is treated as the cause of most accidents, the systems in which people work are safe, the main threat to system safety is human unreliability, and safety progress is achieved by protecting systems from human unreliability through automation, training, discipline, selection, and proceduralization. In the new view, however, human error is treated as a symptom of problems within the system, safety is not inherent within systems, and human error is linked to the tools used, tasks performed, and operating environment. A brief description of the person approach (or old view) and the systems approach (or new view) is presented below.

14.3.1 PERSON APPROACH

The person approach focuses on the errors that operators make at the sharp end of system operation (Reason, 2000) and views error occurrence as the result of psychological factors within an individual. According to the person approach, errors arise from aberrant mental processes such as forgetfulness, inattention, poor motivation, carelessness, negligence, and recklessness. Research related to the person approach typically attempts to identify the nature and frequency of the errors made by operators within complex systems, the ultimate aim being to propose strategies, remedial measures, and countermeasures designed to prevent future error occurrence. Although person-based research is worthwhile for these reasons, it is often criticized for its contribution to individualistic blame cultures within organizational systems (e.g., operator X made these errors so the incident was operator X's fault), and error countermeasures are ultimately focused on reducing the variability in human

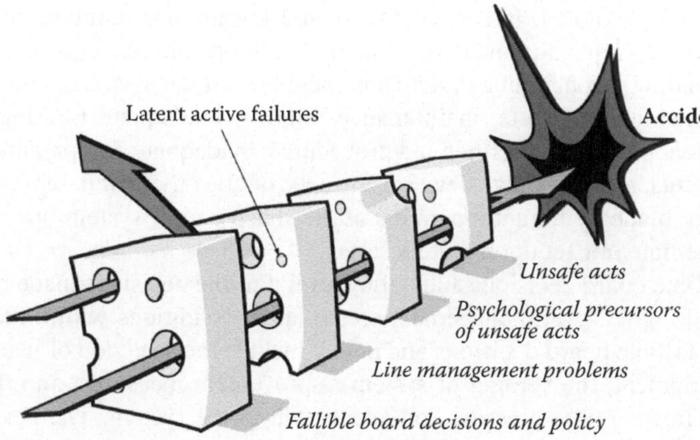

FIGURE 14.2 Systems perspective model of accident causation in complex systems. (Adapted from Reason, J., *British Medical Journal*, 320, 768–770, 2000.)

behavior. Typical person-approach error countermeasures include poster campaigns, additional procedures, disciplinary measures, threat of litigation, retraining and naming, and blaming and shaming. Despite the obvious failings of the person approach to human error (e.g., failure to consider contributory factors in error causation, promotion of blame culture within systems), it still remains the dominant approach in a number of domains, including the healthcare (Reason, 2000) and road transport (Salmon, Regan, and Johnston, 2005) domains. Person-based models of human error include Rasmussen's Skill, Rule, and Knowledge (SRK) framework (Rasmussen, 1983; Vicente, 1999), Reason's (1990) Generic Error Modeling System (GEMS), and Rasmussen's (1982) model of human malfunction.

14.3.2 Systems Approach

Reason's (1990) systems perspective, or Swiss cheese model, as it is more commonly referred to, is the most commonly embraced systems perspective approach to human error and organizational accidents. Reason's model focuses on the interaction between latent conditions and errors and their contribution to organizational accidents. According to the model, complex systems consist of various levels, each of which contributes to production (e.g., decision makers, line management, productive activities, and defenses). Each layer has defenses that are designed to prevent the occurrence of occupational accidents. Examples of defenses include protective equipment, rules and regulations, training, checklists, and engineered safety features. It is notable that effective supervision is a critical factor in the success or failure of these defenses. Holes or weaknesses in these defenses, caused by the latent conditions and unsafe acts, create windows of opportunity for accident trajectories to breach the defenses and cause an accident. Organizational accidents occur when the holes in the system's defenses line up in such a way that the accident trajectory can breach each of the different layers. Reason's systems perspective model is presented in Figure 14.2.

Within Reason's model, latent conditions and unsafe acts combine to produce organizational accidents and incidents. Latent conditions are inadequate conditions or failures residing throughout a system and include poor designs, inadequate supervision, manufacturing defects, maintenance failures, inadequate training, clumsy automation, inappropriate or ill-defined procedures, inadequate equipment, and procedural shortcuts, to name only a few. Unsafe acts, on the other hand, represent those errors that are made by human operators at the sharp end of system operation that have an immediate impact upon system safety.

At the fallible board decisions and policy level, fallible decisions made by system designers and higher level management create latent conditions within the system. Examples of fallible board decisions and policy include the selection of inadequately designed equipment, the vetoing of system improvement measures, and the use of policies that impose time pressures on agents within the system. The next level of Reason's model is the line management level. Line management problems arise from incompetent management and the fallible board decisions and policy from the preceding level in the model. Line management problems represent those instances where management is either inadequate or inappropriate. Examples of line management problems include inadequate management or supervision and the use of inadequate or inappropriate training and procedures. The precise nature of unsafe acts is defined through the complex combination of a number of factors, including the task being performed, the environmental conditions, and the presence of hazards. Each precursor can contribute to a great number of unsafe acts, depending on the associated conditions. Examples of these precursors include poor motivation, negative attitudes, and a failure to perceive hazards.

14.4 GUIDANCE AND RECOMMENDATIONS

Cacciabue (2005) pointed out that it is impossible to conceive of a completely human-error-free plant; rather, system safety is improved through measures designed to prevent errors being made, provide strategies for recovering from human errors, or to mitigate the consequences of errors that are made. Error management programs are employed for this purpose in safety-critical systems and use formal methods to develop a deeper understanding of the nature of, and factors surrounding, error occurrence in a particular system. The ultimate goals of such programs—achieved through a variety of means—are the eradication, reduction, management, or mitigation of errors and their consequences within the system in question. Helmreich (2000) offered the following definition of error management. "Error management is based on understanding the nature and extent of error, changing the conditions that induce error, determining the behaviors that prevent and mitigate error, and training personnel in their use."

According to Reason (2005), error management has two components: error prevention and error containment. Error prevention involves the use of strategies designed to identify and prevent the incidence of errors. Error prevention strategies include the identification and removal of error traps, user-friendly equipment, training, briefings, and debriefings. Error containment involves the use of strategies designed to limit the consequences of those errors that do occur. Error containment acknowledges that errors will continue to occur, and containment strategies include error detection and recovery training and the design of error-tolerant systems. According to Reason (1997), error management includes measures to:

- Minimize the error liability of individuals and teams.
- Reduce the error vulnerability of tasks or task elements.
- Determine, assess, and eliminate error-producing factors within the workplace.
- Identify organizational factors that create error-producing factors within the individual, team, task, and workplace.
- Enhance error detection.
- Increase the error tolerance of the system.
- Make latent conditions visible to those concerned with system operation and management.
- Improve the organization's intrinsic resistance to human fallibility.

Reason emphasized that a basic principle of error management is that the best people can sometimes make the worst errors. He described the following basic elements of human nature and error (Reason, 1997):

- Human actions are almost always constrained by factors beyond an individual's immediate control.
- People cannot easily avoid actions that they did not intend to perform in the first place.
- Errors have multiple causes: personal, task-related, situational, and organizational factors.
- Within a skilled, experienced, and largely well-intentioned workforce, situations are more amenable to improvement than people are.

Reason (1990) suggested that comprehensive error management should be directed at the various levels of an organization, including the team, task, workplace, and organizational processes.

Despite the widespread implementation of error management programs in different domains, Reason (1997) observed that implementations of principled and comprehensive error management programs are very rare, and most error management efforts are typically piecemeal, reactive, and event driven (rather than planned, proactive, and principle driven, as they should be). Additionally, typical error management programs ignore developments in the behavioral sciences that have occurred over the past three decades and have furthered our understanding of the nature, variety, and causes of human error. Reason identified the following problems associated with existing error management strategies:

- They focus on retrospective firefighting of errors that have already occurred rather than the anticipation and prevention of errors.
- They focus on active errors rather than the latent conditions residing in the system.

- They focus on the individual rather than organizational, contextual, and environmental factors.
- They rely heavily on exhortations and disciplinary sanctions.
- They use blame-laden terms such as "carelessness" and "irresponsibility."
- They do not distinguish between random and systematic error-causing factors.
- They are not informed by current human factors knowledge regarding error and accident causation.

Error management programs employ a variety of different approaches, methodologies, and techniques to address the problem of error. A review of the literature identifies a wide variety of techniques related to error management in complex, dynamic systems. Reason (1997), for example, cited a wide selection of error management techniques, including selection, training, licensing and certification, skill checks, human resource management, quality monitoring and auditing, technical safety audits, unsafe act auditing, hazard management systems, procedures, checklists, rules and regulations, administrative controls, Total Quality Management, probabilistic safety assessment, human reliability analysis, human error identification, and Crew Resource Management. A key feature of an effective error management program is the collection of specific information on the different types of human error made, including:

- The nature of the errors committed
- The factors contributing to and causing them
- The tasks and equipment involved
- Their consequences
- How they were recovered from

Such information is then used, among other things, to inform the design and development of error-tolerant systems, as well as the development of countermeasures, remedial measures, and strategies designed to eradicate or reduce error in systems. Methods for collecting and classifying human error data include the use of incident reporting systems, accident investigation and analysis procedures, traditional data collection techniques (observational study, interviews, questionnaires), human error identification, human reliability analysis (HRA) techniques, and error modeling. Due to the importance of collecting human error data, techniques used for such purposes are included here in our review of error management techniques.

The types of error management approaches employed within a particular system depend on the domain, system, and organization involved. Different regulatory bodies and organizations all have their own approaches to error management. Normally, a mixture of approaches is used, but, as Reason (2005) pointed out, there is no single best approach to error management, and the mixture of practices, techniques, and measures should be determined on the basis of the culture in question. Error management techniques are required for each of the different components within a particular system, including the person, team, task, workplace, organization, and the system itself (Reason, 2005).

14.5 ASSESSMENT AND EVALUATION

Human error identification (HEI), or error prediction, offers a proactive strategy for investigating human error in complex sociotechnical systems. The prediction of human error is used within risk assessments to identify potential error occurrence and determine the causal factors, consequences, and recovery strategies associated with the errors identified. The error information derived is then typically used to highlight system design flaws, propose remedial design measures, identify procedural deficiencies, and quantify error incidence probabilities. HEI works on the premise that an understanding of an employee's work task and the characteristics of the technology being used allows us to identify potential errors that may arise from the resulting interaction (Stanton and Baber, 1996). The concept of error prediction was first investigated in response to a number of high-profile catastrophes attributed to human error in the nuclear and chemical processing domains (e.g., Three Mile Island, Bhopal, and Chernobyl). The use of HEI techniques is now widespread, and it has been applied in a wide range of domains, including the nuclear power and petrochemical processing industries (Kirwan, 1996), air-traffic control (Shorrock and Kirwan, 2002), aviation (Stanton et al., 2006), naval operations, military systems, space operations (Nelson, Haney, Ostrom, and Richards, 1998), medicine, and public technology (Baber and Stanton, 1996). The diverse HEI techniques available range from simple error-taxonomy-based techniques based on error modes linked to operator behaviors to more sophisticated error quantification and computational error simulation techniques. HEI techniques can be also be broadly classified into two groups: qualitative and quantitative. Qualitative HEI techniques are used to predict the different types of errors that may occur, whereas quantitative HEI techniques are used to predict the numerical probability of the different errors occurring. A review of the literature identified over 50 HEI techniques, which were classified into the following categories:

- Taxonomy-based techniques
- Error identifier prompt techniques
- Error quantification techniques
- Cognitive modeling techniques
- Cognitive simulation techniques

Taxonomy-based HEI techniques use external error mode (EEM) taxonomies to identify potential errors within complex sociotechnical systems. Typically, EEMs are considered for each component step in a particular task or scenario to determine credible errors that may arise during human–machine interaction. Techniques such as the Systematic Human Error Reduction and Prediction Approach (SHERPA) (Embrey, 1986), the Human Error Template (HET) (Stanton et al., 2006), the Technique for the Retrospective and Predictive Analysis of Cognitive Errors (TRACEr) (Shorrock and Kirwan, 2002), and the Cognitive Reliability and Error Analysis Method (CREAM) (Hollnagel, 1998) all use domain-specific EEM taxonomies. Taxonomic approaches to HEI are typically the most successful in terms of sensitivity and are also the least expensive, quickest, and easiest to use; however, these techniques depend greatly on the judgment of the analyst, and their reliability and validity may at times be questionable. Different analysts with different experience may make different error predictions for the same task (inter-analyst reliability). Similarly, the same analyst may make different judgments on different occasions (intra-analyst reliability).

Error identifier techniques use prompts or questions to aid the analyst in identifying potential errors. Examples of error identifier prompts include: "Could the operator fail to carry out the act in time?" or "Could the operator carry out the task too early?" or "Could the operator carry out the task inadequately?" (Kirwan, 1994). The prompts are linked to a set of error modes and reduction strategies. Although these techniques attempt to remove the reliability problems associated with taxonomy-based approaches, they add considerable time to the analysis because each prompt must be considered. One example of an error identifier HEI technique is the Human Error Identification in Systems Tool (HEIST) approach (Kirwan, 1994).

Error quantification techniques are used to determine numerical probabilities of error occurrence. Potential errors are identified, and a numerical probability value that represents the probability of the error occurring is assigned to each error. Performance shaping factors (PSFs) are typically used to aid the analyst in the identification of potential errors. Error quantification techniques are typically used in the probabilistic safety assessment (PSA) of nuclear processing plants. Kirwan (1996) reported use of the Justification of Human Error Data Information (JHEDI) system in a HRA risk assessment for the BNFL thermal oxide reprocessing plant at Sellafield and the Human Error Assessment and Reduction Technique (HEART) in a HRA risk assessment of the Sizewell B pressurized water reactor. HEART (Williams, 1986) was developed specifically for the nuclear power and chemical processing domains and offers a structured approach for the prediction and quantification of human error. The HEART approach utilizes a set of generic task categories, each with an associated human error probability value; a set of error-producing conditions (EPCs); and a taxonomy of remedial measures.

14.5.1 SHERPA

The Systematic Human Error Reduction and Prediction Approach (SHERPA) (Embrey, 1986) was originally developed for use in the nuclear reprocessing industry and is now the most commonly used HEI approach, with applications in a number of domains, including aviation (Stanton et al., 2006), health care (Lane, Stanton, and Harrison, 2006), nuclear power (Embrey, 1986), ticket machines (Baber and Stanton, 1996), vending machines (Stanton and Stevenage, 1998), and in-car entertainment systems (Stanton and Young, 1999). SHERPA uses a generic external error mode (EEM) taxonomy linked to a behavioral taxonomy and is applied to a hierarchical task analysis (HTA) (Stanton, 2006) of the task or scenario under analysis. The behavioral and EEM taxonomies are used to identify credible errors that are likely to occur during each step in the HTA; for example, each bottom-level task step from the HTA is first classified as one of the five following behavior types from the SHERPA behavior taxonomy:

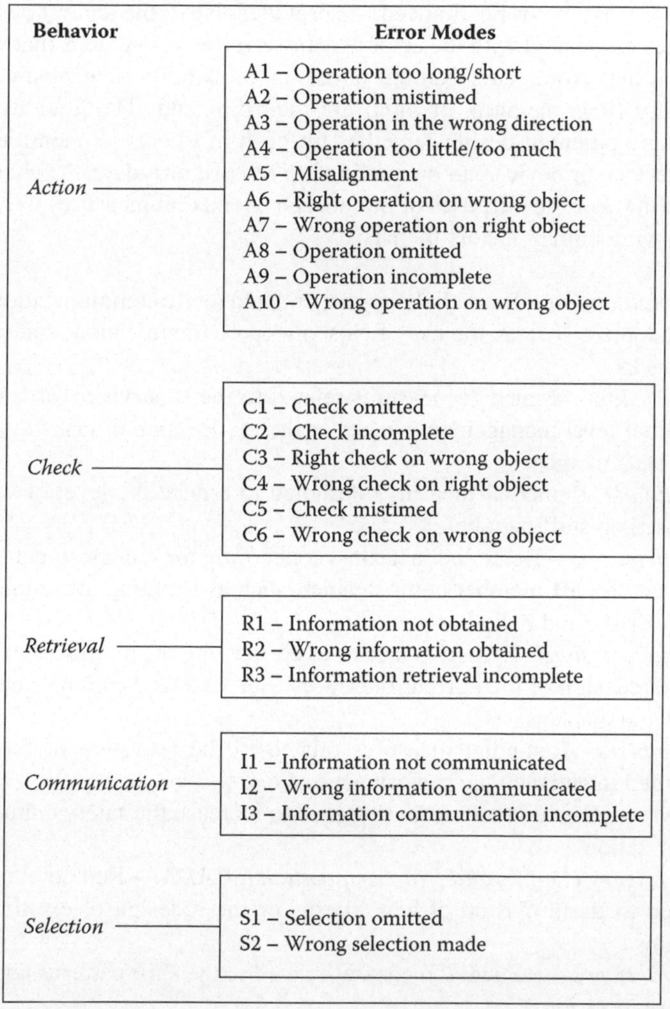

FIGURE 14.3 SHERPA external error mode (EEM) taxonomy.

- Action (e.g., pressing a button, typing in data)
- Information retrieval (e.g., from a display or document)
- Check (e.g., making a procedural check)
- Selection (e.g., selecting one alternative over another)
- Information communication (e.g., talking to a colleague or coworker)

Each SHERPA behavior classification has a set of associated external error modes. The SHERPA EEM taxonomy is presented in Figure 14.3.

The EEM taxonomy and domain expertise are then used to identify, based on the analyst's subjective judgment, any credible error modes for the task step in question. For each credible error identified, a description of the form that the error would take, such as "pilot dials in wrong airspeed," is provided. Next, the analyst describes any consequences associated with the error and any error recovery steps that would have to be taken if the error occurs. Ratings of ordinal probability (low, medium, or high) and criticality (low, medium, or high) are then provided. The final step involves specifying any potential design remedies for each of the errors identified, such as how the interface or device can be modified to remove or reduce the chances of the error occurring. For the purposes of our analysis, remedial measures were proposed for the following human factors themes:

- *Communication*—Remedial measures related to the communication strategies adopted, such as the use of closed loop communication strategies and readbacks
- *Supervision*—Remedial measures related to the supervision of staff, such as higher level management checks or the reallocation of tasks to a higher level staff member
- *Workload*—Remedial measures designed to reduce the level of workload imposed on staff members
- *Competencies*—Remedial measures concerning the enhancement or maintenance of staff member competencies, such as training, retraining, emergency drills, and refresher workshops
- *Interface design*—Remedial measures concerning the interfaces used, such as the redesign of the current interfaces used or specification of new interfaces and displays
- *Procedures*—Remedial measures related to the provision of procedures designed to enhance the performance of tasks
- *Culture*—Remedial measures designed to increase the safety culture of the organization
- *Supervisory Control and Data Acquisition (SCADA)*—Remedial measures related to the provision of new alarms or the redesign of existing alarm systems
- *Shift patterns*—Remedial measures related to the shift patterns adopted by the organization
- *Environment*—Remedial measures related to the enhancement of the environment in which the work under analysis is undertaken

A flowchart depicting the SHERPA procedure is presented in Figure 14.4.

14.7 EXAMPLE

The SHERPA technique was used to predict errors for three control room roles (Network Manager, Principal Operations Engineer, and Operations Engineer) and a range of alarm handling tasks from the control room operations. The purpose of the analysis was to predict the errors for each of these roles that are likely to occur under the current system. The HTAs constructed for each role or scenario detailed the goals, subgoals, and task steps involved. For example purposes, the main goal of the Network Manager was defined as the "safe and efficient operation of the distribution

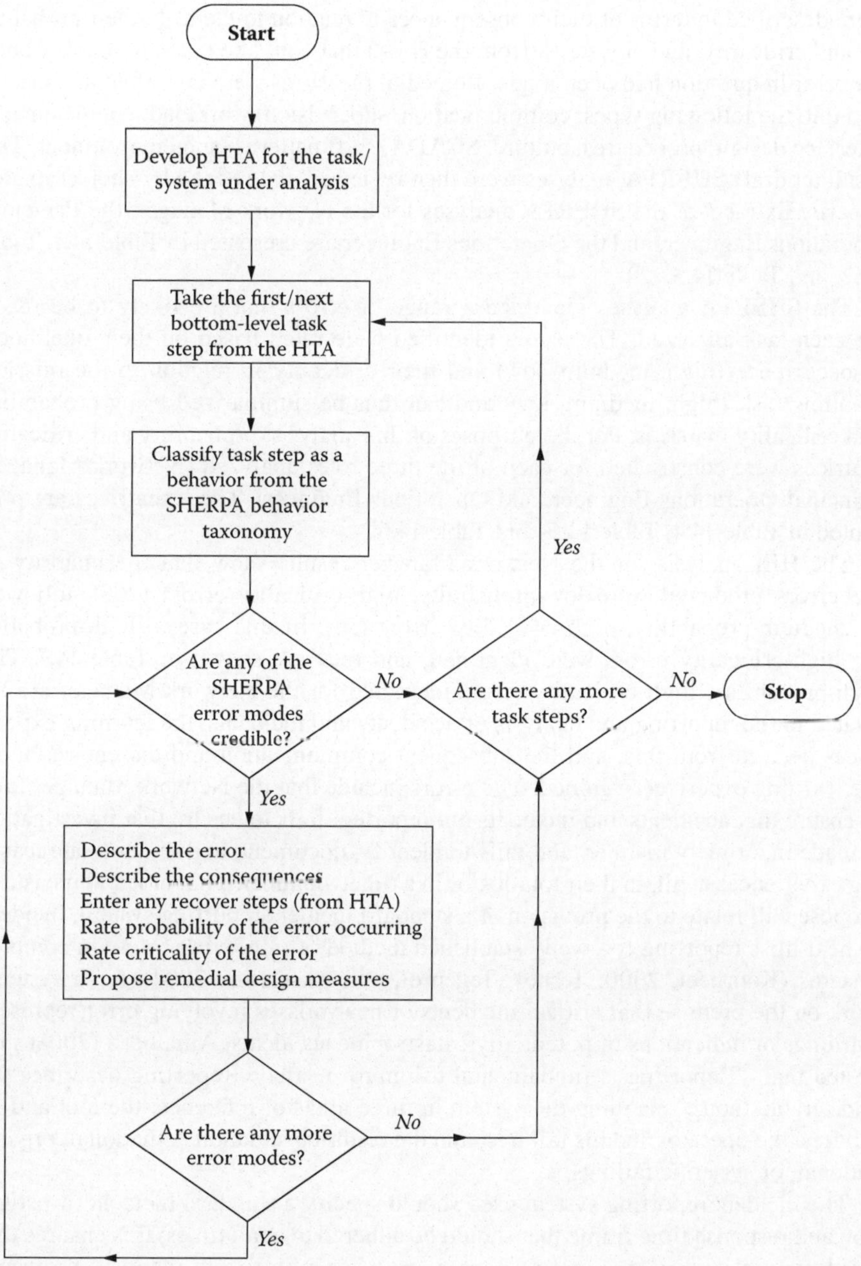

FIGURE 14.4 SHERPA flowchart.

networks while maximizing commercial return." The SHERPA was then applied to each HTA for the error analysis. The analysts involved had significant experience in the use of HEI approaches in a variety of complex domains. The errors identified were described in terms of their consequences in relation to the task, their probability and criticality, and any steps (from the HTA) that could be taken to recover once the error in question had been made. Remedial measures were identified and classified into the following types: communication, supervision, workload, competencies, interface design, procedures, culture, SCADA, shift patterns, and environment. The resultant draft SHERPA analyses were then reviewed and refined by subject matter experts. Extracts of the SHERPA analyses for the Network Manager, the Principal Operations Engineer, and the Operations Engineer are presented in Table 14.1, Table 14.2, and Table 14.3.

The SHERPA analyses identified a range of errors that are likely to be made for each task analyzed. The errors identified were rated based on their likelihood of occurrence (high, medium, low) and their criticality in relation to the mission planning task (high, medium, low) and can thus be summarized using probability and criticality matrices. For the purposes of this analysis, probability and criticality matrices were constructed for each of the three roles analyzed (Network Manager, Principal Operations Engineer, and Operations Engineer). These matrices are presented in Table 14.4, Table 14.5, and Table 14.6.

The HEI analysis for the Network Manager results show that the majority of the errors predicted were low-probability, high-criticality errors (268), followed by medium-probability, high-criticality errors (56). In this case, 6 high-probability, high-criticality errors were identified, and they are shown in Table 14.7. The high-probability, high-criticality errors identified for the Network Manager are all related to the reporting and analysis of accidents and incidents, the learning experiences derived from this, and the subsequent communication and dissemination of the learning experiences gained. The errors include that the Network Manager fails to ensure that accidents and incidents are reported, fails to ensure that investigation is made in a timely manner, and fails to identify, document, and disseminate learning experiences at all, in their totality, or in a timely manner. The remedial measures proposed all relate to the provision of a standard incident reporting system; incident or near-miss reporting is a well-established method of improving safety in complex systems (Koorneef, 2000; Ternov, Tegenrot, and Akselsson, 2004). Such systems work on the premise that critical incidents or near misses involving error represent warnings or indications of potentially catastrophic accidents. Amalberti (2001) suggested that, "Reporting is fundamental to improve safety. Reporting describes the undesirable 'noise' plaguing the system in three areas of reference: the tool and its failures, the operator and his failures, and the resulting situation (situational, organizational, or systemic failings)."

The incident reporting system used should specify a standard incident investigation and response time frame that should be adhered to at all times; this ensures that incidents are investigated and responses are taken in a timely manner. To ensure that all incidents and accidents are reported, the incident reporting system should be well known to all staff members (e.g., poster campaigns, monthly magazine), and all staff members should be given access to training in the system (e.g., how to report incidents). One problem associated with such systems is that staff members do not really understand what constitutes an incident worth reporting and how to report an incident appropriately. To further encourage the reporting of incidents, the system should be confidential, anonymous, blame free, and nonpunitive (where appropriate). Reason (1997) suggested that potential informants often fail to see the value in making reports, or they may not wish to admit that they have made an error or contributed to an incident that compromised system safety in any way. Informants may fear reprisals that may be taken against them as a result of the incidents they report. Staff members often do not report incidents because they are skeptical of the value of incident reporting systems, particularly of the data actually being acted upon, and the extra work involved in creating and filing the report puts them off ever doing so. To counteract this, the outcomes of reports should also be communicated throughout the organization. Typically, successful systems have a magazine of some sort that details the responses to reports, measures taken, and statistics regarding safety and incidents reported.

The HEI analysis of the Principal Operations Engineer showed that the majority of the errors predicted were low-probability, high-criticality errors (149), followed by medium-probability, high-criticality errors (146), low-probability, medium-criticality errors (88), and medium-probability, medium-criticality errors (73). In this case, 18 high-probability, high-criticality errors were identified for the Principal Operations Engineer role (see Table 14.8). For Task 3.1.6 ("Notify appropriate system support personnel/help desk of problem"), the high-probability, high-criticality errors involved the Principal Operations Engineer giving the help desk the wrong information (e.g., inaccurate or incomplete description of the problem) or the help desk misunderstanding or misinterpreting the problem being described. Remedial measures specified for these errors included the use of a standard readback at the end of any communications between the Principal Operations Engineer and the help desk, which allows both parties to verify that the correct information has been given and that the parties fully understand the content of the communication.

For the task of taking on the role of demand controller/incident controller and setting up an incident committee (Task 6.1.4.), the analysis indicated that the process is likely to take too long. Remedial measures proposed to ensure that the process does not take too long included the provision of clear and concise procedural guidelines on how to set up emergency response groups, the provision of an up-to-date emergency group contact list, and an example emergency group setup included within the emergency procedures (e.g., what personnel should be involved). Further remedial measures applicable to this error included the provision of regular emergency scenario training and refresher courses.

Transferring control of an emergency to the emergency management team (Tasks 6.1.13 and 6.2.17) is also likely to produce high-probability, high-criticality errors, including the communication of wrong information (the details provided are incorrect), communication of incomplete information, and misunderstanding of the situation by the emergency management team. Remedial measures proposed in this case include regular emergency scenario training and procedure refresher courses, as well as the provision of a transfer of control checklist detailing the tasks required and the information that should be passed between the parties involved.

Task 6.2.6 involves declaring the incident as severe and faxing forms to a number of different parties (emergency bureaus). The analysis indicated that the information may not be sent to all parties, which in turn means that all parties may not be fully

TABLE 14.1
Network Manager SHERPA Analysis Extract

Task	Error Mode	Error Description	Consequence	Recovery	Probability	Criticality	Remedial Measures
1.1.1. Notified by POE/OE of any significant or abnormal issues regarding demand, storage, forecasting and intake	R2—Wrong information obtained	Network Manager (NM) misunderstands or misinterprets information provided by the Principal Operations Engineer (POE) or Operations Engineer (OE).	NM's understanding of significant or abnormal issues is incorrect.	1.1.3	L	M	*Procedures*. Readback for clarification *Procedures*. Consistent terminology *Competencies*. Training issue
	I2—Wrong information communicated	POE/OE description of issues is incorrect.	NM's understanding of significant or abnormal issues is incorrect.	1.1.3	L	M	*Competencies and Procedures*. Training and procedural issue
	I3—Information communication incomplete R3—Information retrieval incomplete	POE/OE description of issues is incomplete; significant details are missing.	NM's understanding of significant or abnormal issues is incomplete.	1.1.3	M	M	*Procedures*. Checklist of issue items to include in comms with NM
	I1—Information not communicated	POE/OE fails to notify NM of issues	NM is not notified of issues	1.1.3	L	H	*Competencies and Procedures*. Training and procedural issue
	A2—Operation mistimed	POE/OE takes too long to notify NM of issues.	NM is not notified of issues in a timely manner; response is delayed.	NM requests information	L	M	*Competencies and Procedures*. Training and procedural issue

1.1.2. Question POE/OE about any significant or abnormal issues picked out from GTMS where identified	R2—Wrong information obtained	NM misunderstands or misinterprets information provided by POE/OE.	NM's understanding of significant or abnormal issues is incorrect.	1.1.3	L	M	*Procedures*. Readback for clarification *Procedures*. Consistent terminology
	I2—Wrong information communicated	POE/OE description of issues is incorrect.	NM's understanding of significant or abnormal issues is incorrect.	1.1.3	L	M	*Competencies and Procedures*. Training/procedural issue
	I3—Information communication incomplete R3—Information retrieval incomplete	POE/OE description of issues is incomplete; significant details are missing.	NM's understanding of significant or abnormal issues is incomplete.	1.1.3	M	M	*Procedures*. Checklist of issue items to include in comms with NM *Competencies*. Training issue *Procedures*. Frequent audits of quality of information logging
	I1—Information not communicated A8—Operation omitted	NM fails to question POE/OE.	NM is not notified of issues; issues are not interrogated by NM as required.	1.1.3	L	H	*Competencies and Procedures*. Training/procedural issue

TABLE 14.2
Principal Operations Engineer SHERPA Analysis Extract

Task	Error Mode	Error Description	Consequence	Recovery	Probability	Criticality	Remedial Measures
2.1.6. Check that forecast time deadlines are met	C1—Check omitted	Principal Operations Engineer (POE) fails to check that forecast time deadlines are met.	Forecast time deadlines may not be met; POE does not have knowledge of whether or not they are met.	2.1.6	L	M	*Competencies and Procedures.* Training and procedural issue
2.1.7. Check that the networks figures are entered into the system	C1—Check omitted	POE fails to check that the networks figures are entered into the system.	All figures may not be entered into the system; POE does not have knowledge of whether or not they are met.	2.1.7	L	M	*Competencies and Procedures.* Training and procedural issue
	C2—Check incomplete	POE fails to check that all networks figures are entered into the system.	All figures may not be entered into the system; POE does not have knowledge of whether or not they are met.	None	M	M	*Competencies and Procedures.* Training and procedural issue *Interface design.* System presents summary info detailing networks whose figures have not been entered into the system
	R2—Wrong information obtained	POE misunderstands or misinterprets the information checked.	POE's understanding of networks figures entered into the system is incorrect.	None	M	M	*Interface design.* System presents summary info detailing networks whose figures have not been entered into the system

2.1.8. Check that the forecast figures are logged into the system	C1—Check omitted	POE fails to check that the forecast figures are logged into the system.	All forecast figures may not be logged into the system.	2.1.8	L	M	*Competencies and Procedures.* Training and procedural issue
	C2—Check incomplete	POE fails to check that all of forecast figures are logged into the system.	All forecast figures may not be logged into the system.	None	M	M	*Competencies and Procedures.* Training and procedural issue *Interface design.* System presents summary info detailing forecast figures not entered
	R2—Wrong information obtained	POE misunderstands or misinterprets the information checked.	POE's understanding of forecast figures entered is incorrect.	None	M	M	*Interface design.* System presents summary info detailing forecast figures not entered
2.1.9. Send intake figures and demand attribution figures into the system to inform GNCC and shippers	A7—Wrong operation on right object I2—Wrong information obtained	Operations Engineer misreads the data and enters the wrong intake figures into the system	Wrong data are entered and the shippers receive erroneous intake figure data.	None	L	H	*Automation.* System recognizes erroneous values or interrogates questionable values

TABLE 14.3
Operations Engineer SHERPA Analysis Extract

Task	Error Mode	Error Description	Consequence	Recovery	Probability	Criticality	Remedial Measures
2.3.2. Accept alarm	A8—Operation omitted	Operations Engineer (OE) fails to accept alarm.	Alarm is not accepted.	2.3.3	L	H	*Communication/SCADA*. Accept alarm prompts
	A1—Operation too long	OE is distracted and takes too long to accept the alarm.	Problem is not resolved and may deteriorate.	2.2.3	M	H	*Procedures*. Procedural issue *Communication/SCADA*. Alarm design
	A9—Operation incomplete	OE is distracted while accepting alarm.	Alarm is not accepted; process is delayed.	2.2.3	M	H	*Procedures*. Procedural issue
2.3.3. Interpret alarm meaning and its consequence on system and prepare appropriate responses	R2—Wrong information obtained	OE misinterprets alarm meaning.	OE has an incorrect understanding of the alarm and its consequences or implications for system.	2.3.4	M	H	*Communication/SCADA*. Alarm meaning and consequences presented by system *Communication/SCADA*. Improved alarm design *Automation*. Intelligent system interprets alarms and identifies meaning, consequences and required tasks

	A1—Operation too long	OE takes too long to interpret alarm.	OE takes too long to understand alarm and its consequences for the system.	None	M	H	*Communication/SCADA*. Alarm meaning and consequences presented by system *Automation*. Intelligent system interprets alarms and identifies meaning, consequences, and required tasks
	A8—Operation omitted	OE fails to interpret alarm.	OE does not understand what alarm is and what its consequences are.	2.3.5	M	H	*Communication/SCADA*. Alarm meaning and consequences presented by system *Automation*. Intelligent system interprets alarms and identifies meaning, consequences, and required tasks
2.3.4. Interrogate alarm (bring up the appropriate display and check alarm history and association with ongoing maintenance and activities on site)	A8—Operation omitted	OE fails to interrogate alarm.	OE does not understand what alarm is and what its consequences are.	2.3.5	L	H	*Communication*. System prompts user to interrogate alarm via the appropriate databases

TABLE 14.4
Network Manager SHERPA Error Probability and Criticality Matrix

		Criticality		
		Low	Medium	High
Probability	Low	9	14	268
	Medium	2	16	56
	High	—	—	7

TABLE 14.5
Principal Operations Engineer SHERPA Error Probability and Criticality Matrix

		Criticality		
		Low	Medium	High
Probability	Low	35	88	149
	Medium	7	73	146
	High	3	28	18

TABLE 14.6
Operations Engineer SHERPA Error Probability and Criticality Matrix

		Criticality		
		Low	Medium	High
Probability	Low	12	38	90
	Medium	2	76	115
	High	—	3	2

cognizant of the unfolding situation and its status. To remedy this potential error, it is suggested that the appropriate procedures contain a checklist detailing which parties should be sent the information in question. When using procedures, the engineer can then quickly check who should be sent the information.

For the task of quickly reviewing available data (Task 6.3.4), the analysis indicated that the Principal Operations Engineer may either take too long to review the data or may not review all available data as required. To remedy this, it was proposed that regular emergency scenario/blackout procedure training should be provided and that the blackout procedures contain a checklist of all useful data that should be reviewed as part of this task.

The analysis suggested that when logging out of all work stations the Principal Operations Engineer may switch the machines off. The requirement not to switch off the machines can be reinforced by regular emergency scenario/blackout procedure training, the provision of procedures that clearly detail the requirement not to switch off the machines, and the use of a system prompt message that asks users if they are sure they want to turn off the work station (including a warning that the work stations should not be turned off in the event of a blackout).

Task 6.4.19 involves transferring to the secondary route immediately. The analysis indicated that high-probability, high-criticality errors during this task might include that the task takes too long or is mistimed, the corollary of which is that the transfer is not initiated in a timely manner. The high-probability, high-criticality error identified for Task 6.5.4 (evacuation notice) was that all parties are not informed of the impending evacuation; for Task 6.6.3 (monitor restoration), it was that the Principal Operations Engineer fails to completely monitor the restoration. Remedial measures for these errors included the provision of regular emergency scenario and evacuation procedure training, procedure refresher workshops, and a list of contacts detailing all contact requirements within the emergency and evacuation procedures.

The HEI analysis for the Operations Engineer showed that the majority of the errors predicted were medium-probability, high-criticality errors (115 in total) and low-probability, high-criticality errors (90 in total), followed by medium-probability, medium-criticality errors (76) and low-probability, medium-criticality errors (38). In this case, 2 high-probability, high criticality errors were identified and are presented in Table 14.9. Remedial measures for the first error, failing to monitor the hourly prediction against hourly demand, involve ensuring that training, regular retraining, and procedures clearly highlight the requirement for monitoring hourly prediction against hourly demand. Also, in the event that hourly prediction against hourly demand reaches a threshold value of some sort, it is recommended that alarms and prompts are used to notify the engineer. For the second error, failing to take into account constraints or all constraints when obtaining minimum and maximum stock predictions, it is recommended that the system prompt the engineer to take into account constraints (including presenting those constraints that should be considered) and that the importance of taking into account constraints and a list of the constraints to be taken into account be detailed in the procedures for this task. Alternatively, the use of an intelligent system that automatically takes into account the relevant constraints and calculates the minimum and maximum stock predictions automatically could be used.

The analysis of error potential for the roles in the control room identified 7 high-probability, high-criticality errors for the Network Manager, 18 high-probability, high-criticality errors for the Principal Operations Engineer, and 2 high-probability, high-criticality errors for the Operations Engineer. It is recommended that all of these be addressed in the redesign of the control operations.

TABLE 14.7
Network Manager High-Probability, High-Criticality Errors

Task	Error Mode	Error Description	Consequence	Recovery	Probability	Criticality	Remedial Measures
3.1.2. Ensure that procedures for accidents, incidents, and near misses are reported and that timely investigation is carried out in accordance with HSE	A2—Operation mistimed	Action is not taken in a timely manner.	Action is not taken in a timely manner; incident continues to occur.	Information requested	H	H	*Procedures*. Make incident investigation and response standard and within a specified timeframe
	A8—Operation omitted A9—Operation incomplete	Network Manager (NM) fails to ensure that accidents, incidents, and near misses are reported.	All accidents, incidents, and near misses are not reported; error reduction and mitigation strategies cannot be put in place.	None	H	H	*Procedures*. Standardized error reporting system *Competencies and Procedures*. All members of staff given access to and training in incident reporting system *Procedures*. Incident reporting system should be confidential, anonymous, blame free, and non-punishable (where appropriate) Have an incident reporting system monthly magazine detailing incidents, measures taken, and stats *Procedures*. SER process

3.1.3. Identify any learning experience that occur and document them in the appropriate place; manage and maintain the use of the learning experience database in the system.	R1—Information not obtained	NM fails to identify learning experience.	Learning experiences are not identified; learning experiences are not learned from.	None	H	H	*Procedures.* Group reviews to identify, discuss, and document learning experiences
	R3—Information retrieval incomplete	NM fails to identify and document all learning experiences gained.	All learning experiences are not identified and documented; learning experiences are not learned from.	None	H	H	*Procedures.* Group reviews to identify, discuss, and document learning experiences
3.1.4. Ensure all learning experiences are disseminated to and understood by shift team	A2—Operation mistimed	Learning experiences are not disseminated in a timely manner.	Learning experiences are not disseminated in a timely manner; learning experiences are not initially learned from.	None	H	H	*Procedures.* Group reviews to identify, discuss, and document learning experiences
3.1.5. Ensure that learning experiences and related information are well communicated to all concerned giving benefit to the system including other shift teams.	A2—Operation mistimed	Learning experiences are not disseminated in a timely manner.	Learning experiences are not disseminated in a timely manner; learning experiences are not initially learned from.	None	H	H	*Procedures.* Group reviews to identify, discuss, and document learning experiences *Procedures.* Organizational circular and publications

TABLE 14.8
Principal Operations Engineer High-Probability, High-Criticality Errors

Task	Error Mode	Error Description	Consequence	Recovery	Probability	Criticality	Remedial Measures
3.1.6. Notify appropriate system support personnel/ help desk of problem	I2—Wrong information communicated	Principal Operations Engineer (POE) gives system support personnel or help desk the wrong information.	System support personnel or help desk's understanding of the problem is incorrect.	None	H	H	*Procedures*. Readback for confirmation of correct understanding
	R2—Wrong information obtained	System support personnel or help desk misunderstand or misinterpret the information given to them by the POE.	System support personnel or help desk's understanding of the problem is incorrect.	None	H	H	*Procedures*. Readback for confirmation of correct understanding
6.1.4. Take the role as demand controller/ incident controller and set up committee of OE, communication route, and district personnel immediately	A1—Operation too long	Process takes too long.	Process takes too long and response to incident is delayed.	None	H	H	*Procedures*. Clear concise guidelines on how to set up emergency response groups *Procedures*. Emergency group contact list *Procedures*. Example emergency group provided in emergency procedures
6.1.13. Transfer control of emergency to emergency management team	I2—Wrong information communicated	Details given are incorrect.	Emergency management team does not have full picture of the situation.	None	H	H	*Competencies and Procedures*. Training and procedural issue *Procedures*. Transfer of control checklist

	R2—Wrong information obtained	Emergency management team misunderstands the situation.	Emergency management team does not have correct picture of the situation.	None	H	H	*Competencies and Procedures*. Training and procedural issue *Procedures*. Transfer of control checklist
	I3—Information communication incomplete	Emergency management team is not given the full details of the situation.	Emergency management team does not have full picture of the situation.	None	H	H	*Competencies and Procedures*. Training and procedural issue *Procedures*. Transfer of control checklist
6.2.4. Take the role of demand controller and set up committee	A1—Operation too long	Process takes too long.	Process takes too long and response to incident is delayed.	None	H	H	*Procedures*. Clear, concise guidelines on how to set up emergency response groups *Procedures*. Emergency group contact list *Procedures*. Example emergency group provided in emergency procedures
6.2.6. Declare the event major/severe and fax forms to the system emergency bureaus, etc.	A9—Operation incomplete I3—Information communication incomplete	Information is not sent to all parties.	All parties are not aware of the situation	Party request information	H	H	*Procedures*. Include in emergency procedures a checklist of parties to send information to
6.2.13. Activate emergency interruption	A8—Operation omitted	POE fails to activate emergency interruption.	Emergency interruption is not activated.	None	H	H	*Competencies and Procedures*. Training and procedural issue
6.2.17. Transfer control to alternative center	I2—Wrong information communicated	Details given are incorrect.	Emergency management team does not have full picture of the situation.	None	H	H	*Competencies and Procedures*. Training and procedural issue *Procedures*. Transfer of control checklist

(continued)

TABLE 14.8 (continued)
Principal Operations Engineer High-Probability, High-Criticality Errors

Task	Error Mode	Error Description	Consequence	Recovery	Probability	Criticality	Remedial Measures
6.2.17. Transfer control to alternative center (cont.)	R2—Wrong information obtained	Emergency management team misunderstands the situation.	Emergency management team does not have correct picture of the situation.	None	H	H	*Competencies and Procedures.* Training and procedural issue *Procedures.* Transfer of control checklist
	I3—Information communication incomplete	Emergency management team is not given the full details of the situation.	Emergency management team does not have full picture of the situation	None	H	H	*Competencies and Procedures.* Training and procedural issue *Procedures.* Transfer of control checklist
6.3.4. Quickly review available data	A1—Operation too long	Review of data takes too long.	Process takes too long; other tasks are delayed as a result (e.g., contacting other parties).	None	H	H	*Competencies and Procedures.* Regular emergency scenario and blackout procedure training
	A9—Operation incomplete	Not all useful data are consulted.	Picture is incomplete.	None	H	H	*Competencies and Procedures.* Regular emergency scenario and blackout procedure training *Procedures.* Include in emergency procedures a checklist of all useful data to review

6.4.8. Log out of all work stations, but don't switch them off	A7—Wrong operation on right object	POE turns stations off.	Work stations are turned off.	None	H	H	*Competencies and Procedures.* Regular emergency scenario and evacuation procedure training *Procedures.* Procedure detailed in emergency procedures *Communication.* System asks users if they are sure they want to turn work station off
6.4.19. Transfer to secondary route immediately	A1—Operation too long A2—Operation mistimed	POE does not initiate transfer immediately.	Transfer is not undertaken quickly enough.	None	H	H	*Competencies and Procedures.* Regular emergency scenario and evacuation procedure training
6.5.4. Notify other security and network owners that the control room will be evacuating with date time and duration	A9—Operation incomplete I3—Information communication incomplete	All parties are not informed.	All parties are not informed.	None	H	H	*Competencies and Procedures.* Regular emergency scenario and evacuation procedure training *Procedures.* List of contacts and contact requirements in emergency procedures
6.6.3. Monitor restoration	A9—Operation incomplete	POE fails to completely monitor restoration.	All restoration is not monitored.	None	H	H	*Competencies and Procedures.* Training and procedural issue

TABLE 14.9
Operations Engineer High-Probability, High-Criticality Errors

Task	Error Mode	Error Description	Consequence	Recovery	Probability	Criticality	Remedial Measures
2.2.1.1. Monitor hourly prediction against actual hourly demand using the system, forecasting tools, weather, Internet information source (BBC), and local knowledge to see whether demand is/will be balanced	A8—Operation omitted	Operations Engineer (OE) fails to monitor hourly prediction against actual hourly demand.	Hourly prediction against actual hourly demand is not monitored; demand may or may not be balanced without OE knowing.	2.2.1.2	H	H	*Competencies and Procedures*. Training and procedural issue *Communication*. Monitor prompts *Communication/SCADA*. Alarms when levels are close to being insufficient
2.2.2.4. Obtain min/max stock predictions from the daily program, taking account of any constraints (e.g., maintenance)	A8—Operation omitted	OE fails to take into account constraints (e.g., maintenance).	The stock prediction data may be incorrect.	2.2.2.6	H	H	*Communication*. System prompts engineer to take into account constraints *Automation*. System automatically takes into account constraints in figures *Procedures*. Checklists of constraints to take into account

BIBLIOGRAPHY

Amalberti, R. (2001). The paradoxes of almost totally safe transportation systems. *Safety Science*, 37, 109–126.

Baber, C. and Stanton, N.A. (1996). Human error identification techniques applied to public technology: predictions compared with observed use. *Applied Ergonomics*, 27, 119–131.

Cacciabue, C. (2005). Human error risk management methodology for safety audit of a large railway organization. *Applied Ergonomics*, 36(6), 709–718.

Dekker, S.W.A. (2002). Reconstructing human contributions to accidents: the new view on human error and performance. *Journal of Safety Research*, 33, 371–385.

Embrey, D.E. (1986). SHERPA: A Systematic Human Error Reduction and Prediction Approach. Paper presented at the International Congress on Advances in Nuclear Power Plants, Knoxville, TN, April 1986.

Helmreich, R.L. (2000). On error management: lessons from aviation. *British Medical Journal*, 320, 781–785.

Hollnagel, E. (1993). *Human Reliability Analysis: Context and Control.* San Diego, CA: Academic Press.

Hollnagel, E. (1998). *Cognitive Reliability and Error Analysis Method—CREAM.* Amsterdam: Elsevier.

Isaac, A., Shorrock, S.T., Kennedy, R., Kirwan, B., Anderson, H., and Bove, T. (2002). *Technical Review of Human Performance Models and Taxonomies of Error in Air Traffic Management (HERA)*, Eurocontrol Project Report, HRS/HSP-002-REP-01. Brussels: European Organisation for the Safety of Air Navigation.

Johnson, C. (1999). Why human error modelling has failed to help systems development. *Interacting with Computers*, 11, 517–524.

Kirwan, B. (1994). *A Guide to Practical Human Reliability Assessment.* London: Taylor & Francis.

Kirwan, B. (1996). The validation of three human reliability quantification techniques: THERP, HEART, and JHEDI. Part 1. Technique descriptions and validation issues. *Applied Ergonomics*, 27, 359–373.

Koorneef, F. (2000). *Organizational Learning from Small-Scale Incidents.* Amsterdam: Delft University Press.

Lane, R., Stanton, N.A., and Harrison, D. (2006). Applying hierarchical task analysis to medication administration errors. *Applied Ergonomics*, 37(5), 669–679.

Nelson, W.R., Haney, L.N., Ostrom, L.T., and Richards, R.E. (1998). Structured methods for identifying and correcting potential human errors in space operations. *Acta Astronautica*, 43, 211–222.

Nyssen, A. and Blavier, A. (2006). Error detection: a study in anaesthesia. *Ergonomics*, 49, 517–525.

Payne, D. and Altman, J.W. (1992). *An Index of Electronic Equipment Operability, Report of Development*, Report AIR-C-43-1/62-FR. Pittsburgh, PA: American Institute for Research.

Rasmussen, J. (1982). Human error: a taxonomy for describing human malfunction in industrial installations. *Journal of Occupational Accidents*, 4, 311–333.

Rasmussen, J. (1983). Skills, rules, knowledge; signals, signs, and symbols; and other distinctions in human performance models. *IEEE Transactions in Systems, Man, and Cybernetics*, SMC-13, 257–266.

Reason, J. (1990). *Human Error.* Cambridge, U.K.: Cambridge University Press.

Reason, J. (1997). *Managing the Risks of Organizational Accidents.* Burlington, VT: Ashgate.

Reason, J. (2000). Human error: models and management. *British Medical Journal*, 320, 768–770.

Reason, J. (2005). Managing Error Management Workshop, Brisbane, Australia, May 26, 2005.

Salmon, P.M., Regan, M., and Johnston, I. (2005). *Human Error and Road Transport: Literature Review*. Clayton, Victoria, Australia: Monash University Accident Research Centre.

Senders, J. and Moray, N. (1991). *Human Error: Cause, Prediction and Reduction*. Mahwah, NJ: Lawrence Erlbaum.

Shorrock, S.T. and Kirwan, B. (2002). Development and application of a human error identification tool for air traffic control. *Applied Ergonomics*, 33, 319–336.

Stanton, N.A. (2006). Hierarchical task analysis: developments, applications, and extensions. *Applied Ergonomics*, 37, 55–79.

Stanton, N.A. and Baber, C. (1996). A systems approach to human error identification. *Safety Science*, 22, 215–228.

Stanton, N.A. and Stevenage, S.V. (1998). Learning to predict human error: issues of acceptability, reliability and validity. *Ergonomics*, 41(11), 1737–1756.

Stanton, N.A. and Young, M.S. (1999). *A Guide to Methodology in Ergonomics: Designing for Human Use*. London: Taylor & Francis.

Stanton, N.A., Harris, D., Salmon, P.M., Demagalski, J.M., Marshall, A., Young, M.S., Dekker, S.W.A., and Waldmann, T. (2006). Predicting design-induced pilot error using HET (human error template): a new formal human error identification method for flight decks. *Journal of Aeronautical Sciences*, 110, 107–115.

Strauch, B. (2005). *Investigating Human Error: Incidents, Accidents and Complex Systems*. Burlington, VT: Ashgate.

Ternov, S., Tegenrot, G., and Akselsson, R. (2004). Operator-centred local error management in air traffic control. *Safety Science*, 42(10), pp. 907–920.

Vicente, K.J. (1999). *Cognitive Work Analysis: Toward Safe, Productive, and Healthy Computer-Based Work*. Mahwah, NJ: Lawrence Erlbaum.

Williams, J.C. (1986). HEART: a proposed method for assessing and reducing human error. In *Proceedings of the 9th Advances in Reliability Technology Symposium (ARTS)*. Bradford, U.K.: University of Bradford.

15 Safety Culture

15.1 OUTLINE

A necessity in the safety-critical domains (Olive, O'Connor, and Mannan, 2006), an organization's safety culture is perhaps the most important factor involved in safe system operation. Without one, no amount of safety measures and defenses will guarantee system safety. The importance of this point is such that there is a huge body of literature surrounding it and yet, notwithstanding this, safety culture remains both ill defined and misunderstood (Haukelid, 2008; Reason, 1997) and something of an elusive commodity for organizations. Rao (2007) suggested that one of the most challenging aspects facing organizations in the safety-critical domains is to create and preserve a self-sustaining safety culture; Reason (1997, p. 191) also observed that "few things are so sought after and yet so little understood."

An organization's safety culture was formally defined by the Advisory Committee on the Safety of Nuclear Installations (ACSNI, 1993, p. 23) as the "product of individual and group values, attitudes, perceptions, competencies and patterns of behavior that determine the commitment to, and the style and proficiency of, an organization's health and safety management. Organizations with a positive safety culture are characterized by communications founded on mutual trust, by shared perceptions of the importance of safety, and by confidence in the efficacy of preventive measures."

The importance of a safety culture cannot be understated. Safety culture has been identified as a contributory factor in many catastrophic incidents occurring within the hazardous industries in recent times, including the Ladbroke Grove and Southall rail crashes, Texas BP refinery explosion, the Challenger and Columbia space shuttle disasters, and the Piper Alpha oil platform disaster (Hendershot, 2007).

15.2 KEY ISSUES

The key issues underpinning safety culture relate to what it actually is, what it is comprised of, and how organizations can measure, develop, and sustain one. Pertinent lines of inquiry, therefore, include the investigation and identification of what measures organizations should take to develop and sustain an appropriate safety culture and how they can measure the quality of their own safety culture. A summary of the key issues is presented below:

- *Development and maintenance of an appropriate safety culture*—An organization needs to identify what measures it must take to develop and maintain an appropriate safety culture. These typically include making a commitment at all levels to the development of a safety culture, increasing communication between all levels of the organization, collecting and analyzing data regarding the current performance of the system in relation to safety, and developing appropriate safety management systems.

- *Measurement of safety culture*—It is critical that organizations have reliable and valid ways in which to measure their safety culture if they are to develop and sustain it appropriately. Organizations need to continuously know and understand the current status of their safety culture. This involves the continuous assessment of their safety culture, which typically involves assessing the safety-related attitudes and perceptions of their workers at all levels of the organization.

15.3 STANDARDS AND GUIDELINES

Perhaps due to its multidimensional nature, no standards and guidelines specifically focus on the safety culture; however, some guidelines and standards are available that relate to an organization's occupational health and safety. A summary of these is presented below.

15.3.1 *Guide to Occupational Health and Safety Management Systems*

The Occupational Health and Safety Advisory Services (OHSAS, 2004) standard BS/OHSAS 8800 was developed to aid organizations in the development of a framework for managing occupational health and safety. The standard helps organizations:

- Minimize risk to employees and others by aiding the development of efficient working practices to prevent accidents and work-related ill health.
- Improve business performance and enable organizations to establish a responsible image within the marketplace.
- Assist an organization's continuous performance improvement beyond legal compliance.
- Help organizations to achieve compliance with occupational policies and objectives.

15.3.2 *Occupational Health and Safety Zone*

The Occupational Health and Safety Advisory Services (OHSAS, 2007) standard BS/OHSAS 18001 provides assessment specification for occupational health and safety management systems. The standard can be adopted by any organization wishing to implement a formal procedure to reduce the risks associated with health and safety in the working environment for employees, customers, and the general public. BS/OHSAS 18001 addresses the following areas:

- Planning for hazard identification, risk assessment, and risk control
- OHSAS management program
- Structure and responsibility
- Training, awareness, and competence

- Consultation and communication
- Operational control
- Emergency preparedness and response
- Performance measuring, monitoring, and improvement

15.4 HUMAN FACTORS LITERATURE

15.4.1 Defining Safety Culture

The concept first emerged during the aftermath of the Chernobyl disaster in 1986, when the International Nuclear Safety Advisory Group (INSAG) introduced the term *safety culture* to denote the management and organizational factors important to system safety (Sorensen, 2002). The errors and violations made during the incident were seen by many as evidence of a poor safety culture at the Chernobyl plant (Fleming, 1999). Following its initial emergence, the concept has become something of a buzzword within the safety-critical industries and has received considerable attention over the past two decades; consequently, a huge body of literature surrounding the concept exists.

Immediately apparent when looking at the literature are both the great number of definitions available and the contention among these definitions; the concept remains both ill defined and ill understood. The Health and Safety Executive (HSE) defined safety culture as "individual and group values, attitudes, competencies, and patterns of behavior that determine the commitment to, and the style and proficiency of, an organization's health and safety programs" (HSE, 2005a). Richter and Koch (2004, p. 705) defined the concept as "the shared and learned meanings, experiences, and interpretations of work and safety—expressed partially symbolically—which guide people's actions toward risks, accidents, and prevention. Safety culture is shaped by people in the structures and social relations within and outside the organization." Reason (1997) defined safety culture more simply as "the way we do things around here," and Guldenmund (2000) defined it as "those aspects of the organizational culture which will impact on attitudes and behavior related to increasing or decreasing risk."

The most widely cited definition was presented by the Advisory Committee on the Safety of Nuclear Installations (ACSNI). They defined the safety culture of an organization as the "product of individual and group values, attitudes, perceptions, competencies, and patterns of behavior that determine the commitment to, and the style and proficiency of, an organization's health and safety management. Organizations with a positive safety culture are characterized by communications founded on mutual trust, by shared perceptions of the importance of safety, and by confidence in the efficacy of preventive measures" (ACSNI, 1993, p. 23).

From the definitions presented here, an organization's safety culture represents everything about the organization's approach to safety; it concerns an organization's and its workers' safety-related values, attitudes, behaviors, policies, and commitment to safety, and these factors impact what they do on a day-to-day basis. Its definition has proved problematic, and as yet there appears to be a lack of consensus within the literature.

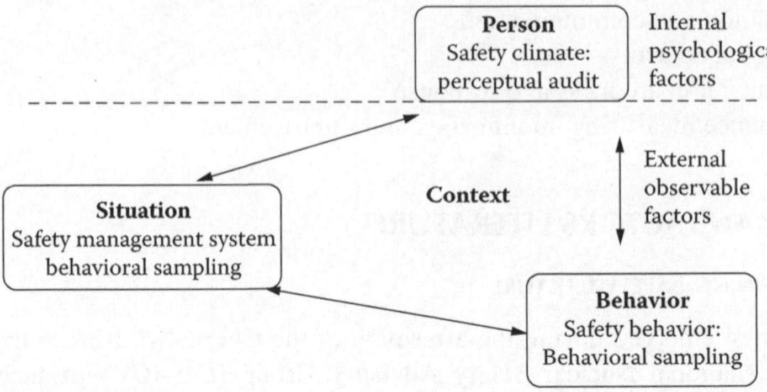

FIGURE 15.1 Cooper's safety culture model. (Adapted from Cooper, M.D., *Safety Science*, 35, 111–136, 2000.)

Much of the literature devotes significant attention to the apparent confusion between safety *culture* and safety *climate*. Olive et al. (2005) suggested that, although both can be used to describe the underlying safety attitude of an organization, they are fundamentally different concepts. Safety culture refers to an organization's overarching policies and goals that are set in relation to the overall safety level of their environment, whereas safety climate refers simply to the workers' attitudes toward safety. Safety culture, then, can be viewed as the background influence on an organization, while safety climate is at the forefront of the organization.

15.4.2 Models of Safety Culture

A number of models of safety culture are presented in the literature. Cooper (2000), for example, presented a model that suggests that safety culture is comprised of three interrelated aspects: psychological, behavioral, and situational. The model encompasses internal psychological factors, observable ongoing safety-related behaviors, and objective situational features. Cooper's model is presented in Figure 15.1.

The HSE (2005a) has provided a succinct summary of Cooper's model. The psychological component refers to how people within organizations feel about safety and safety management systems. This includes the beliefs, attitudes, values, and perceptions of individuals and teams across all organizational levels. The HSE pointed out that this component of safety culture represents the safety climate of an organization. The behavioral component is concerned with what people do within the organization. This includes all of the safety-related activities, actions, and behaviors undertaken by the workers at all levels within the organization. Finally, the situational component refers to what policies, procedures, and measures the organization has in place, or what the organization has, as the HSE put it. This includes the organization's safety-related policies, procedures, management systems, control systems, communication flows, and work flow systems (HSE, 2005a).

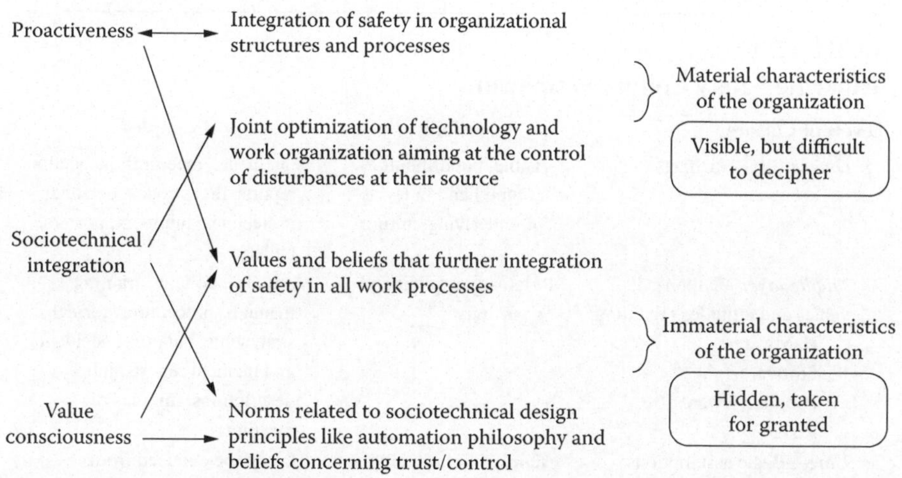

FIGURE 15.2 Sociotechnical model of safety culture. (Adapted from Grote, G. and Kunzler, C., *Safety Science*, 34, 131–150, 2000.)

Grote and Kunzler (2000) developed a sociotechnical model of safety culture that suggests that, to achieve a safety culture, organizations must be proactive in integrating safety into organizational structure and processes and in optimizing the organization of technology and work while also taking into account the material and immaterial characteristics of the organization. Grote and Kunzler's model is presented in Figure 15.2. Geller (1994) distinguished between the following three interactive features of an organization's safety culture: person factors (i.e., knowledge, skills, abilities, intelligence, motives, and personality), behavioral factors (i.e., compliance, coaching, recognition, communication, demonstration, and active caring), and environmental factors (i.e., equipment, tools, machinery, housekeeping, heat and cold, engineering).

In the conclusion to an exhaustive review of the safety culture literature (including models and measures), Guldenmund (2000) observed that both safety culture and safety climate remain ill defined and that the relationship between the two remains unclear. Guldenmund brought attention to the fact that there is significant confusion over the exact cause, content, and consequences of a safety culture and subsequently proposed a framework for safety culture (see Table 15.1). In this framework, safety culture has three separate layers: an outer layer consisting of artifacts; a middle layer comprised of values and attitudes regarding hardware, software, and risks; and a core layer comprised of the basic assumptions regarding aspects of time, space, human nature, activity, and relationships.

Other models presented in the literature attempt to define the factors affecting an organization's safety culture (e.g., Ball and Scotney, 1998; Wright, Brabazon, Tipping, and Talwalkar, 1999) or to provide guidance on how an organization should

TABLE 15.1
Tripartite Safety Culture Framework

Levels of Culture	Visibility	Examples
1. *Outer layer*—Artifacts	Visible, but difficult to comprehend in terms of underlying culture	Statements, meetings, inspection reports, dress codes, personal protective equipment, posters, bulletins
2. *Middle layer*—Espoused values and attitudes regarding: Hardware, Software, People/liveware, Risks	Relatively explicit and conscious	Attitudes, policies, training manuals, procedures, formal statements, bulletins, accident and incident reports, job descriptions, minutes of meetings
3. *Core*—Basic assumptions regarding the nature of: Reality and truth, Time, Space, Human nature, Human activity, Human relationships	Mainly implicit; obvious for the members, invisible, preconscious	Have to be deduced from artifacts and espoused values as well as through observation

Source: Adapted from Guldenmund, F.W., *Safety Science*, 34, 215–257, 2000.

go about developing and sustaining a safety culture (e.g., Fleming, 1999; Wright et al., 1999). These models are addressed in the following sections.

In summary, many models of safety culture are presented in the literature. Analogous to the concept's definition, it is apparent, too, that there is a lack of a universally accepted model (Guldenmund, 2000; HSE, 2005a). Based on a review of safety culture models and measures, Guldenmund (2000) concluded that there was no satisfying model of safety culture (or climate). In the time since Guldenmund's review, no satisfactory models have yet been presented; however, from the models proposed in the literature, it is possible to make the following conclusions:

- Safety culture is a complex, multidimensional concept comprised of organizational, personal, psychological, behavioral, and environmental factors.
- Safety culture therefore concerns how organizations and workers behave, what they think about safety (e.g., perceptions and attitudes), and what an organization has and does with regard to safety (e.g., equipment, procedures, safety-related policies and strategies).

Although much has been written about the concept, it remains largely ill defined and misunderstood. Davies, Ross, Wallace, and Wright (2003, p. 14), for example, pointed out that "the term *safety culture* appears to be bandied about and is a common figure of speech in many firms and organizations with scant reference to, or even lack of knowledge of, any of its various definitions."

15.4.3 Characteristics of an Efficient Safety Culture

Many in the field have attempted to articulate exactly what it is that characterizes a safety culture. Reason (1997), for example, outlined the following four critical components of an organization's safety culture: a *reporting* culture, a *just* culture, a *flexible* culture, and a *learning* culture. He suggested that together these components combine to produce an informed culture in which managers and operators of the system have up-to-date knowledge regarding the human, technical, organizational, and environmental factors that determine the safety of the system (Reason, 1997). The reporting culture refers to the provision of incident reporting systems that allow the workforce to report their errors, safety-compromising incidents, and near-miss incidents. The just culture relates to how an organization handles blame and punishment. A no-blame culture is neither feasible nor desirable; rather, it is preferable to strive for a just culture where the workforce is encouraged, even rewarded, for providing safety-related information but where they are also well aware of where the line between acceptable and unacceptable behavior lies. The flexible culture relates to the flexibility of the organization in the face of high-tempo or dangerous situations; it involves the ability to reconfigure from conventional hierarchical structures to flatter professional structures. Finally, the concept of a learning culture relates to the conclusions that are drawn from the safety-related data gathered and analyzed. Reason (1997) described this as the willingness and competence of an organization to draw the right conclusions from its safety information system and the organization's will to implement major reforms when needed.

Based on a comprehensive review of safety culture studies, Muniz, Peon, and Ordas (2007) identified three key components of an organization's safety culture. These were the manager's commitment to safety, the involvement of employees, and the safety management system that the organization has in place. Muniz et al. (2007) suggested that an organization can be deemed to possess a safety culture when it has adopted an adequate safety management system, such that the organization:

- Has defined a safety policy reflecting the organization's safety-related principles and values
- Has established incentives for its workers to become involved in safety activities
- Provides employees with continuous training that enables them to undertake their jobs in the most efficient and safest way possible
- Provides information about the risks that workers are exposed to and how to combat them
- Plans strategies to undertake in order to avoid accidents (preventive planning) and is able to react quickly in emergencies (emergency planning)
- Provides feedback based on analysis of the working conditions and events occurring inside the organization (internal control) and by comparing itself with other organizations (benchmarking)
- Ensures that managers have a strong commitment to safety, show a personal interest in employee working conditions, and are personally involved in safety activities

FIGURE 15.3 Ball and Scotney's safety culture model detailing 13 factors that influence an organization's safety culture. (Adapted from Wright, M.S. et al., *Development of a Business Excellence Model of Safety Culture*, Health and Safety Executive, London, 1999. Reproduced under the terms of the Click-Use Licence.)

- Ensures that workers are conscious of the importance of working safely so they comply with regulations and procedures, participating in safety meetings, and offering suggestions on how to improve safety in their workplace

In its exhaustive review of the safety culture literature, the HSE (2005a) concluded that there are five key indicators that influence an organization's safety culture: leadership, two-way communication, employee involvement, learning culture, and attitude toward blame. In addition to these five indicators, the HSE also identified six negative evidence points that represent aspects inhibiting the development of a safety culture: lack of ownership, isolationism, lack of learning, unwillingness to share information or cooperate, failure to deal with independent external safety review findings, and lack of management commitment.

Ball and Scotney (1998) presented a safety culture model detailing 13 factors that influence an organization's safety culture (see Figure 15.3). Pidgeon and O'Leary (1994) argued that a good safety culture can be promoted in four ways: (1) a commitment to safety from senior management; (2) a shared care and concern for hazards and solicitude over their impacts upon people; (3) realistic and flexible norms and rules regarding hazards; and (4) a continual reflection upon organizational practice via monitoring, analysis, and feedback systems (labeled *organizational learning*). Olive et al. (2005) identified the following characteristics of a good safety culture:

- *Commitment*—There should be a commitment to the improvement of safety behaviors and attitudes at all levels of the organization.
- *Communication*—The structure of the organization should be such that it promotes open and clear communication where people feel free from intimidation or retribution in raising issues and are encouraged to ask questions at all times.
- *Resilience and flexibility*—The organization should possess a propensity for resilience and flexibility to adapt effectively and safely to new situations.
- *Vigilance*—There should be a prevailing attitude toward constant vigilance throughout the organization.

Hale and Hovden (1998) reviewed 34 programs of research related to safety culture and identified the following common features of a safety culture:

- The availability of resources, a problem-solving approach, a stable workforce, appropriate safety training systems, and efficient communication channels
- Learning systems in design, maintenance, and construction operations
- Good-quality communication and leadership style that demonstrates concern for the group and its dynamics
- Openness to criticism, good labor relations, low stress, and low grievance rates
- Management commitment, along with supervisor and individual commitment and safety attitudes of coworkers

In summary, many articles presented in the literature attempt to define the components of an effective safety culture. Although total consensus has yet to be achieved, it is possible to identify a series of commonly cited general characteristics of an effective safety culture. These are presented in Figure 15.4.

15.4.4 Building a Safety Culture

Safety managers, professionals, and organizations should be under no illusion that building a safety culture can be quickly and easily achieved. Developing an efficient safety culture is an intensive, long, drawn-out process that requires the full dedication of all members of an organization at all levels of the organization (Olive et al., 2005). Fleming (1999) presented a safety culture maturity model (see Figure 15.5) that can be used by organizations to aid in the development of safety culture. According to the model, the safety culture maturity of an organization is a function of the following factors:

- Management commitment and visibility
- Communication
- Productivity vs. safety
- Learning organization
- Safety resources

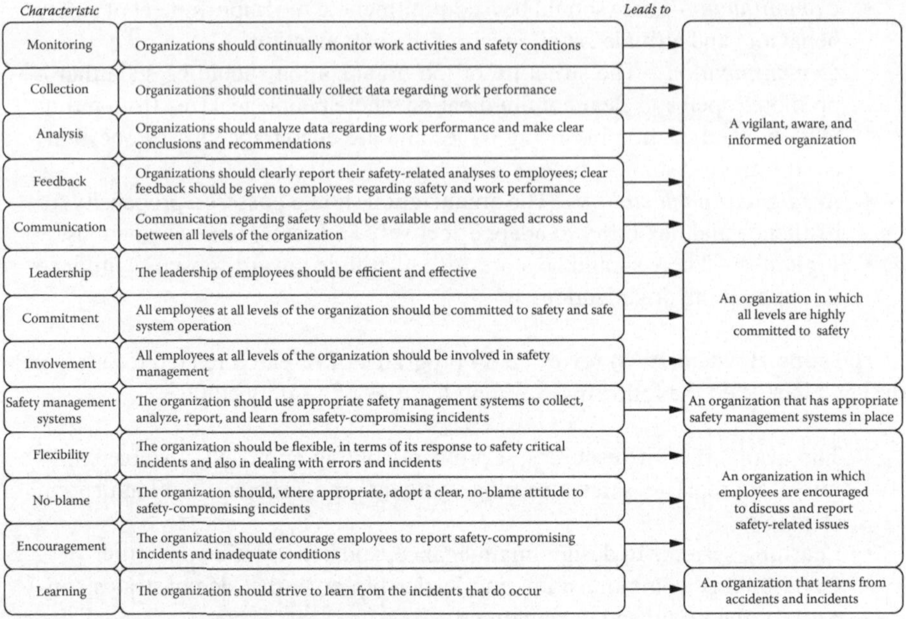

FIGURE 15.4 Characteristics of safety culture.

- Participation
- Shared perceptions about safety
- Trust
- Industrial relations and job satisfaction
- Training.

Based on a comprehensive review of the literature, Wright et al. (1999) developed a business model of safety culture. This model (see Figure 15.6) attempts to specify what it is that an organization must do to develop and sustain an effective safety culture. Wright et al. (1999) examined each of the components of the model in order to specify a series of micro subelements. Examples are presented in Table 15.2; for the full set of subelements, the reader is referred to Wright et al. (1999).

15.4.5 Collecting Safety-Related Data

Reason's (1997) descriptions of reporting and learning cultures relate to how organizations can gather and analyze safety-related data to identify what measures are needed to enhance system safety. One of the key commodities involved in developing and sustaining a safety culture is the collection and analysis of safety-related data (i.e., an informed culture) such that organizations can draw appropriate conclusions. Organizations can collect safety-related data in various ways. Two of the more commonly used approaches are accident investigation and analysis and incident reporting systems. An overview of both approaches is presented below.

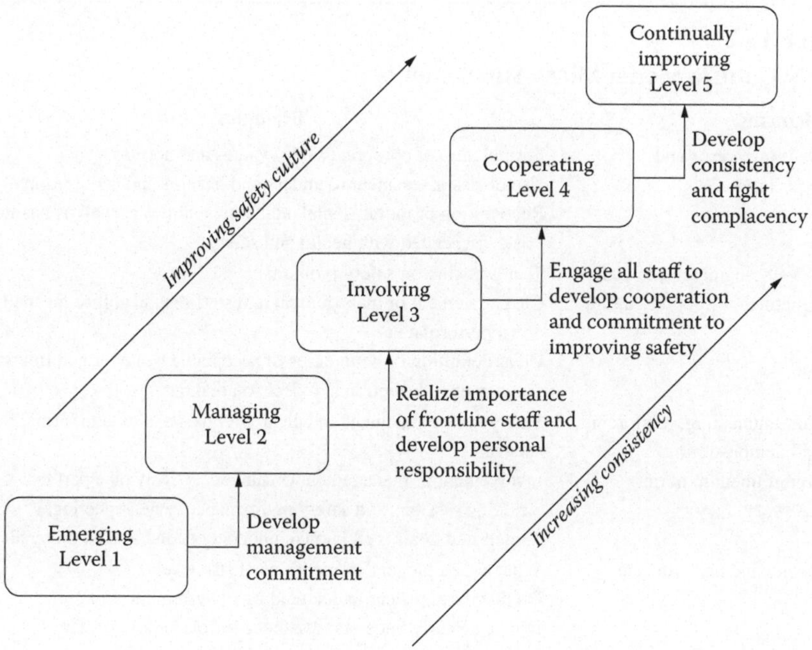

FIGURE 15.5 Safety culture maturity model. (Adapted from Fleming, M., *Safety Culture Maturity Model*, Offshore Technology Report 2000/049, HSE Books, London, 1999. Reproduced under the terms of the Click-Use Licence.)

15.4.5.1 Accident Investigation and Analysis

One of the most obvious means of collecting and analyzing error data in complex, safety-critical systems is through the retrospective investigation and analysis of accidents and incidents. Accident investigation is used to reconstruct accidents and

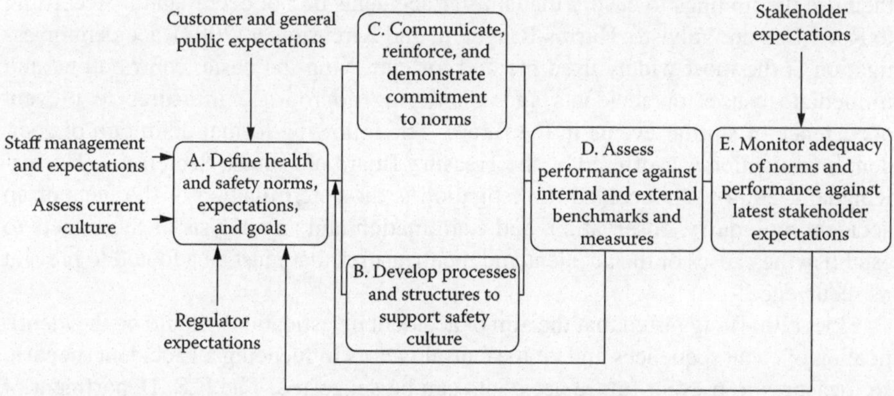

FIGURE 15.6 Composite safety culture model. (Adapted from Wright, M.S. et al., *Development of a Business Excellence Model of Safety Culture*, Health and Safety Executive, London, 1999. Reproduced under the terms of the Click-Use Licence.)

TABLE 15.2
Safety Culture Model Micro Subelements

Subelements	Examples
A. Defining norms and expectations	Formulation of corporate safety values and norms Use of risk assessment to understand hazards and set of norms Recognition of moral, social, and ethical duties as well as business risks associated with health and safety
B. Developing processes and structures to support culture	Teamworking on safety problems Clear statement of management and staff responsibility for their own performance Clear definition of boundaries of acceptable behavior and limits on personal discretion and decision making
C. Communicating, reinforcing, and demonstrating commitment to norms	Safety policy statement and clear-cut goals and targets formally stated Involvement of management (managers actively take part in training sessions and safety performance review meetings) Targets and goals well known, understood, and agreed on by all
D. Reviewing performance	What are employers' perceptions of risk levels? Do they have a clear understanding of system hazards? Do they clearly understand safety controls and procedures?
E. Recognizing and responding to changes in expectations or poor internal performance	Open mindedness Proactive seeking out lessons learned No-blame (appropriate blame) culture

Source: Wright, M.S. et al., *Development of a Business Excellence Model of Safety Culture*, Health and Safety Executive, London, 1999. With permission.

identify the human and system contributions, including error, to a particular accident or incident. It allows practitioners to specify exactly what happened and why and then use the findings to ensure that similar accidents do not occur again. According to Roed-Larsen, Valvisto, Harms-Ringdahl, and Kirchsteiger (2004), accident investigation is the most widely used method for clarifying the basic, contributing, and immediate causes of accidents and identifying appropriate measures to prevent occurrence of similar events in the future. The following formal definition of accident investigation was offered by the Treasury Board of Canada Secretariat (Harms-Ringdahl, 2004): "An accident investigation is the determination of the facts of an accident by inquiry, observation, and examination and an analysis of these facts to establish the causes of the accident and the measures that must be adopted to prevent its recurrence."

Sklet (2004) suggested that the aim of accident investigations should be the identification of event sequences and all the causal factors influencing an accident scenario so measures to prevent future accidents can be suggested. The U.S. Department of Energy (DOE) divides the accident investigation process into three main phases: collection of evidence and facts, analysis of evidence and facts and development of conclusions, and the development of judgments and need in writing the report. The

Centre for Chemical Process Safety (CCPS) (Sklet, 2004) described three main purposes of accident investigation techniques as follows:

- Organization of information regarding the accident under analysis
- Describing accident causation and developing hypotheses for further examination
- The assessment of proposed corrective actions

Numerous accident investigation and analysis techniques are available; for example, in a review and comparison of accident investigation methods, Sklet (2004) selected the following accident investigation methods for comparison: events and causal factors charting and analysis; barrier analysis; change analysis; root cause analysis; fault-tree analysis; influence diagrams; event-tree analysis; Management Oversight Risk Tree (MORT); Systematic Cause Analysis Technique (SCAT); Sequentially Timed Events Plotting (STEP); Man, Technology, and Organization (MTO) analysis; Accident Evolution and Barrier (AEB) model; Tripod Accident Causation Kit (TRACK); and AcciMaps.

The process of accident analysis and investigation is attractive for a number of reasons:

- Hindsight ensures that investigators know more about the accident than those who were caught up in it (Dekker, 2002).
- The entire sequence of events is exposed, including triggering conditions, outcome, and the various twists and turns involved (Dekker, 2002).
- It permits an exhaustive analysis of the errors involved, including the contributory factors, nature, and associated consequences.
- It permits the identification of the human and systemic causal factors involved in a particular accident.
- It permits the identification of system failures or inadequacies, such as bad design, inadequate training, inadequate equipment, poor management, etc.
- The identification of such factors leads to the development of appropriate countermeasures designed to prevent similar errors and accidents from occurring in the future.

Despite its obvious utility, a number of problems are associated with accident investigation and analysis:

- Typical accident analyses lead to the apportioning of blame to the individuals involved.
- Hindsight can potentially lead to oversimplified causality and counterfactual reasoning (Dekker, 2002).
- Countermeasures are typically aimed at the individual (e.g., increased training) and thus ignore the systemwide contributory factors that may have been involved in the accident in question.
- Accurate accident analysis and investigation are entirely dependent on the availability of accurate and comprehensive data. Such data are not always available, so much of the investigation is based on assumptions, domain knowledge, and expertise.

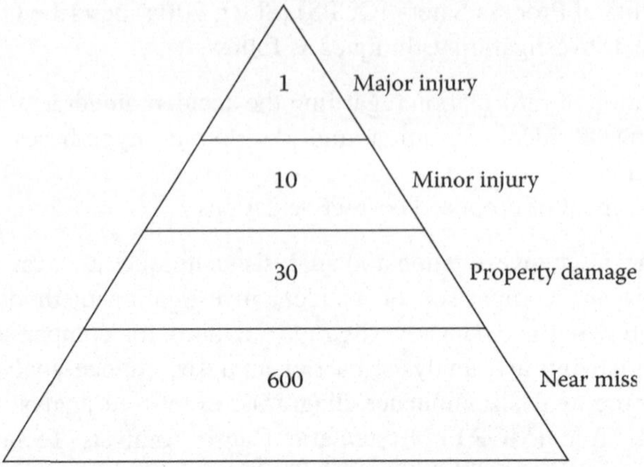

FIGURE 15.7 Bird's accident triangle. (Adapted from Jones, S. et al., *Journal of Loss Prevention in the Process Industries*, 12, 59–67, 1999.)

15.4.5.2 Incident Reporting Systems

Incident or near-miss reporting is a well-established method of improving safety in complex systems (Koorneef, 2000; Ternov, Tegenrot, and Akselsson, 2004). Incident reporting systems are used to collect pertinent information regarding critical incidents (or near misses), errors, safety-compromising incidents, and safety concerns within complex, safety-critical systems. The incident data gathered are used for a number of purposes, including enhancing system safety, increasing understanding of safety-related issues, and facilitating safety-related research. Incident reporting systems work on the premise that critical incidents, or near misses, involving error represent warnings or indications of potentially catastrophic accidents; the concept first emerged within the aviation domain, with the introduction of the Aviation Safety Reporting System (ASRS) in 1975. Incident reporting systems now exist in most safety-critical domains, including aviation, air-traffic control, healthcare, rail systems, and the nuclear power and petrochemical domains. Amalberti (2001) suggested that, "Reporting is fundamental to improve safety. Reporting describes the undesirable 'noise' plaguing the system in three areas of reference: the tool and its failures; the operator and his failures; and the resulting situation (situational, organizational, or systemic failings)."

Jones, Kirchsteiger, and Bjerke (1999) recommended that the internal investigation of near misses should be an integral part of safety management systems and should aim to prevent accidents and the occurrence of similar events in the near future. In the past, researchers have attempted to quantify the link between near-miss incidents and major accidents within complex, dynamic systems. Bird's near-miss triangle (Figure 15.7), for example, demonstrates that for each major accident there are a greater number of associated minor accidents and an even greater number of near misses. Research has also indicated that reducing the number of near misses reduces the number of full-blown accidents by reducing the number of near misses that evolve into actual accidents (Jones et al., 1999).

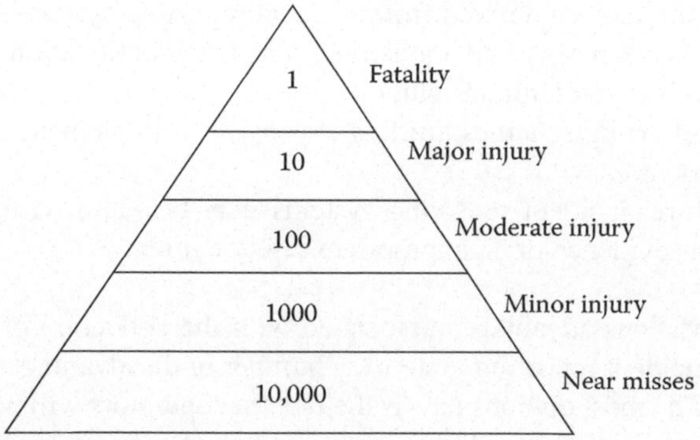

FIGURE 15.8 Critical incident triangle.

Heinrich, Petersen, and Roos (1980) also developed a triangle that demonstrates the relationship between near misses and fatal accidents in industrial settings. They estimated that, for every fatal accident, there are 10,000 associated near accidents or near-miss scenarios. The triangle proposed by Heinrich et al. is presented in Figure 15.8.

A number of researchers have highlighted the potential advantages associated with the collection and analysis of near-miss data. According to Reason (1997), the collection and analysis of near-miss data offer a number of advantages, including:

- When the right conclusions are drawn and acted upon, they work like vaccines to mobilize a system's defenses against more serious future occurrences.
- They provide a qualitative insight into how small defensive failures can line up to create large disasters.
- Because they occur more frequently than actual accidents, they provide the numbers required for more pertinent quantitative analysis.
- They provide a powerful reminder of the hazards that reside within a particular system.

According to Van der Schaaf (1995), near misses are more numerous than actual accidents; contain valuable information on system functioning, such as why things did not go wrong in the end; and contain references to safety rules, training, and safety equipment defenses in action. According to Johnson (2003), incident reporting systems are attractive for the following reasons:

- Incident reports help to identify why accidents do not occur.
- A high frequency of incident reports allows for quantitative analysis of the data.
- Incident reports provide a reminder of systemwide hazards.
- Feedback derived from incident reports keeps system actors in the loop and encourages agent participation in safety improvement. Operators can see that their concerns are being followed up effectively.

- The data and lessons derived from incident reporting systems can be compared across domains and industries, and the identification of common causes can lead to common solutions.
- Incident reporting schemes are less expensive to implement than the cost of an accident.
- In the future, incident reporting systems may be required by regulatory agencies as evidence of an appropriate safety culture.

Despite the various advantages associated with the collection of near-miss data and the use of incident reporting systems, a number of disadvantages may affect the data collected. The most obvious flaw is the fact that operators will not always report near-miss or critical incidents. There are a number of reasons for this. Informants may not wish to admit that they have made an error or contributed to an incident that compromised system safety in any way. According to Reason (1997), persuading people to file critical incident and near-miss reports is not an easy task, particularly when it involves divulging their own errors. Informants may fear that action will be taken against them as a result of the incidents they report. To counteract this, most incident reporting schemes are confidential, anonymous, and nonpunitive. Reason (1997) pointed out that potential informants often fail to see the value in making reports. Often people are skeptical of such systems, particularly the data actually being acted upon, and the extra work involved in creating and filing the report puts them off ever doing so. Another flaw concerns the quality of the data collected. The retrospective description of incidents is fraught with problems, including memory degradation (Klein and Armstrong, 2004) and a potential glorification of events. Further, informants may be unable to provide a sufficiently accurate and detailed account of an incident and its associated contributory events (Reason, 1997). Finally, incident reporting systems may suffer from problems associated with the understanding of what it is that actually constitutes a near-miss or critical incident. Although most pilots would report a gross error such as failing to lower the landing gear, they may not report errors that they perceive to be minor incidents, such as entering the wrong altitude or airspeed into the flight management computer. Such incidents could potentially lead to catastrophes yet may not be reported due to their (wrongly) perceived insignificance. In addition to the issues described above, Johnson (2002, 2003) observed the following disadvantages associated with the use of incident reporting systems:

- It is difficult to elicit information from actors who are involved in adverse incidents.
- The data obtained are subject to a number of biases, including confidence bias, hindsight bias, judgment bias, political bias, and recency bias.
- Incident reporting systems are typically expensive to set up and run. The ASRS, for example, spends approximately $3 million per year analyzing around 30,000 reports.
- Only limited sharing of data occurs between established incident reporting systems.
- Incident reporting systems may fail to keep actors in the loop.

- Incident reporting systems may serve merely as reminders of failures that operators within the system already know exist but do not have the incentive to address.
- Incident reporting systems typically recommend only short-term fixes or countermeasures that fail to address the various underlying causes of incidents.
- Only limited attempts have been made to ensure consistency in responses to incidents (e.g., different countermeasures may be proposed for the same incident in different cultures or countries).
- Incident reporting systems have a tendency to remind actors of their failures rather than address the associated root causes.

Despite these problems, the success of a number of schemes indicates that the advantages greatly outweigh the flaws (Reason, 1997).

15.5 GUIDELINES AND RECOMMENDATIONS

Based on the literature reviewed, we can postulate the following guidance for developing and sustaining an effective safety culture:

1. *Monitor and assess safety culture*. Organizations should continually monitor and assess the state of their own safety cultures. This involves monitoring work activities, assessing the safety conditions within the organizations, and assessing workers' perceptions and attitudes toward safety.
2. *Monitor and assess safety performance*. Organizations should continually monitor, assess, and learn from their own performance with regard to safety. Incident and accident data should be collected and analyzed accordingly, and appropriate lessons should be drawn from such endeavors.
3. *Communicate to all and provide feedback throughout the organization*. Organizations should clearly report their analyses of safety culture and safety-related performance to all workers at all levels of the organization. Measures taken to enhance safety or the safety culture should also be clearly communicated.
4. *Make a clear commitment to safety*. Commitment at all levels of the organization is essential for an efficient safety culture. All employees at all levels should therefore be committed to safety; this begins with a clear commitment to safety from managers within the organization.
5. *Involve employees at all levels*. Involving all employees is key, not only for improving the safety culture but also for improving workers' perceptions and attitudes toward safety and safety measures. It is therefore recommended that employees at all levels of the organization should be actively encouraged to become involved with safety-related endeavors.
6. *Implement safety management systems*. Organizations should develop and implement appropriate safety management systems designed to collect, analyze, report, and learn from safety-compromising behavior or incidents.

7. *Be flexible.* It is recommended that organizations should be flexible in their response to safety-compromising incidents. Organizations should, where appropriate, adopt a clear no-blame attitude toward employees involved in safety-compromising incidents.
8. *Encourage employees to be involved in system safety.* Employees should be actively encouraged to become involved in measures designed to increase system safety, including the reporting of safety-compromising incidents and inadequate working conditions and suggesting measures for enhancing system safety.
9. *Learn from safety-compromising incidents.* It is vital that organizations learn from any safety-compromising incidents that do occur. Organizations should therefore collect and analyze data related to such incidents.

15.6 ASSESSMENT AND EVALUATION

Various safety culture measures have been developed and applied in the hazardous industries. Most approaches involve the use of questionnaires or interviews to canvass employee opinions on the current state of the organization's safety culture. In its conclusion to a review of existing measures, the HSE (2005a) reported that the majority of tools available are questionnaires that use rating scales, that the majority of approaches are domain specific (i.e., they are developed specifically for a certain domain), and that most of the approaches assess safety climate rather than safety culture. The HSE (2005b) recently developed a safety culture assessment toolkit for the U.K. rail domain that focuses on five indicators found to influence safety culture (leadership, two-way communication, employee involvement, learning culture, and attitude toward blame). The tool consists of a question set, a pocket card, an overview diagram, and a guide for inspectors.

The general consensus within the literature is that assessing employee attitudes alone is not satisfactory when wishing to measure an organization's safety culture, as they can only elicit safety perceptions and attitudes (HSE, 2005a). Other methods that can be used in addition to safety culture questionnaires include observational studies, focus groups, interviews, case studies, surveys, and questionnaires.

It is also recommended that safety culture assessments focus on all levels of the organization in question rather than merely frontline operators; the HSE (2005a) pointed out that culture is likely to vary widely across an organization so assessments should be made across the entire organization.

15.7 EXAMPLE

Glendon, Stanton, and Harrison (1994) developed the Safety Culture Questionnaire (SCQ) for assessing safety cultures across the following eight dimensions:

1. *Work pressure* (the degree to which employees feel under pressure to complete work, the amount of time there is to plan and carry out work, and the balance of workload)

2. *Incident investigation and development of procedures* (the degree to which people are involved in the development of procedures, the extent to which incident investigations get to the underlying causes of incidents, and the effectiveness of procedures)
3. *Adequacy of procedures* (the accuracy, completeness, comprehensiveness, clarity, and appropriateness of procedures as well as the ease of selection and use of procedures)
4. *Communication and training* (the degree of openness and extent to which communication transgresses all levels in the organization, the extent to which training incorporates all aspects of the job, and the relevance and effectiveness of training)
5. *Relationships* (the degree of trust and support within the organization, confidence people have in the future of the organization, working relationships with others, and general morale)
6. *Personal protective equipment* (the training, use, and enforcement of personal protective equipment, as well as user group involvement in the identification of problems with and design of personal protective equipment)
7. *Spares* (the availability, timeliness, and fitting of spares and the appropriateness and availability of back-up systems)
8. *Safety* (the degree to which safety is a priority, the extent to which people are consulted on safety matters, and the practicality of implementing safety policies and procedures)

It is worth noting that the SCQ is not intended to be used without other methods and approaches. In the current study, the SCQ is being used in the context of a much larger assessment of the human factors of control operations (namely, competencies, training, procedures, communication processes, manning and workload, supervision, shift patterns, control room layout, and environment) and an investigation of human error potential in control room operations, tasks, and roles. By way of example, the topic of procedures is covered by various approaches, as shown in Table 15.3, which indicates that some of the human factors areas were addressed by up to three of the approaches, whereas others are addressed by one or two. Where more than one approach addressed an area, it would be prudent to cross reference between the findings. The important point to be made is that it is worth checking the correspondence between the objective and subjective data.

The SCQ was developed over a period of 2 years with over 350 participants. The original version of the questionnaire had over 300 items, but this number was reduced through factor analytic studies to 58 items and 8 factors. An extract of the SCQ is presented in Figure 15.9. This questionnaire provides a baseline judgment about how good or poor a particular organization's safety culture is; they propose that a score below 5 (based on summing the responses for each dimension and calculating the mean score) for any of the 8 dimensions is of particular concern.

The current safety culture of the example organization was assessed using Glendon et al.'s (1994) questionnaire. A summary of the study and results is presented below. The personal protective equipment factor was dropped from this study as it was not relevant to the control room staff.

TABLE 15.3
Different Approaches to the Human Factors Study

Human Factors Topic Area (Chapter in Book)	Analysis and Review	Human Error	Safety Culture	EEMUA 191 and 201[a]
Competencies (2)	✓	✓		
Training (3)	✓	✓	✓	
Procedures (4)	✓	✓	✓	
Communication (5)	✓	✓	✓	
Manning (6)	✓		✓	
Workload (6)	✓	✓	✓	
Automation (7)	✓	✓	✓	
Supervision (8)	✓	✓	✓	
Shift pattern (9)	✓	✓		
Control room layout (10)	✓			
Interface design (11 and 12)	✓	✓		✓
Environment (13)	✓			
Error (14)	✓	✓	✓	
Culture (15)	✓	✓	✓	

[a] Engineering Equipment and Materials Users' Association (EEMUA) standards: EEMUA 191 (Alarm Systems: A Guide to Design, Management and Procurement) and EEMUA 201 (Human–Computer Interfaces).

A total of 28 questionnaires were completed by employees of the organization in question; the respondent population included 11 Operations Engineers, 9 Principal Operations Engineers, and 3 Network Managers. Five respondents did not report their current role within the organization. All respondents were male. The SCQ was distributed in either a paper or electronic (Microsoft® Word) format based on the preference of the participants. Participants completed the questionnaire using pen or paper or electronically via a desktop computer with Word. Microsoft® Excel was used to record and analyze the data collected. The mean ratings derived from the questionnaire sample are presented in Figure 15.10. The questionnaire data were also analyzed based on the job role of the respondents—for this purpose, the mean responses for each safety culture dimension from each job role group (Network Manager, Principal Operations Engineer, Operations Engineer, and unknown respondent group). The mean overall ratings per job role group are presented in Figure 15.11.

The SCQ allows judgments to be made on how good employees feel an organization's safety culture is. Glendon et al. (1994) suggested that, as a rule of thumb, organizations should be concerned about any dimension that receives a mean score of 5 or below. The mean results for each dimension show that the relationship and spares dimensions achieved mean scores of below 5 (4.48 and 4.74, respectively). The mean score on the relationship dimension suggests that there may be problems

1. Safety rules are adhered to even under production pressures.

Never *Sometimes* *Always*

1 2 3 4 5 6 7 8 9

2. Safety rules can be implemented without conflicting with established work practices.

Never *Sometimes* *Always*

1 2 3 4 5 6 7 8 9

3. Safety rules are practical to apply in all situations.

Never *Sometimes* *Always*

1 2 3 4 5 6 7 8 9

4. There are adequate opportunities for staff to express their views about operational problems.

Never *Sometimes* *Always*

1 2 3 4 5 6 7 8 9

5. There are adequate opportunities to discuss important policy issues.

Never *Sometimes* *Always*

1 2 3 4 5 6 7 8 9

6. Consultation is adequate when changes in working practices are proposed.

Never *Sometimes* *Always*

1 2 3 4 5 6 7 8 9

7. Meetings take place where causes of operational problems are openly discussed between engineers and management.

Never *Sometimes* *Always*

1 2 3 4 5 6 7 8 9

8. An effective system exists for communicating the nature of plant changes and their implications for safety to operating personnel.

Never *Sometimes* *Always*

1 2 3 4 5 6 7 8 9

9. An adequate system exists for transmitting critical information regarding the state of the system during shift changeover.

Never *Sometimes* *Always*

1 2 3 4 5 6 7 8 9

10. Users are involved in developing the incident investigation system.

Never *Sometimes* *Always*

1 2 3 4 5 6 7 8 9

11. Members of investigation teams are trained to identify factors that influence the causes of errors.

Never *Sometimes* *Always*

1 2 3 4 5 6 7 8 9

12. The investigation system considers management and policy influences on the causes of incidents.

Never *Sometimes* *Always*

1 2 3 4 5 6 7 8 9

FIGURE 15.9 Safety culture questionnaire extract. (From Stanton, N.A. and Glendon, A.I., *Safety Science*, 22(1–3), 1–13, 1996. With permission.)

regarding the relationships between the different levels of the organization, such as the nature of relationships in general within the organization, between the operators and management in terms of the level of trust between them, the support provided by management to lower level staff, and employee confidence regarding their own

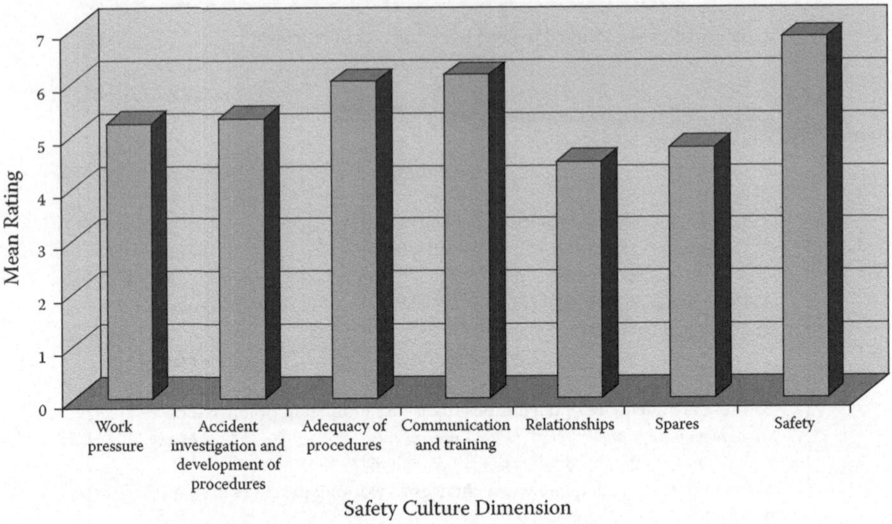

FIGURE 15.10 Mean safety culture dimension ratings.

future within the company. The low score obtained for this dimension is of concern because of the importance placed on relationships between management and lower level staff in developing and maintaining an efficient safety culture within the research literature. The low score on the spares dimension suggests that the respondents perceived problems with the availability of spare parts (the ratings provided suggest that critical spare parts are not always available), time taken to obtain critical

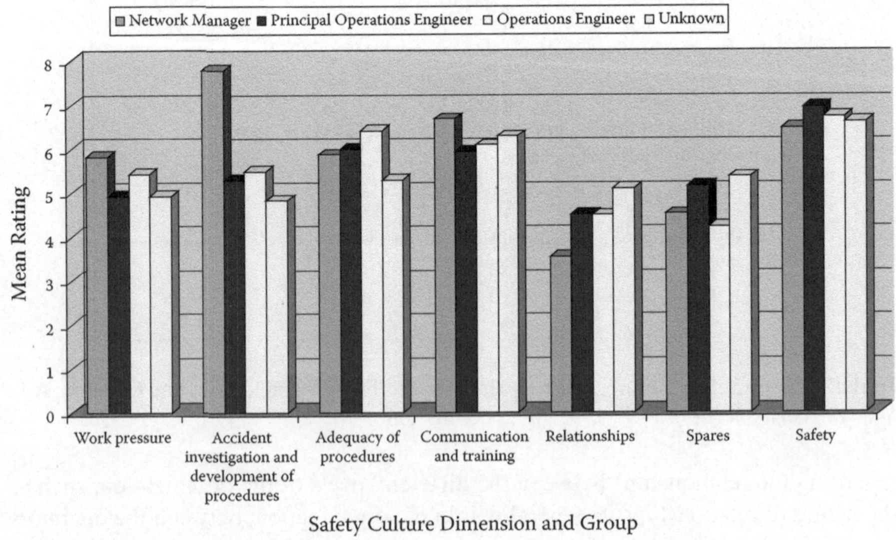

FIGURE 15.11 Mean safety culture dimension ratings per job role group.

spare parts (the ratings provided indicate that the time taken to obtain spare parts is not always acceptable), and that inappropriate spare parts may also be fitted due to poor availability of appropriate spare parts (the ratings provided suggest that the correct spare parts are not always fitted).

The remainder of the dimensions assessed achieved acceptable ratings, although the work pressure dimension received a low mean score of 5.2. This suggests that, although satisfactory ratings were achieved overall, there may be some problems regarding the level of workload imposed on staff, the number of staff available, and the time available to plan and undertake tasks. Similarly, accident investigation and the development of procedures received a satisfactory but low mean score (5.3). This suggests that the participants felt that the accident investigation and procedure development processes in place within the organization may not always be sufficient. Factors here include the provision of sufficient levels of training to allow workers to identify the causal factors involved in accidents and incidents, the use of clear and well-documented procedures, the use of an efficient and well-documented procedure development process, the use of error analysis, and the provision of human factors guidance during procedure development. All of the other dimensions assessed (adequacy of procedures, communication, and safety) achieved mean scores of 6 or more, which suggests that factors related to these dimensions are currently satisfactory within the organization.

The responses were also analyzed in terms of the participants' current job roles within the organization (Network Manager, Principal Operations Engineer, or Operations Engineer). For work pressure, the Network Managers gave the highest mean rating (5.8), whereas the Principal Operations Engineers gave the lowest mean rating (4.9), which is below the threshold proposed by Stanton and Glendon. This suggests that, of the three groups, the Principal Operations Engineers feel under the most pressure in terms of time available to plan and undertake tasks, frustration levels, and the level of workload imposed on them. For the dimension of accident investigation and development of procedures, the Network Managers provided the highest ratings (mean of 7.77), whereas the Principal Operations Engineers provided the lowest ratings (mean of 5.27). For the dimension of adequacy of procedures, the Operations Engineers provided the highest ratings overall, with a mean of 6.40, followed by the Principal Operations Engineers with a mean rating of 5.9. The Network Managers provided the lowest mean rating (5.8) for adequacy of procedures.

For communication and training, the Network Managers provided the highest mean rating (6.66), followed by the Operations Engineers (6.08) and the Principal Operations Engineers (5.9). All three of the participant groups provided a mean rating of below 5 for the relationships dimension. The Network Managers provided the lowest mean rating (3.53), followed by the Operations Engineers (4.47) and the Principal Operations Engineers (4.48). The results suggest that all three participant groups felt that there were significant problems relating to the relationships within the organization in question. For the spares dimension, the Network Managers and Operations Engineers provided mean ratings of 4.5 and 4.2, respectively, which suggests that both groups perceive significant problems regarding the availability, time taken to obtain, and use of critical spare parts. In terms of overall safety, the groups provided similar mean rating scores, with the Principal Operations Engineers

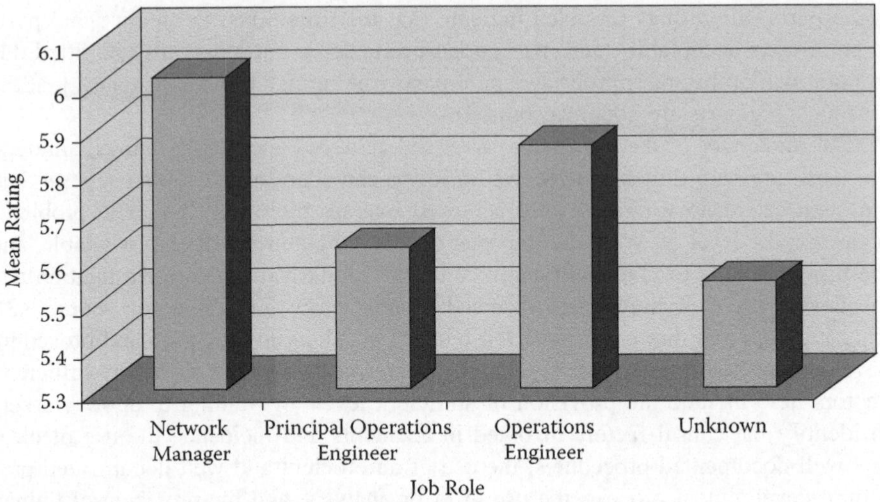

FIGURE 15.12 Mean overall safety culture questionnaire ratings per job role group.

achieving a mean rating of 6.94, followed by the Operations Engineers (6.71) and then the Network Managers (6.45).

The overall ratings given by each participant group (Figure 15.12) suggest that the Principal Operations Engineers have the worst view of the organization's safety culture (mean overall response rating of 5.62), followed by the Operations Engineers (mean overall response rating of 5.85) and then the Network Managers (mean overall response rating of 6.01), who view the organization's safety culture the most favorably.

BIBLIOGRAPHY

Advisory Committee on the Safety of Nuclear Installations (ACSNI). (1993). *Human Factors Study Group Third Report: Organising for Safety*. London: HMSO.

Amalberti, R. (2001). The paradoxes of almost totally safe transportation systems. *Safety Science*, 37, 109–126.

Ball, P.W. and Scotney, V. (1998). *Approaches to Safety Culture Enhancement*. Daresbury, U.K.: British Nuclear Fuels, Ltd.

Cooper, M.D. (2000). Toward a model of safety culture. *Safety Science*, 35, 111–136.

Davies, J., Ross, A., Wallace, B., and Wright, L. (2003). *Safety Management: A Qualitative Systems Approach*. London: Taylor & Francis.

Dekker, S.W.A. (2002). Reconstructing human contributions to accidents: the new view on human error and performance. *Journal of Safety Research*, 33, 371–385.

Fleming, M. (1999). *Safety Culture Maturity Model*, Offshore Technology Report 2000/049. London: HSE Books.

Geller, E.S. (1994). Ten principles for achieving a total safety culture. *Professional Safety*, 39(9), 18.

Glendon, A.I., Stanton, N.A., and Harrison, D. (1994). Factor analysing a performance shaping concepts questionnaire. In S. Robertson (Ed.), *Contemporary Ergonomics* (pp. 340–345). London: Taylor & Francis.

Grote, G. and Kunzler, C. (2000). Diagnosis of safety culture in safety management audits. *Safety Science*, 34, 131–150.

Guldenmund, F.W. (2000). The nature of safety culture: a review of theory and research. *Safety Science*, 34, 215–257.

Hale, A. and Hovden, J. (1998). Management and culture: the third age of safety. A review of approaches to organisational aspects of safety, health, and environment. In A. Feyer and A. Williamson (Eds.), *Occupational Injury: Risk Prevention, and Intervention* (pp. 129–166). London: Taylor & Francis.

Harms-Ringdahl, L. (2004). Relationship between investigations, risk analysis, and safety management. *Journal of Hazardous Materials*, 111, 13–19.

Haukelid, K. (2008). Theories of (safety) culture revisited: an anthropological approach. *Safety Science*, 46, 413–426.

Health and Safety Executive (HSE). (2005a). *A Review of the Safety Culture and Safety Climate Literature for the Development of the Safety Culture Inspection Toolkit*, HSE Research Report 367. London: HSE Books.

Health and Safety Executive (HSE). (2005b). *Development and Validation of the HMRI Safety Culture Inspection Toolkit*, HSE Research Report 365. London: HSE Books.

Health and Safety Laboratory. (2002). *Safety Culture: A Review of the Literature*, Research Report HSL/2002/25. Derbyshire, U.K.: Health and Safety Laboratory.

Heinrich, H.W., Petersen, D., and Roos, N. (1980). *Industrial Accident Prevention: A Safety Management Approach*, 5th ed. New York: McGraw-Hill.

Hendershot, D. (2007). Process safety culture. *Journal of Chemical Health and Safety*, 14(3), 39–40.

Johnson, C. (2002). Software tools to support incident reporting in safety-critical systems. *Safety Science*, 40, 765–780.

Johnson, C. (2003). *Failure in Safety-Critical Systems: A Handbook of Accident and Incident Reporting*. Glasgow, Scotland: University of Glasgow Press.

Jones, S., Kirchsteiger, C., and Bjerke, W. (1999). The importance of near miss reporting to further improve safety performance. *Journal of Loss Prevention in the Process Industries*, 12, 59–67.

Klein, G. and Armstrong, A.A. (2004). Critical decision method. In N.A. Stanton, A. Hedge, K. Brookhuis, E. Salas, and H. Hendrick (Eds.), *Handbook of Human Factors and Ergonomics Methods* (pp. 58-1–58-6). Boca Raton, FL: CRC Press.

Koorneef, F. (2000). *Organizational Learning from Small-Scale Incidents*. Amsterdam: Delft University Press.

Muniz, B.F., Peon, J.M.M., and Ordas, C.J.V. (2007). Safety culture: analysis of the causal relationships between its key dimensions. *Journal of Safety Research*, 38, 627–641.

Occupational Health and Safety Advisory Services. (2004). *BS/OHSAS 8800: Guide to Occupational Health and Safety Management Systems*. London: OHSAS.

Occupational Health and Safety Advisory Services. (2007). *BS/OHSAS 18001: Occupational Health and Safety Zone*. London: OHSAS.

Olive, C., O'Connor, M.T., and Mannan, S.M. (2006). Relationship of safety culture and process safety. *Journal of Hazardous Materials*, 130, 133–140.

Pidgeon, N.F. and O'Leary, M. (1994). Organizational safety culture: implications for aviation practice. In N. Johnston, N. McDonald, and R. Fuller (Eds.), *Aviation Psychology in Practice* (pp. 21–43). Aldershot: Avebury.

Pidgeon, N.F. and O'Leary, M. (2000). Man-made disasters: why technology and organizations (sometimes) fail. *Safety Science*, 34, 15–30.

Rao, S. (2007). Safety culture and accident analysis: a socio-management approach based on organizational safety social capital. *Journal of Hazardous Materials*, 142, 730–740.

Reason, J. (1997). *Managing the Risks of Organizational Accidents*. Burlington, VT: Ashgate.

Richter, A. and Koch, C. (2004). Integration, differentiation and ambiguity in safety cultures. *Safety Science*, 42(8), 703–722.

Roed-Larsen, S., Valvisto, T., Harms-Ringdahl, L., and Kirchsteiger, C. (2004). Accident investigation practices in Europe: main responses from a recent study of accidents in industry and transport. *Journal of Hazardous Materials*, 111, 7–12.

Sklet, S. (2004). Comparison of some selected methods for accident investigation. *Journal of Hazardous Materials*, 111, 29–37.

Sorensen, J.N. (2002). Safety culture: a survey of the state-of-the-art. *Reliability Engineering and System Safety*, 76, 189–204.

Stanton, N.A. and Glendon, A.I. (1996). Risk homeostasis and risk assessment. *Safety Science*, 22(1–3), 1–13.

Ternov, S., Tegenrot, G., and Akselsson, R. (2004). Operator-centred local error management in air traffic control. *Safety Science*, 42(10), 907–920.

Van der Schaaf, T.W. (1995). Near miss reporting in the chemical process industry: an overview. *Microelectronics and Reliability*, 35, 1233–1243.

Wierwille, W.W., Hanowski, R.J., Hankey, J.M., Kieliszewski, C.A., Lee, S.E., Medina, A., Keisler, A.S., and Dingus, T.A. (2002). *Identification and Evaluation of Driver Errors: Overview and Recommendations*, Report No. FHWA-RD-02-003. Washington, D.C.: U.S. Department of Transportation.

Wright, M.S., Brabazon, P., Tipping, A., and Talwalkar, M. (1999). *Development of a Business Excellence Model of Safety Culture*. London: Health and Safety Executive.

Author Index

D

E

F

G

H

I

J

K

T

V

W

X

Y

Z

Subject Index

A

B

C

D

E

F

G

H

I

M

N

O

P

Q

R

S

T

U

V

W